Canadian Mathematical Society
Société mathématique du Canada

CMS Books in Mathematics
Ouvrages de mathématiques de la SMC

1 HERMAN/KUČERA/ŠIMŠA Equations and Inequalities

2 ARNOLD Abelian Groups and Representations of Finite Partially Ordered Sets

3 BORWEIN/LEWIS Convex Analysis and Nonlinear Optimization, Second Edition

4 LEVIN/LUBINSKY Orthogonal Polynomials for Exponential Weights

5 KANE Reflection Groups and Invariant Theory

6 PHILLIPS Two Millennia of Mathematics

7 DEUTSCH Best Approximation in Inner Product Spaces

8 FABIAN ET AL. Functional Analysis and Infinite-Dimensional Geometry

9 KŘÍŽEK/LUCA/SOMER 17 Lectures on Fermat Numbers

10 BORWEIN Computational Excursions in Analysis and Number Theory

11 REED/SALES (Editors) Recent Advances in Algorithms and Combinatorics

12 HERMAN/KUČERA/ŠIMŠA Counting and Configurations

13 NAZARETH Differentiable Optimization and Equation Solving

14 PHILLIPS Interpolation and Approximation by Polynomials

15 BEN-ISRAEL/GREVILLE Generalized Inverses, Second Edition

16 ZHAO Dynamical Systems in Population Biology

17 GÖPFERT ET AL. Variational Methods in Partially Ordered Spaces

18 AKIVIS/GOLDBERG Differential Geometry of Varieties with Degenerate Gauss Maps

19 MIKHALEV/SHPILRAIN/YU Combinatorial Methods

20 BORWEIN/ZHU Techniques of Variational Analysis

21 VAN BRUMMELEN/KINYON Mathematics and the Historian's Craft: The Kenneth O. May Lectures

22 LUCCHETTI Convexity and Well-Posed Problems

23 NICULESCU/PERSSON Convex Functions and Their Applications

24 SINGER Duality for Nonconvex Approximation and Optimization

25 HIGGINSON/PIMM/SINCLAIR Mathematics and the Aesthetic

Jonathan M. Borwein
Adrian S. Lewis

Convex Analysis and Nonlinear Optimization

Theory and Examples

Second Edition

Jonathan Borwein
Faculty of Computer Science
Dalhousie University
6050 University Avenue
Halifax B3H 1W5
NS, Canada
jborwein@cs.dal.ca

Adrian Lewis
School of Operations Research
and Industrial Engineering
Cornell University
Ithaca, NY 14853
USA
aslewis@orie.cornell.edu

Editors-in-Chief
Rédacteurs-en-chef
Jonathan Borwein
Karl Dilcher
Department of Mathematics and Statistics
Dalhousie University
Halifax, Nova Scotia B3H 3J5
Canada
cbs-editors@cms.math.ca

Mathematics Subject Classification (2000): 90–01 49–01 90C25 49J53 52A41 46N10 47H10

ISBN 978-1-4419-2127-7

e-ISBN 978-0-387-31256-9

Printed on acid-free paper.

Printed in the United States of America. (MVY)

9 8 7 6 5 4 3 2 1

springeronline.com

To our families

Preface

Optimization is a rich and thriving mathematical discipline. Properties of minimizers and maximizers of functions rely intimately on a wealth of techniques from mathematical analysis, including tools from calculus and its generalizations, topological notions, and more geometric ideas. The theory underlying current computational optimization techniques grows ever more sophisticated—duality-based algorithms, interior point methods, and control-theoretic applications are typical examples. The powerful and elegant language of convex analysis unifies much of this theory. Hence our aim of writing a concise, accessible account of convex analysis and its applications and extensions, for a broad audience.

For students of optimization and analysis, there is great benefit to blurring the distinction between the two disciplines. Many important analytic problems have illuminating optimization formulations and hence can be approached through our main variational tools: subgradients and optimality conditions, the many guises of duality, metric regularity and so forth. More generally, the idea of convexity is central to the transition from classical analysis to various branches of modern analysis: from linear to nonlinear analysis, from smooth to nonsmooth, and from the study of functions to multifunctions. Thus, although we use certain optimization models repeatedly to illustrate the main results (models such as linear and semidefinite programming duality and cone polarity), we constantly emphasize the power of abstract models and notation.

Good reference works on finite-dimensional convex analysis already exist. Rockafellar's classic *Convex Analysis* [167] has been indispensable and ubiquitous since the 1970s, and a more general sequel with Wets, *Variational Analysis* [168], appeared recently. Hiriart–Urruty and Lemaréchal's *Convex Analysis and Minimization Algorithms* [97] is a comprehensive but gentler introduction. Our goal is not to supplant these works, but on the contrary to promote them, and thereby to motivate future researchers. This book aims to make converts.

We try to be succinct rather than systematic, avoiding becoming bogged down in technical details. Our style is relatively informal; for example, the text of each section creates the context for many of the result statements. We value the variety of independent, self-contained approaches over a single, unified, sequential development. We hope to showcase a few memorable principles rather than to develop the theory to its limits. We discuss no algorithms. We point out a few important references as we go, but we make no attempt at comprehensive historical surveys.

Optimization in infinite dimensions lies beyond our immediate scope. This is for reasons of space and accessibility rather than history or application: convex analysis developed historically from the calculus of variations, and has important applications in optimal control, mathematical economics, and other areas of infinite-dimensional optimization. However, rather like Halmos's *Finite Dimensional Vector Spaces* [90], ease of extension beyond finite dimensions substantially motivates our choice of approach. Where possible, we have chosen a proof technique permitting those readers familiar with functional analysis to discover for themselves how a result extends. We would, in part, like this book to be an entrée for mathematicians to a valuable and intrinsic part of modern analysis. The final chapter illustrates some of the challenges arising in infinite dimensions.

This book can (and does) serve as a teaching text, at roughly the level of first year graduate students. In principle we assume no knowledge of real analysis, although in practice we expect a certain mathematical maturity. While the main body of the text is self-contained, each section concludes with an often extensive set of optional exercises. These exercises fall into three categories, marked with zero, one, or two asterisks, respectively, as follows: examples that illustrate the ideas in the text or easy expansions of sketched proofs; important pieces of additional theory or more testing examples; longer, harder examples or peripheral theory.

We are grateful to the Natural Sciences and Engineering Research Council of Canada for their support during this project. Many people have helped improve the presentation of this material. We would like to thank all of them, but in particular Patrick Combettes, Guillaume Haberer, Claude Lemaréchal, Olivier Ley, Yves Lucet, Hristo Sendov, Mike Todd, Xianfu Wang, and especially Heinz Bauschke.

Jonathan M. Borwein
Adrian S. Lewis

Gargnano, Italy
September 1999

Preface to the Second Edition

Since the publication of the First Edition of this book, convex analysis and nonlinear optimization has continued to flourish. The "interior point revolution" in algorithms for convex optimization, fired by Nesterov and Nemirovski's seminal 1994 work [148], and the growing interplay between convex optimization and engineering exemplified by Boyd and Vandenberghe's recent monograph [47], have fuelled a renaissance of interest in the fundamentals of convex analysis. At the same time, the broad success of key monographs on general variational analysis by Clarke, Ledyaev, Stern and Wolenski [56] and Rockafellar and Wets [168] over the last decade testify to a ripening interest in nonconvex techniques, as does the appearance of [43].

The Second Edition both corrects a few vagaries in the original and contains a new chapter emphasizing the rich applicability of variational analysis to concrete examples. After a new sequence of exercises ending Chapter 8 with a concise approach to monotone operator theory via convex analysis, the new Chapter 9 begins with a presentation of Rademacher's fundamental theorem on differentiability of Lipschitz functions. The subsequent sections describe the appealing geometry of proximal normals, four approaches to the convexity of Chebyshev sets, and two rich concrete models of nonsmoothness known as "amenability" and "partial smoothness". As in the First Edition, we develop and illustrate the material through extensive exercises.

Convex analysis has maintained a Canadian thread ever since Fenchel's original 1949 work on the subject in Volume 1 of the *Canadian Journal of Mathematics* [76]. We are grateful to the continuing support of the Canadian academic community in this project, and in particular to the Canadian Mathematical Society, for their sponsorship of this book series, and to the Canadian Natural Sciences and Engineering Research Council for their support of our research endeavours.

JONATHAN M. BORWEIN
ADRIAN S. LEWIS
September 2005

Preface to the Second Edition

Contents

Chapter 1

Background

1.1 Euclidean Spaces

We begin by reviewing some of the fundamental algebraic, geometric and analytic ideas we use throughout the book. **Our setting, for most of the book, is an arbitrary Euclidean space $\mathbf{E}$**, by which we mean a finite-dimensional vector space over the reals $\mathbf{R}$, equipped with an inner product $\langle \cdot, \cdot \rangle$. We would lose no generality if we considered only the space $\mathbf{R}^n$ of real (column) n-vectors (with its standard inner product), but a more abstract, coordinate-free notation is often more flexible and elegant.

We define the *norm* of any point x in $\mathbf{E}$ by $\|x\| = \sqrt{\langle x, x \rangle}$, and the *unit ball* is the set

$$B = \{x \in \mathbf{E} \mid \|x\| \leq 1\}.$$

Any two points x and y in $\mathbf{E}$ satisfy the *Cauchy–Schwarz inequality*

$$|\langle x, y \rangle| \leq \|x\| \|y\|.$$

We define the sum of two sets C and D in $\mathbf{E}$ by

$$C + D = \{x + y \mid x \in C,\ y \in D\}.$$

The definition of $C - D$ is analogous, and for a subset Λ of $\mathbf{R}$ we define

$$\Lambda C = \{\lambda x \mid \lambda \in \Lambda,\ x \in C\}.$$

Given another Euclidean space $\mathbf{Y}$, we can consider the Cartesian product Euclidean space $\mathbf{E} \times \mathbf{Y}$, with inner product defined by $\langle (e, x), (f, y) \rangle = \langle e, f \rangle + \langle x, y \rangle$.

We denote the nonnegative reals by $\mathbf{R}_+$. If C is nonempty and satisfies $\mathbf{R}_+ C = C$ we call it a *cone*. (Notice we require that cones contain the origin.) Examples are the positive orthant

$$\mathbf{R}^n_+ = \{x \in \mathbf{R}^n \mid \text{each } x_i \geq 0\},$$

and the cone of vectors with nonincreasing components

$$\mathbf{R}^n_{\geq} = \{x \in \mathbf{R}^n \mid x_1 \geq x_2 \geq \cdots \geq x_n\}.$$

The smallest cone containing a given set $D \subset \mathbf{E}$ is clearly $\mathbf{R}_+ D$.

The fundamental geometric idea of this book is *convexity*. A set C in $\mathbf{E}$ is *convex* if the line segment joining any two points x and y in C is contained in C: algebraically, $\lambda x + (1-\lambda)y \in C$ whenever $0 \leq \lambda \leq 1$. An easy exercise shows that intersections of convex sets are convex.

Given any set $D \subset \mathbf{E}$, the *linear span* of D, denoted span(D), is the smallest linear subspace containing D. It consists exactly of all linear combinations of elements of D. Analogously, the *convex hull* of D, denoted conv(D), is the smallest convex set containing D. It consists exactly of all *convex combinations* of elements of D, that is to say points of the form $\sum_{i=1}^m \lambda_i x^i$, where $\lambda_i \in \mathbf{R}_+$ and $x^i \in D$ for each i, and $\sum \lambda_i = 1$ (see Exercise 2).

The language of elementary point-set topology is fundamental in optimization. A point x lies in the *interior* of the set $D \subset \mathbf{E}$ (denoted $\operatorname{int} D$) if there is a real $\delta > 0$ satisfying $x + \delta B \subset D$. In this case we say D is a *neighbourhood* of x. For example, the interior of $\mathbf{R}^n_+$ is

$$\mathbf{R}^n_{++} = \{x \in \mathbf{R}^n \mid \text{each } x_i > 0\}.$$

We say the point x in $\mathbf{E}$ is the *limit* of the sequence of points $x^1, x^2, \ldots$ in $\mathbf{E}$, written $x^j \to x$ as $j \to \infty$ (or $\lim_{j\to\infty} x^j = x$), if $\|x^j - x\| \to 0$. The *closure* of D is the set of limits of sequences of points in D, written $\operatorname{cl} D$, and the *boundary* of D is $\operatorname{cl} D \setminus \operatorname{int} D$, written $\operatorname{bd} D$. The set D is *open* if $D = \operatorname{int} D$, and is *closed* if $D = \operatorname{cl} D$. Linear subspaces of $\mathbf{E}$ are important examples of closed sets. Easy exercises show that D is open exactly when its complement D^c is closed, and that arbitrary unions and finite intersections of open sets are open. The interior of D is just the largest open set contained in D, while $\operatorname{cl} D$ is the smallest closed set containing D. Finally, a subset G of D is *open in* D if there is an open set $U \subset \mathbf{E}$ with $G = D \cap U$.

Much of the beauty of convexity comes from *duality* ideas, interweaving geometry and topology. The following result, which we prove a little later, is both typical and fundamental.

Theorem 1.1.1 (Basic separation) *Suppose that the set $C \subset \mathbf{E}$ is closed and convex, and that the point y does not lie in C. Then there exist real b and a nonzero element a of $\mathbf{E}$ satisfying $\langle a, y\rangle > b \geq \langle a, x\rangle$ for all points x in C.*

Sets in $\mathbf{E}$ of the form $\{x \mid \langle a, x\rangle = b\}$ and $\{x \mid \langle a, x\rangle \leq b\}$ (for a nonzero element a of $\mathbf{E}$ and real b) are called *hyperplanes* and *closed halfspaces*, respectively. In this language the above result states that the point y is *separated* from the set C by a hyperplane. In other words, C is contained in a certain closed halfspace whereas y is not. Thus there is a "dual" representation of C as the intersection of all closed halfspaces containing it.

The set D is *bounded* if there is a real k satisfying $kB \supset D$, and it is *compact* if it is closed and bounded. The following result is a central tool in real analysis.

Theorem 1.1.2 (Bolzano–Weierstrass) *Bounded sequences in* $\mathbf{E}$ *have convergent subsequences.*

Just as for sets, geometric and topological ideas also intermingle for the functions we study. Given a set D in $\mathbf{E}$, we call a function $f : D \to \mathbf{R}$ *continuous* (on D) if $f(x^i) \to f(x)$ for any sequence $x^i \to x$ in D. In this case it easy to check, for example, that for any real α the *level set* $\{x \in D \mid f(x) \leq \alpha\}$ is closed providing D is closed.

Given another Euclidean space $\mathbf{Y}$, we call a map $A : \mathbf{E} \to \mathbf{Y}$ *linear* if any points x and z in $\mathbf{E}$ and any reals λ and μ satisfy $A(\lambda x + \mu z) = \lambda Ax + \mu Az$. In fact any linear function from $\mathbf{E}$ to $\mathbf{R}$ has the form $\langle a, \cdot \rangle$ for some element a of $\mathbf{E}$. Linear maps and *affine* functions (linear functions plus constants) are continuous. Thus, for example, closed halfspaces are indeed closed. A *polyhedron* is a finite intersection of closed halfspaces, and is therefore both closed and convex. The *adjoint* of the map A above is the linear map $A^* : \mathbf{Y} \to \mathbf{E}$ defined by the property

$$\langle A^*y, x\rangle = \langle y, Ax\rangle \text{ for all points } x \text{ in } \mathbf{E} \text{ and } y \text{ in } \mathbf{Y}$$

(whence $A^{**} = A$). The *null space* of A is $N(A) = \{x \in \mathbf{E} \mid Ax = 0\}$. The *inverse image* of a set $H \subset \mathbf{Y}$ is the set $A^{-1}H = \{x \in \mathbf{E} \mid Ax \in H\}$ (so for example $N(A) = A^{-1}\{0\}$). Given a subspace G of $\mathbf{E}$, the *orthogonal complement* of G is the subspace

$$G^{\perp} = \{y \in \mathbf{E} \mid \langle x, y\rangle = 0 \text{ for all } x \in G\},$$

so called because we can write $\mathbf{E}$ as a direct sum $G \oplus G^{\perp}$. (In other words, any element of $\mathbf{E}$ can be written uniquely as the sum of an element of G and an element of $G^{\perp}$.) Any subspace G satisfies $G^{\perp\perp} = G$. The range of any linear map A coincides with $N(A^*)^{\perp}$.

Optimization studies properties of minimizers and maximizers of functions. Given a set $\Lambda \subset \mathbf{R}$, the *infimum* of Λ (written $\inf \Lambda$) is the greatest lower bound on Λ, and the *supremum* (written $\sup \Lambda$) is the least upper bound. To ensure these are always defined, it is natural to append $-\infty$ and $+\infty$ to the real numbers, and allow their use in the usual notation for open and closed intervals. Hence, $\inf \emptyset = +\infty$ and $\sup \emptyset = -\infty$, and for example

$(-\infty, +\infty]$ denotes the interval $\mathbf{R} \cup \{+\infty\}$. We try to avoid the appearance of $+\infty - \infty$, but when necessary we use the convention $+\infty - \infty = +\infty$, so that any two sets C and D in $\mathbf{R}$ satisfy $\inf C + \inf D = \inf(C + D)$. We also adopt the conventions $0 \cdot (\pm\infty) = (\pm\infty) \cdot 0 = 0$. A *(global) minimizer* of a function $f : D \to \mathbf{R}$ is a point $\bar{x}$ in D at which f attains its infimum

$$\inf_D f = \inf f(D) = \inf\{f(x) \mid x \in D\}.$$

In this case we refer to $\bar{x}$ as an *optimal solution* of the *optimization problem* $\inf_D f$.

For a positive real δ and a function $g : (0, \delta) \to \mathbf{R}$, we define

$$\liminf_{t \downarrow 0} g(t) = \lim_{t \downarrow 0} \inf_{(0,t)} g$$

and

$$\limsup_{t \downarrow 0} g(t) = \lim_{t \downarrow 0} \sup_{(0,t)} g.$$

The limit $\lim_{t \downarrow 0} g(t)$ exists if and only if the above expressions are equal.

The question of *attainment*, or in other words the *existence* of an optimal solution for an optimization problem is typically topological. The following result is a prototype. The proof is a standard application of the Bolzano–Weierstrass theorem above.

Proposition 1.1.3 (Weierstrass) *Suppose that the set $D \subset \mathbf{E}$ is nonempty and closed, and that all the level sets of the continuous function $f : D \to \mathbf{R}$ are bounded. Then f has a global minimizer.*

Just as for sets, convexity of functions will be crucial for us. Given a convex set $C \subset \mathbf{E}$, we say that the function $f : C \to \mathbf{R}$ is *convex* if

$$f(\lambda x + (1 - \lambda)y) \le \lambda f(x) + (1 - \lambda) f(y)$$

for all points x and y in C and $0 \le \lambda \le 1$. The function f is *strictly convex* if the inequality holds strictly whenever x and y are distinct in C and $0 < \lambda < 1$. It is easy to see that a strictly convex function can have at most one minimizer.

Requiring the function f to have bounded level sets is a "growth condition". Another example is the stronger condition

$$\liminf_{\|x\| \to \infty} \frac{f(x)}{\|x\|} > 0, \tag{1.1.4}$$

where we define

$$\liminf_{\|x\| \to \infty} \frac{f(x)}{\|x\|} = \lim_{r \to +\infty} \inf\left\{ \frac{f(x)}{\|x\|} \;\middle|\; x \in C \cap rB^c \right\}.$$

Surprisingly, for *convex* functions these two growth conditions are equivalent.

Proposition 1.1.5 *For a convex set $C \subset \mathbf{E}$, a convex function $f : C \to \mathbf{R}$ has bounded level sets if and only if it satisfies the growth condition (1.1.4).*

The proof is outlined in Exercise 10.

Exercises and Commentary

Good general references are [177] for elementary real analysis and [1] for linear algebra. Separation theorems for convex sets originate with Minkowski [142]. The theory of the relative interior (Exercises 11, 12, and 13) is developed extensively in [167] (which is also a good reference for the recession cone, Exercise 6).

1. Prove the intersection of an arbitrary collection of convex sets is convex. Deduce that the convex hull of a set $D \subset \mathbf{E}$ is well-defined as the intersection of all convex sets containing D.

2. (a) Prove that if the set $C \subset \mathbf{E}$ is convex and if

$$x^1, x^2, \ldots, x^m \in C, \ 0 \le \lambda_1, \lambda_2, \ldots, \lambda_m \in \mathbf{R},$$

and $\sum \lambda_i = 1$ then $\sum \lambda_i x^i \in C$. Prove, furthermore, that if $f : C \to \mathbf{R}$ is a convex function then $f(\sum \lambda_i x^i) \le \sum \lambda_i f(x^i)$.

(b) We see later (Theorem 3.1.11) that the function $-\log$ is convex on the strictly positive reals. Deduce, for any strictly positive reals $x^1, x^2, \ldots, x^m$, and any nonnegative reals $\lambda_1, \lambda_2, \ldots, \lambda_m$ with sum 1, the *arithmetic-geometric mean inequality*

$$\sum_i \lambda_i x^i \ge \prod_i (x^i)^{\lambda_i}.$$

(c) Prove that for any set $D \subset \mathbf{E}$, $\mathrm{conv}\, D$ is the set of all convex combinations of elements of D.

3. Prove that a convex set $D \subset \mathbf{E}$ has convex closure, and deduce that $\mathrm{cl}\,(\mathrm{conv}\, D)$ is the smallest closed convex set containing D.

4. **(Radstrom cancellation)** Suppose sets $A, B, C \subset \mathbf{E}$ satisfy

$$A + C \subset B + C.$$

(a) If A and B are convex, B is closed, and C is bounded, prove

$$A \subset B.$$

(Hint: Observe $2A + C = A + (A + C) \subset 2B + C$.)

(b) Show this result can fail if B is not convex.

5. * **(Strong separation)** Suppose that the set $C \subset \mathbf{E}$ is closed and convex, and that the set $D \subset \mathbf{E}$ is compact and convex.

   (a) Prove the set $D - C$ is closed and convex.

   (b) Deduce that if in addition D and C are disjoint then there exists a nonzero element a in $\mathbf{E}$ with $\inf_{x \in D} \langle a, x \rangle > \sup_{y \in C} \langle a, y \rangle$. Interpret geometrically.

   (c) Show part (b) fails for the closed convex sets in $\mathbf{R}^2$,

$$\begin{aligned} D &= \{x \mid x_1 > 0,\ x_1 x_2 \geq 1\}, \\ C &= \{x \mid x_2 = 0\}. \end{aligned}$$

6. ** **(Recession cones)** Consider a nonempty closed convex set $C \subset \mathbf{E}$. We define the *recession cone* of C by

$$0^+(C) = \{d \in \mathbf{E} \mid C + \mathbf{R}_+ d \subset C\}.$$

   (a) Prove $0^+(C)$ is a closed convex cone.

   (b) Prove $d \in 0^+(C)$ if and only if $x + \mathbf{R}_+ d \subset C$ for some point x in C. Show this equivalence can fail if C is not closed.

   (c) Consider a family of closed convex sets C_γ $(\gamma \in \Gamma)$ with nonempty intersection. Prove $0^+(\cap C_\gamma) = \cap 0^+(C_\gamma)$.

   (d) For a unit vector u in $\mathbf{E}$, prove $u \in 0^+(C)$ if and only if there is a sequence (x^r) in C satisfying $\|x^r\| \to \infty$ and $\|x^r\|^{-1} x^r \to u$. Deduce C is unbounded if and only if $0^+(C)$ is nontrivial.

   (e) If $\mathbf{Y}$ is a Euclidean space, the map $A : \mathbf{E} \to \mathbf{Y}$ is linear, and $N(A) \cap 0^+(C)$ is a linear subspace, prove AC is closed. Show this result can fail without the last assumption.

   (f) Consider another nonempty closed convex set $D \subset \mathbf{E}$ such that $0^+(C) \cap 0^+(D)$ is a linear subspace. Prove $C - D$ is closed.

7. For any set of vectors $a^1, a^2, \ldots, a^m$ in $\mathbf{E}$, prove the function $f(x) = \max_i \langle a^i, x \rangle$ is convex on $\mathbf{E}$.

8. Prove Proposition 1.1.3 (Weierstrass).

9. **(Composing convex functions)** Suppose that the set $C \subset \mathbf{E}$ is convex and that the functions $f_1, f_2, \ldots, f_n : C \to \mathbf{R}$ are convex, and define a function $f : C \to \mathbf{R}^n$ with components f_i. Suppose further that $f(C)$ is convex and that the function $g : f(C) \to \mathbf{R}$ is convex and *isotone*: any points $y \leq z$ in $f(C)$ satisfy $g(y) \leq g(z)$. Prove the composition $g \circ f$ is convex.

10. * **(Convex growth conditions)**

    (a) Find a function with bounded level sets which does not satisfy the growth condition (1.1.4).

    (b) Prove that any function satisfying (1.1.4) has bounded level sets.

    (c) Suppose the convex function $f : C \to \mathbf{R}$ has bounded level sets but that (1.1.4) fails. Deduce the existence of a sequence (x^m) in C with $f(x^m) \leq \|x^m\|/m \to +\infty$. For a fixed point $\bar{x}$ in C, derive a contradiction by considering the sequence

    $$\bar{x} + \frac{m}{\|x^m\|}(x^m - \bar{x}).$$

    Hence complete the proof of Proposition 1.1.5.

The relative interior

Some arguments about finite-dimensional convex sets C simplify and lose no generality if we assume C contains 0 and spans $\mathbf{E}$. The following exercises outline this idea.

11. ** **(Accessibility lemma)** Suppose C is a convex set in $\mathbf{E}$.

    (a) Prove $\mathrm{cl}\, C \subset C + \epsilon B$ for any real $\epsilon > 0$.

    (b) For sets D and F in $\mathbf{E}$ with D open, prove $D + F$ is open.

    (c) For x in $\mathrm{int}\, C$ and $0 < \lambda \leq 1$, prove $\lambda x + (1 - \lambda)\mathrm{cl}\, C \subset C$. Deduce $\lambda\, \mathrm{int}\, C + (1 - \lambda)\mathrm{cl}\, C \subset \mathrm{int}\, C$.

    (d) Deduce $\mathrm{int}\, C$ is convex.

    (e) Deduce further that if $\mathrm{int}\, C$ is nonempty then $\mathrm{cl}\,(\mathrm{int}\, C) = \mathrm{cl}\, C$. Is convexity necessary?

12. ** **(Affine sets)** A set L in $\mathbf{E}$ is *affine* if the entire line through any distinct points x and y in L lies in L: algebraically, $\lambda x + (1 - \lambda)y \in L$ for *any* real λ. The *affine hull* of a set D in $\mathbf{E}$, denoted $\mathrm{aff}\, D$, is the smallest affine set containing D. An *affine combination* of points $x^1, x^2, \ldots, x^m$ is a point of the form $\sum_1^m \lambda_i x^i$, for reals λ_i summing to one.

    (a) Prove the intersection of an arbitrary collection of affine sets is affine.

    (b) Prove that a set is affine if and only if it is a translate of a linear subspace.

    (c) Prove $\mathrm{aff}\, D$ is the set of all affine combinations of elements of D.

    (d) Prove $\mathrm{cl}\, D \subset \mathrm{aff}\, D$ and deduce $\mathrm{aff}\, D = \mathrm{aff}\,(\mathrm{cl}\, D)$.

(e) For any point x in D, prove $\operatorname{aff} D = x + \operatorname{span}(D - x)$, and deduce the linear subspace $\operatorname{span}(D - x)$ is independent of x.

13. ** **(The relative interior)** (We use Exercises 11 and 12.) The *relative interior* of a convex set C in $\mathbf{E}$, denoted $\operatorname{ri} C$, is its interior relative to its affine hull. In other words, a point x lies in $\operatorname{ri} C$ if there is a real $\delta > 0$ with $(x + \delta B) \cap \operatorname{aff} C \subset C$.

    (a) Find convex sets $C_1 \subset C_2$ with $\operatorname{ri} C_1 \not\subset \operatorname{ri} C_2$.

    (b) Suppose $\dim \mathbf{E} > 0$, $0 \in C$ and $\operatorname{aff} C = \mathbf{E}$. Prove C contains a basis $\{x^1, x^2, \ldots, x^n\}$ of $\mathbf{E}$. Deduce $(1/(n+1)) \sum_1^n x^i \in \operatorname{int} C$. Hence deduce that any nonempty convex set in $\mathbf{E}$ has nonempty relative interior.

    (c) Prove that for $0 < \lambda \leq 1$ we have $\lambda \operatorname{ri} C + (1 - \lambda) \operatorname{cl} C \subset \operatorname{ri} C$, and hence $\operatorname{ri} C$ is convex with $\operatorname{cl}(\operatorname{ri} C) = \operatorname{cl} C$.

    (d) Prove that for a point x in C, the following are equivalent:

        (i) $x \in \operatorname{ri} C$.

        (ii) For any point y in C there exists a real $\epsilon > 0$ with $x + \epsilon(x - y)$ in C.

        (iii) $\mathbf{R}_+(C - x)$ is a linear subspace.

    (e) If $\mathbf{F}$ is another Euclidean space and the map $A : \mathbf{E} \to \mathbf{F}$ is linear, prove $\operatorname{ri} AC \supset A \operatorname{ri} C$.

1.2 Symmetric Matrices

Throughout most of this book our setting is an abstract Euclidean space $\mathbf{E}$. This has a number of advantages over always working in $\mathbf{R}^n$: the basis-independent notation is more elegant and often clearer, and it encourages techniques which extend beyond finite dimensions. But more concretely, identifying $\mathbf{E}$ with $\mathbf{R}^n$ may obscure properties of a space beyond its simple Euclidean structure. As an example, in this short section we describe a Euclidean space which "feels" very different from $\mathbf{R}^n$: the space $\mathbf{S}^n$ of $n \times n$ real symmetric matrices.

The nonnegative orthant $\mathbf{R}^n_+$ is a cone in $\mathbf{R}^n$ which plays a central role in our development. In a variety of contexts the analogous role in $\mathbf{S}^n$ is played by the cone of positive semidefinite matrices, $\mathbf{S}^n_+$. (We call a matrix X in $\mathbf{S}^n$ *positive semidefinite* if $x^T X x \geq 0$ for all vectors x in $\mathbf{R}^n$, and *positive definite* if the inequality is strict whenever x is nonzero.) These two cones have some important differences; in particular, $\mathbf{R}^n_+$ is a polyhedron, whereas the cone of positive semidefinite matrices $\mathbf{S}^n_+$ is not, even for $n = 2$. The cones $\mathbf{R}^n_+$ and $\mathbf{S}^n_+$ are important largely because of the orderings they induce. (The latter is sometimes called the *Loewner ordering*.) For points x and y in $\mathbf{R}^n$ we write $x \leq y$ if $y - x \in \mathbf{R}^n_+$, and $x < y$ if $y - x \in \mathbf{R}^n_{++}$ (with analogous definitions for $\geq$ and $>$). The cone $\mathbf{R}^n_+$ is a *lattice cone*: for any points x and y in $\mathbf{R}^n$ there is a point z satisfying

$$w \geq x \text{ and } w \geq y \Leftrightarrow w \geq z.$$

(The point z is just the componentwise maximum of x and y.) Analogously, for matrices X and Y in $\mathbf{S}^n$ we write $X \preceq Y$ if $Y - X \in \mathbf{S}^n_+$, and $X \prec Y$ if $Y - X$ lies in $\mathbf{S}^n_{++}$, the set of positive definite matrices (with analogous definitions for $\succeq$ and $\succ$). By contrast, it is straightforward to see $\mathbf{S}^n_+$ is *not* a lattice cone (Exercise 4).

We denote the identity matrix by I. The *trace* of a square matrix Z is the sum of the diagonal entries, written $\operatorname{tr} Z$. It has the important property $\operatorname{tr}(VW) = \operatorname{tr}(WV)$ for any matrices V and W for which VW is well-defined and square. We make the vector space $\mathbf{S}^n$ into a Euclidean space by defining the inner product

$$\langle X, Y \rangle = \operatorname{tr}(XY) \quad \text{for } X, Y \in \mathbf{S}^n.$$

Any matrix X in $\mathbf{S}^n$ has n real eigenvalues (counted by multiplicity), which we write in nonincreasing order $\lambda_1(X) \geq \lambda_2(X) \geq \ldots \geq \lambda_n(X)$. In this way we define a function $\lambda : \mathbf{S}^n \to \mathbf{R}^n$. We also define a linear map $\operatorname{Diag} : \mathbf{R}^n \to \mathbf{S}^n$, where for a vector x in $\mathbf{R}^n$, $\operatorname{Diag} x$ is an $n \times n$ diagonal matrix with diagonal entries x_i. This map embeds $\mathbf{R}^n$ as a subspace of $\mathbf{S}^n$ and the cone $\mathbf{R}^n_+$ as a subcone of $\mathbf{S}^n_+$. The determinant of a square matrix Z is written $\det Z$.

We write $\mathbf{O}^n$ for the group of $n \times n$ *orthogonal* matrices (those matrices U satisfying $U^T U = I$). Then any matrix X in $\mathbf{S}^n$ has an *ordered spectral decomposition* $X = U^T(\operatorname{Diag}\lambda(X))U$, for some matrix U in $\mathbf{O}^n$. This shows, for example, that the function λ is *norm-preserving*: $\|X\| = \|\lambda(X)\|$ for all X in $\mathbf{S}^n$. For any X in $\mathbf{S}^n_+$, the spectral decomposition also shows there is a unique matrix $X^{1/2}$ in $\mathbf{S}^n_+$ whose square is X.

The Cauchy–Schwarz inequality has an interesting refinement in $\mathbf{S}^n$ which is crucial for variational properties of eigenvalues, as we shall see.

Theorem 1.2.1 (Fan) *Any matrices X and Y in $\mathbf{S}^n$ satisfy the inequality*

$$\operatorname{tr}(XY) \leq \lambda(X)^T \lambda(Y). \tag{1.2.2}$$

Equality holds if and only if X and Y have a **simultaneous ordered spectral decomposition***: there is a matrix U in $\mathbf{O}^n$ with*

$$X = U^T(\operatorname{Diag}\lambda(X))U \quad \textit{and} \quad Y = U^T(\operatorname{Diag}\lambda(Y))U. \tag{1.2.3}$$

A standard result in linear algebra states that matrices X and Y have a simultaneous (*unordered*) spectral decomposition if and only if they commute. Notice condition (1.2.3) is a stronger property.

The special case of Fan's inequality where both matrices are diagonal gives the following classical inequality. For a vector x in $\mathbf{R}^n$, we denote by $[x]$ the vector with the same components permuted into nonincreasing order. We leave the proof of this result as an exercise.

Proposition 1.2.4 (Hardy–Littlewood–Pólya) *Any vectors x and y in $\mathbf{R}^n$ satisfy the inequality*

$$x^T y \leq [x]^T [y].$$

We describe a proof of Fan's theorem in the exercises, using the above proposition and the following classical relationship between the set $\mathbf{\Gamma}^n$ of *doubly stochastic* matrices (square matrices with all nonnegative entries, and each row and column summing to one) and the set $\mathbf{P}^n$ of *permutation* matrices (square matrices with all entries zero or one, and with exactly one entry of one in each row and in each column).

Theorem 1.2.5 (Birkhoff) *Doubly stochastic matrices are convex combinations of permutation matrices.*

We defer the proof to a later section (Section 4.1, Exercise 22).

Exercises and Commentary

Fan's inequality (1.2.2) appeared in [73], but is closely related to earlier work of von Neumann [184]. The condition for equality is due to [180]. The Hardy–Littlewood–Pólya inequality may be found in [91]. Birkhoff's theorem [15] was in fact proved earlier by König [115].

1. Prove $\mathbf{S}^n_+$ is a closed convex cone with interior $\mathbf{S}^n_{++}$.

2. Explain why $\mathbf{S}^2_+$ is not a polyhedron.

3. **($\mathbf{S}^3_+$ is not strictly convex)** Find nonzero matrices X and Y in $\mathbf{S}^3_+$ such that $\mathbf{R}_+X \neq \mathbf{R}_+Y$ and $(X+Y)/2 \notin \mathbf{S}^3_{++}$.

4. **(A nonlattice ordering)** Suppose the matrix Z in $\mathbf{S}^2$ satisfies
$$W \succeq \begin{bmatrix} 1 & 0 \\ 0 & 0 \end{bmatrix} \text{ and } W \succeq \begin{bmatrix} 0 & 0 \\ 0 & 1 \end{bmatrix} \Leftrightarrow W \succeq Z.$$
   (a) By considering diagonal W, prove
$$Z = \begin{bmatrix} 1 & a \\ a & 1 \end{bmatrix}$$
   for some real a.

   (b) By considering $W = I$, prove $Z = I$.

   (c) Derive a contradiction by considering
$$W = \tfrac{2}{3}\begin{bmatrix} 2 & 1 \\ 1 & 2 \end{bmatrix}.$$

5. **(Order preservation)**

   (a) Prove any matrix X in $\mathbf{S}^n$ satisfies $(X^2)^{1/2} \succeq X$.

   (b) Find matrices $X \succeq Y$ in $\mathbf{S}^2_+$ such that $X^2 \not\succeq Y^2$.

   (c) For matrices $X \succeq Y$ in $\mathbf{S}^n_+$, prove $X^{1/2} \succeq Y^{1/2}$. (Hint: Consider the relationship
$$\langle (X^{1/2}+Y^{1/2})x, (X^{1/2}-Y^{1/2})x\rangle = \langle (X-Y)x, x\rangle \geq 0,$$
   for eigenvectors x of $X^{1/2}-Y^{1/2}$.)

6. * **(Square-root iteration)** Suppose a matrix A in $\mathbf{S}^n_+$ satisfies $I \succeq A$. Prove that the iteration
$$Y_0 = 0, \quad Y_{n+1} = \frac{1}{2}(A + Y_n^2) \quad (n = 0, 1, 2, \ldots)$$
   is nondecreasing (that is, $Y_{n+1} \succeq Y_n$ for all n) and converges to the matrix $I - (I-A)^{1/2}$. (Hint: Consider diagonal matrices A.)

7. **(The Fan and Cauchy–Schwarz inequalities)**

   (a) For any matrices X in $\mathbf{S}^n$ and U in $\mathbf{O}^n$, prove $\|U^T XU\| = \|X\|$.

   (b) Prove the function λ is norm-preserving.

   (c) Explain why Fan's inequality is a refinement of the Cauchy–Schwarz inequality.

8. Prove the inequality $\operatorname{tr} Z + \operatorname{tr} Z^{-1} \geq 2n$ for all matrices Z in $\mathbf{S}^n_{++}$, with equality if and only if $Z = I$.

9. Prove the Hardy–Littlewood–Pólya inequality (Proposition 1.2.4) directly.

10. Given a vector x in $\mathbf{R}^n_+$ satisfying $x_1 x_2 \ldots x_n = 1$, define numbers $y_k = 1/x_1 x_2 \ldots x_k$ for each index $k = 1, 2, \ldots, n$. Prove

$$x_1 + x_2 + \ldots + x_n = \frac{y_n}{y_1} + \frac{y_1}{y_2} + \ldots \frac{y_{n-1}}{y_n}.$$

By applying the Hardy–Littlewood–Pólya inequality (1.2.4) to suitable vectors, prove $x_1 + x_2 + \ldots + x_n \geq n$. Deduce the inequality

$$\frac{1}{n} \sum_1^n z_i \geq \Big(\prod_1^n z_i \Big)^{1/n}$$

for any vector z in $\mathbf{R}^n_+$.

11. For a fixed column vector s in $\mathbf{R}^n$, define a linear map $A : \mathbf{S}^n \to \mathbf{R}^n$ by setting $AX = Xs$ for any matrix X in $\mathbf{S}^n$. Calculate the adjoint map A^*.

12. * **(Fan's inequality)** For vectors x and y in $\mathbf{R}^n$ and a matrix U in $\mathbf{O}^n$, define

$$\alpha = \langle \operatorname{Diag} x, U^T (\operatorname{Diag} y) U \rangle.$$

   (a) Prove $\alpha = x^T Z y$ for some doubly stochastic matrix Z.

   (b) Use Birkhoff's theorem and Proposition 1.2.4 to deduce the inequality $\alpha \leq [x]^T [y]$.

   (c) Deduce Fan's inequality (1.2.2).

13. **(A lower bound)** Use Fan's inequality (1.2.2) for two matrices X and Y in $\mathbf{S}^n$ to prove a *lower* bound for $\operatorname{tr}(XY)$ in terms of $\lambda(X)$ and $\lambda(Y)$.

14. * **(Level sets of perturbed log barriers)**

    (a) For δ in $\mathbf{R}_{++}$, prove the function

    $$t \in \mathbf{R}_{++} \mapsto \delta t - \log t$$

    has compact level sets.

    (b) For c in $\mathbf{R}^n_{++}$, prove the function

    $$x \in \mathbf{R}^n_{++} \mapsto c^T x - \sum_{i=1}^{n} \log x_i$$

    has compact level sets.

    (c) For C in $\mathbf{S}^n_{++}$, prove the function

    $$X \in \mathbf{S}^n_{++} \mapsto \langle C, X \rangle - \log \det X$$

    has compact level sets. (Hint: Use Exercise 13.)

15. * **(Theobald's condition)** Assuming Fan's inequality (1.2.2), complete the proof of Fan's theorem (1.2.1) as follows. Suppose equality holds in Fan's inequality (1.2.2), and choose a spectral decomposition

    $$X + Y = U^T (\mathrm{Diag}\, \lambda(X + Y))U$$

    for some matrix U in $\mathbf{O}^n$.

    (a) Prove $\lambda(X)^T \lambda(X + Y) = \langle U^T (\mathrm{Diag}\, \lambda(X))U, X + Y \rangle$.

    (b) Apply Fan's inequality (1.2.2) to the two inner products

    $$\langle X, X + Y \rangle \quad \text{and} \quad \langle U^T (\mathrm{Diag}\, \lambda(X))U, Y \rangle$$

    to deduce $X = U^T (\mathrm{Diag}\, \lambda(X))U$.

    (c) Deduce Fan's theorem.

16. ** **(Generalizing Theobald's condition [122])** Consider a set of matrices $X^1, X^2, \ldots, X^m$ in $\mathbf{S}^n$ satisfying the conditions

    $$\mathrm{tr}\,(X^i X^j) = \lambda(X^i)^T \lambda(X^j) \quad \text{for all } i \text{ and } j.$$

    Generalize the argument of Exercise 15 to prove the entire set of matrices $\{X^1, X^2, \ldots, X^m\}$ has a simultaneous ordered spectral decomposition.

17. ** **(Singular values and von Neumann's lemma)** Let $\mathbf{M}^n$ denote the vector space of $n \times n$ real matrices. For a matrix A in $\mathbf{M}^n$ we define the *singular values* of A by $\sigma_i(A) = \sqrt{\lambda_i(A^T A)}$ for $i = 1, 2, \ldots, n$, and hence define a map $\sigma : \mathbf{M}^n \to \mathbf{R}^n$. (Notice zero may be a singular value.)

(a) Prove

$$\lambda \begin{bmatrix} 0 & A^T \\ A & 0 \end{bmatrix} = \begin{bmatrix} \sigma(A) \\ [-\sigma(A)] \end{bmatrix}$$

(b) For any other matrix B in $\mathbf{M}^n$, use part (a) and Fan's inequality (1.2.2) to prove

$$\operatorname{tr}(A^T B) \leq \sigma(A)^T \sigma(B).$$

(c) If A lies in $\mathbf{S}^n_+$, prove $\lambda(A) = \sigma(A)$.

(d) By considering matrices of the form $A + \alpha I$ and $B + \beta I$, deduce Fan's inequality from von Neumann's lemma (part (b)).

Chapter 2

Inequality Constraints

2.1 Optimality Conditions

Early in multivariate calculus we learn the significance of differentiability in finding minimizers. In this section we begin our study of the interplay between convexity and differentiability in optimality conditions.

For an initial example, consider the problem of minimizing a function $f : C \to \mathbf{R}$ on a set C in $\mathbf{E}$. We say a point $\bar{x}$ in C is a *local minimizer of f on C* if $f(x) \geq f(\bar{x})$ for all points x in C close to $\bar{x}$. The *directional derivative* of a function f at $\bar{x}$ in a direction $d \in \mathbf{E}$ is

$$f'(\bar{x}; d) = \lim_{t \downarrow 0} \frac{f(\bar{x} + td) - f(\bar{x})}{t},$$

when this limit exists. When the directional derivative $f'(\bar{x}; d)$ is actually linear in d (that is, $f'(\bar{x}; d) = \langle a, d \rangle$ for some element a of $\mathbf{E}$) then we say f is *(Gâteaux) differentiable* at $\bar{x}$, with *(Gâteaux) derivative* $\nabla f(\bar{x}) = a$. If f is differentiable at every point in C then we simply say f is differentiable (on C). An example we use quite extensively is the function $X \in \mathbf{S}^n_{++} \mapsto \log \det X$. An exercise shows this function is differentiable on $\mathbf{S}^n_{++}$ with derivative X^{-1}.

A convex cone which arises frequently in optimization is the *normal cone* to a convex set C at a point $\bar{x} \in C$, written $N_C(\bar{x})$. This is the convex cone of *normal vectors*, vectors d in $\mathbf{E}$ such that $\langle d, x - \bar{x} \rangle \leq 0$ for all points x in C.

Proposition 2.1.1 (First order necessary condition) *Suppose that C is a convex set in $\mathbf{E}$ and that the point $\bar{x}$ is a local minimizer of the function $f : C \to \mathbf{R}$. Then for any point x in C, the directional derivative, if it exists, satisfies $f'(\bar{x}; x - \bar{x}) \geq 0$. In particular, if f is differentiable at $\bar{x}$, then the condition $-\nabla f(\bar{x}) \in N_C(\bar{x})$ holds.*

Proof. If some point x in C satisfies $f'(\bar{x}; x - \bar{x}) < 0$, then all small real $t > 0$ satisfy $f(\bar{x} + t(x - \bar{x})) < f(\bar{x})$, but this contradicts the local minimality of $\bar{x}$. □

The case of this result where C is an open set is the canonical introduction to the use of calculus in optimization: local minimizers $\bar{x}$ must be *critical points* (that is, $\nabla f(\bar{x}) = 0$). This book is largely devoted to the study of first order necessary optimality conditions for a local minimizer of a function subject to constraints. In that case local minimizers $\bar{x}$ may not lie in the interior of the set C of interest, so the normal cone $N_C(\bar{x})$ is not simply $\{0\}$.

The next result shows that when f is convex the first order condition above is *sufficient* for $\bar{x}$ to be a global minimizer of f on C.

Proposition 2.1.2 (First order sufficient condition) *Suppose that the set $C \subset \mathbf{E}$ is convex and that the function $f : C \to \mathbf{R}$ is convex. Then for any points $\bar{x}$ and x in C, the directional derivative $f'(\bar{x}; x - \bar{x})$ exists in $[-\infty, +\infty)$. If the condition $f'(\bar{x}; x - \bar{x}) \geq 0$ holds for all x in C, or in particular if the condition $-\nabla f(\bar{x}) \in N_C(\bar{x})$ holds, then $\bar{x}$ is a global minimizer of f on C.*

Proof. A straightforward exercise using the convexity of f shows the function

$$t \in (0, 1] \mapsto \frac{f(\bar{x} + t(x - \bar{x})) - f(\bar{x})}{t}$$

is nondecreasing. The result then follows easily (Exercise 7). □

In particular, any critical point of a convex function is a global minimizer.

The following useful result illustrates what the first order conditions become for a more concrete optimization problem. The proof is outlined in Exercise 4.

Corollary 2.1.3 (First order conditions for linear constraints) *For a convex set $C \subset \mathbf{E}$, a function $f : C \to \mathbf{R}$, a linear map $A : \mathbf{E} \to \mathbf{Y}$ (where $\mathbf{Y}$ is a Euclidean space) and a point b in $\mathbf{Y}$, consider the optimization problem*

$$\inf\{f(x) \mid x \in C, \ Ax = b\}. \tag{2.1.4}$$

Suppose the point $\bar{x} \in \operatorname{int} C$ satisfies $A\bar{x} = b$.

(a) *If $\bar{x}$ is a local minimizer for the problem (2.1.4) and f is differentiable at $\bar{x}$ then $\nabla f(\bar{x}) \in A^*\mathbf{Y}$.*

(b) *Conversely, if $\nabla f(\bar{x}) \in A^*\mathbf{Y}$ and f is convex then $\bar{x}$ is a global minimizer for (2.1.4).*

The element $y \in \mathbf{Y}$ satisfying $\nabla f(\bar{x}) = A^*y$ in the above result is called a *Lagrange multiplier*. This kind of construction recurs in many different forms in our development.

In the absence of convexity, we need second order information to tell us more about minimizers. The following elementary result from multivariate calculus is typical.

Theorem 2.1.5 (Second order conditions) *Suppose the twice continuously differentiable function $f : \mathbf{R}^n \to \mathbf{R}$ has a critical point $\bar{x}$. If $\bar{x}$ is a local minimizer then the Hessian $\nabla^2 f(\bar{x})$ is positive semidefinite. Conversely, if the Hessian is positive definite then $\bar{x}$ is a local minimizer.*

(In fact for $\bar{x}$ to be a local minimizer it is sufficient for the Hessian to be positive semidefinite locally; the function $x \in \mathbf{R} \mapsto x^4$ highlights the distinction.)

To illustrate the effect of constraints on second order conditions, consider the framework of Corollary 2.1.3 (First order conditions for linear constraints) in the case $\mathbf{E} = \mathbf{R}^n$, and suppose $\nabla f(\bar{x}) \in A^*\mathbf{Y}$ and f is twice continuously differentiable near $\bar{x}$. If $\bar{x}$ is a local minimizer then $y^T \nabla^2 f(\bar{x}) y \geq 0$ for all vectors y in $N(A)$. Conversely, if $y^T \nabla^2 f(\bar{x}) y > 0$ for all nonzero y in $N(A)$ then $\bar{x}$ is a local minimizer.

We are already beginning to see the broad interplay between analytic, geometric and topological ideas in optimization theory. A good illustration is the separation result of Section 1.1, which we now prove.

Theorem 2.1.6 (Basic separation) *Suppose that the set $C \subset \mathbf{E}$ is closed and convex, and that the point y does not lie in C. Then there exist a real b and a nonzero element a of $\mathbf{E}$ such that $\langle a, y \rangle > b \geq \langle a, x \rangle$ for all points x in C.*

Proof. We may assume C is nonempty, and define a function $f : \mathbf{E} \to \mathbf{R}$ by $f(x) = \|x - y\|^2/2$. Now by the Weierstrass proposition (1.1.3) there exists a minimizer $\bar{x}$ for f on C, which by the First order necessary condition (2.1.1) satisfies $-\nabla f(\bar{x}) = y - \bar{x} \in N_C(\bar{x})$. Thus $\langle y - \bar{x}, x - \bar{x} \rangle \leq 0$ holds for all points x in C. Now setting $a = y - \bar{x}$ and $b = \langle y - \bar{x}, \bar{x} \rangle$ gives the result. □

We end this section with a rather less standard result, illustrating another idea which is important later, the use of "variational principles" to treat problems where minimizers may not exist, but which nonetheless have "approximate" critical points. This result is a precursor of a principle due to Ekeland, which we develop in Section 7.1.

Proposition 2.1.7 *If the function $f : \mathbf{E} \to \mathbf{R}$ is differentiable and bounded below then there are points where f has small derivative.*

Proof. Fix any real $\epsilon > 0$. The function $f + \epsilon\|\cdot\|$ has bounded level sets, so has a global minimizer x^ϵ by the Weierstrass proposition (1.1.3). If the vector $d = \nabla f(x^\epsilon)$ satisfies $\|d\| > \epsilon$ then, from the inequality

$$\lim_{t \downarrow 0} \frac{f(x^\epsilon - td) - f(x^\epsilon)}{t} = -\langle \nabla f(x^\epsilon), d \rangle = -\|d\|^2 < -\epsilon\|d\|,$$

we would have for small $t > 0$ the contradiction

$$\begin{aligned} -t\epsilon\|d\| &> f(x^\epsilon - td) - f(x^\epsilon) \\ &= (f(x^\epsilon - td) + \epsilon\|x^\epsilon - td\|) \\ &\quad - (f(x^\epsilon) + \epsilon\|x^\epsilon\|) + \epsilon(\|x^\epsilon\| - \|x^\epsilon - td\|) \\ &\geq -\epsilon t\|d\| \end{aligned}$$

by definition of x^ϵ and the triangle inequality. Hence $\|\nabla f(x^\epsilon)\| \leq \epsilon$. □

Notice that the proof relies on consideration of a *nondifferentiable* function, even though the result concerns derivatives.

Exercises and Commentary

The optimality conditions in this section are very standard (see for example [132]). The simple variational principle (Proposition 2.1.7) was suggested by [95].

1. Prove the normal cone is a closed convex cone.

2. **(Examples of normal cones)** For the following sets $C \subset \mathbf{E}$, check C is convex and compute the normal cone $N_C(\bar{x})$ for points $\bar{x}$ in C:

   (a) C a closed interval in $\mathbf{R}$.

   (b) $C = B$, the unit ball.

   (c) C a subspace.

   (d) C a closed halfspace: $\{x \mid \langle a, x \rangle \leq b\}$ where $0 \neq a \in \mathbf{E}$ and $b \in \mathbf{R}$.

   (e) $C = \{x \in \mathbf{R}^n \mid x_j \geq 0 \text{ for all } j \in J\}$ (for $J \subset \{1, 2, \ldots, n\}$).

3. **(Self-dual cones)** Prove each of the following cones K satisfy the relationship $N_K(0) = -K$.

   (a) $\mathbf{R}^n_+$

   (b) $\mathbf{S}^n_+$

   (c) $\{x \in \mathbf{R}^n \mid x_1 \geq 0,\ x_1^2 \geq x_2^2 + x_3^2 + \cdots + x_n^2\}$

4. **(Normals to affine sets)** Given a linear map $A : \mathbf{E} \to \mathbf{Y}$ (where $\mathbf{Y}$ is a Euclidean space) and a point b in $\mathbf{Y}$, prove the normal cone to the set $\{x \in \mathbf{E} \mid Ax = b\}$ at any point in it is A^*Y. Hence deduce Corollary 2.1.3 (First order conditions for linear constraints).

5. Prove that the differentiable function $x_1^2 + x_2^2(1 - x_1)^3$ has a unique critical point in $\mathbf{R}^2$, which is a local minimizer, but has no global minimizer. Can this happen on $\mathbf{R}$?

6. **(The Rayleigh quotient)**

   (a) Let the function $f : \mathbf{R}^n \setminus \{0\} \to \mathbf{R}$ be continuous, satisfying $f(\lambda x) = f(x)$ for all $\lambda > 0$ in $\mathbf{R}$ and nonzero x in $\mathbf{R}^n$. Prove f has a minimizer.

   (b) Given a matrix A in $\mathbf{S}^n$, define a function $g(x) = x^T Ax/\|x\|^2$ for nonzero x in $\mathbf{R}^n$. Prove g has a minimizer.

   (c) Calculate $\nabla g(x)$ for nonzero x.

   (d) Deduce that minimizers of g must be eigenvectors, and calculate the minimum value.

   (e) Find an alternative proof of part (d) by using a spectral decomposition of A.

   (Another approach to this problem is given in Section 7.2, Exercise 6.)

7. Suppose a convex function $g : [0, 1] \to \mathbf{R}$ satisfies $g(0) = 0$. Prove the function $t \in (0, 1] \mapsto g(t)/t$ is nondecreasing. Hence prove that for a convex function $f : C \to \mathbf{R}$ and points $\bar{x}, x \in C \subset \mathbf{E}$, the quotient $(f(\bar{x} + t(x - \bar{x})) - f(\bar{x}))/t$ is nondecreasing as a function of t in $(0, 1]$, and complete the proof of Proposition 2.1.2.

8. * **(Nearest points)**

   (a) Prove that if a function $f : C \to \mathbf{R}$ is strictly convex then it has at most one global minimizer on C.

   (b) Prove the function $f(x) = \|x - y\|^2/2$ is strictly convex on $\mathbf{E}$ for any point y in $\mathbf{E}$.

   (c) Suppose C is a nonempty, closed convex subset of $\mathbf{E}$.

      (i) If y is any point in $\mathbf{E}$, prove there is a unique nearest point (or *best approximation*) $P_C(y)$ to y in C, characterized by

      $$\langle y - P_C(y), x - P_C(y) \rangle \leq 0 \quad \text{for all } x \in C.$$

      (ii) For any point $\bar{x}$ in C, deduce that $d \in N_C(\bar{x})$ holds if and only if $\bar{x}$ is the nearest point in C to $\bar{x} + d$.

(iii) Deduce, furthermore, that any points y and z in $\mathbf{E}$ satisfy

$$\|P_C(y) - P_C(z)\| \leq \|y - z\|,$$

so in particular the *projection* $P_C : \mathbf{E} \to C$ is continuous.

(d) Given a nonzero element a of $\mathbf{E}$, calculate the nearest point in the subspace $\{x \in \mathbf{E} \mid \langle a, x \rangle = 0\}$ to the point $y \in \mathbf{E}$.

(e) **(Projection on $\mathbf{R}^n_+$ and $\mathbf{S}^n_+$)** Prove the nearest point in $\mathbf{R}^n_+$ to a vector y in $\mathbf{R}^n$ is y^+, where $y_i^+ = \max\{y_i, 0\}$ for each i. For a matrix U in $\mathbf{O}^n$ and a vector y in $\mathbf{R}^n$, prove that the nearest positive semidefinite matrix to $U^T \mathrm{Diag}\, y U$ is $U^T \mathrm{Diag}\, y^+ U$.

9. * **(Coercivity)** Suppose that the function $f : \mathbf{E} \to \mathbf{R}$ is differentiable and satisfies the growth condition $\lim_{\|x\| \to \infty} f(x)/\|x\| = +\infty$. Prove that the gradient map ∇f has range $\mathbf{E}$. (Hint: Minimize the function $f(\cdot) - \langle a, \cdot \rangle$ for elements a of $\mathbf{E}$.)

10. (a) Prove the function $f : \mathbf{S}^n_{++} \to \mathbf{R}$ defined by $f(X) = \mathrm{tr}\, X^{-1}$ is differentiable on $\mathbf{S}^n_{++}$. (Hint: Expand the expression $(X + tY)^{-1}$ as a power series.)

    (b) Define a function $f : \mathbf{S}^n_{++} \to \mathbf{R}$ by $f(X) = \log \det X$. Prove $\nabla f(I) = I$. Deduce $\nabla f(X) = X^{-1}$ for any X in $\mathbf{S}^n_{++}$.

11. ** **(Kirchhoff's law [9, Chapter 1])** Consider a finite, undirected, connected graph with vertex set V and edge set E. Suppose that α and β in V are distinct vertices and that each edge ij in E has an associated "resistance" $r_{ij} > 0$ in $\mathbf{R}$. We consider the effect of applying a unit "potential difference" between the vertices α and β. Let $V_0 = V \setminus \{\alpha, \beta\}$, and for "potentials" x in $\mathbf{R}^{V_0}$ we define the "power" $p : \mathbf{R}^{V_0} \to \mathbf{R}$ by

$$p(x) = \sum_{ij \in E} \frac{(x_i - x_j)^2}{2r_{ij}},$$

where we set $x_\alpha = 0$ and $x_\beta = 1$.

(a) Prove the power function p has compact level sets.

(b) Deduce the existence of a solution to the following equations (describing "conservation of current"):

$$\begin{aligned} \sum_{j \,:\, ij \in E} \frac{x_i - x_j}{r_{ij}} &= 0 \text{ for } i \text{ in } V_0 \\ x_\alpha &= 0 \\ x_\beta &= 1. \end{aligned}$$

(c) Prove the power function p is strictly convex.

(d) Use part (a) of Exercise 8 to show that the conservation of current equations in part (b) have a unique solution.

12. ** **(Matrix completion [86])** For a set $\Delta \subset \{(i,j) \,|\, 1 \le i \le j \le n\}$, suppose the subspace $L \subset \mathbf{S}^n$ of matrices with (i,j)th entry of zero for all (i,j) in Δ satisfies $L \cap \mathbf{S}^n_{++} \neq \emptyset$. By considering the problem (for $C \in \mathbf{S}^n_{++}$)

$$\inf\{\langle C, X\rangle - \log\det X \mid X \in L \cap \mathbf{S}^n_{++}\},$$

use Section 1.2, Exercise 14 and Corollary 2.1.3 (First order conditions for linear constraints) to prove there exists a matrix X in $L \cap \mathbf{S}^n_{++}$ with $C - X^{-1}$ having (i,j)th entry of zero for all (i,j) not in Δ.

13. ** **(BFGS update, cf. [80])** Given a matrix C in $\mathbf{S}^n_{++}$ and vectors s and y in $\mathbf{R}^n$ satisfying $s^T y > 0$, consider the problem

$$\inf\{\langle C, X\rangle - \log\det X \mid Xs = y,\ X \in \mathbf{S}^n_{++}\}.$$

(a) Prove that for the problem above, the point

$$X = \frac{(y - \delta s)(y - \delta s)^T}{s^T(y - \delta s)} + \delta I$$

is feasible for small $\delta > 0$.

(b) Prove the problem has an optimal solution using Section 1.2, Exercise 14.

(c) Use Corollary 2.1.3 (First order conditions for linear constraints) to find the solution. (The solution is called the *BFGS update* of C^{-1} under the *secant condition* $Xs = y$.)

(See also [61, p. 205].)

14. ** Suppose intervals $I_1, I_2, \ldots, I_n \subset \mathbf{R}$ are nonempty and closed and the function $f : I_1 \times I_2 \times \ldots \times I_n \to \mathbf{R}$ is differentiable and bounded below. Use the idea of the proof of Proposition 2.1.7 to prove that for any $\epsilon > 0$ there exists a point $x^\epsilon \in I_1 \times I_2 \times \ldots \times I_n$ satisfying

$$(-\nabla f(x^\epsilon))_j \in N_{I_j}(x^\epsilon_j) + [-\epsilon, \epsilon] \quad (j = 1, 2, \ldots, n).$$

15. * **(Nearest polynomial with a given root)** Consider the Euclidean space of complex polynomials of degree no more than n, with inner product

$$\Big\langle \sum_{j=0}^n x_j z^j, \sum_{j=0}^n y_j z^j \Big\rangle = \sum_{j=0}^n \overline{x_j} y_j.$$

Given a polynomial p in this space, calculate the nearest polynomial with a given complex root α, and prove the distance to this polynomial is $(\sum_{j=0}^{n} |\alpha|^{2j})^{(-1/2)}|p(\alpha)|$.

2.2 Theorems of the Alternative

One well-trodden route to the study of first order conditions uses a class of results called "theorems of the alternative", and, in particular, the Farkas lemma (which we derive at the end of this section). Our first approach, however, relies on a different theorem of the alternative.

Theorem 2.2.1 (Gordan) *For any elements* $a^0, a^1, \ldots, a^m$ *of* $\mathbf{E}$, *exactly one of the following systems has a solution:*

$$\sum_{i=0}^{m} \lambda_i a^i = 0, \quad \sum_{i=0}^{m} \lambda_i = 1, \quad 0 \le \lambda_0, \lambda_1, \ldots, \lambda_m \in \mathbf{R} \tag{2.2.2}$$

$$\langle a^i, x \rangle < 0 \quad \text{for } i = 0, 1, \ldots, m, \quad x \in \mathbf{E}. \tag{2.2.3}$$

Geometrically, Gordan's theorem says that the origin does not lie in the convex hull of the set $\{a^0, a^1, \ldots, a^m\}$ if and only if there is an open halfspace $\{y \mid \langle y, x \rangle < 0\}$ containing $\{a^0, a^1, \ldots, a^m\}$ (and hence its convex hull). This is another illustration of the idea of separation (in this case we separate the origin and the convex hull).

Theorems of the alternative like Gordan's theorem may be proved in a variety of ways, including separation and algorithmic approaches. We employ a less standard technique using our earlier analytic ideas and leading to a rather unified treatment. It relies on the relationship between the optimization problem

$$\inf\{f(x) \mid x \in \mathbf{E}\}, \tag{2.2.4}$$

where the function f is defined by

$$f(x) = \log\left(\sum_{i=0}^{m} \exp\langle a^i, x \rangle\right), \tag{2.2.5}$$

and the two systems (2.2.2) and (2.2.3). We return to the surprising function (2.2.5) when we discuss conjugacy in Section 3.3.

Theorem 2.2.6 *The following statements are equivalent:*

(*i*) *The function defined by (2.2.5) is bounded below.*

(*ii*) *System (2.2.2) is solvable.*

(*iii*) *System (2.2.3) is unsolvable.*

Proof. The implications (ii) $\Rightarrow$ (iii) $\Rightarrow$ (i) are easy exercises, so it remains to show (i) $\Rightarrow$ (ii). To see this we apply Proposition 2.1.7. We deduce that for each $k = 1, 2, \ldots$, there is a point x^k in $\mathbf{E}$ satisfying

$$\|\nabla f(x^k)\| = \left\|\sum_{i=0}^{m} \lambda_i^k a^i\right\| < \frac{1}{k},$$

where the scalars

$$\lambda_i^k = \frac{\exp\langle a^i, x^k\rangle}{\sum_{r=0}^m \exp\langle a^r, x^k\rangle} > 0$$

satisfy $\sum_{i=0}^m \lambda_i^k = 1$. Now the limit λ of any convergent subsequence of the bounded sequence (λ^k) solves system (2.2.2). □

The equivalence of (ii) and (iii) gives Gordan's theorem.

We now proceed by using Gordan's theorem to derive the Farkas lemma, one of the cornerstones of many approaches to optimality conditions. The proof uses the idea of the *projection* onto a linear subspace $\mathbf{Y}$ of $\mathbf{E}$. Notice first that $\mathbf{Y}$ becomes a Euclidean space by equipping it with the same inner product. The projection of a point x in $\mathbf{E}$ onto $\mathbf{Y}$, written $P_{\mathbf{Y}}x$, is simply the nearest point to x in $\mathbf{Y}$. This is well-defined (see Exercise 8 in Section 2.1), and is characterized by the fact that $x - P_{\mathbf{Y}}x$ is orthogonal to $\mathbf{Y}$. A standard exercise shows $P_{\mathbf{Y}}$ is a linear map.

Lemma 2.2.7 (Farkas) *For any points $a^1, a^2, \ldots, a^m$ and c in $\mathbf{E}$, exactly one of the following systems has a solution:*

$$\sum_{i=1}^m \mu_i a^i = c, \quad 0 \le \mu_1, \mu_2, \ldots, \mu_m \in \mathbf{R} \tag{2.2.8}$$

$$\langle a^i, x\rangle \le 0 \;\; \text{for} \;\; i = 1, 2, \ldots, m, \quad \langle c, x\rangle > 0, \quad x \in \mathbf{E}. \tag{2.2.9}$$

Proof. Again, it is immediate that if system (2.2.8) has a solution then system (2.2.9) has no solution. Conversely, we assume (2.2.9) has no solution and deduce that (2.2.8) has a solution by using induction on the number of elements m. The result is clear for $m = 0$.

Suppose then that the result holds in any Euclidean space and for any set of $m - 1$ elements and any element c. Define $a^0 = -c$. Applying Gordan's theorem (2.2.1) to the unsolvability of (2.2.9) shows there are scalars $\lambda_0, \lambda_1, \ldots, \lambda_m \ge 0$ in $\mathbf{R}$, not all zero, satisfying $\lambda_0 c = \sum_1^m \lambda_i a^i$. If $\lambda_0 > 0$ the proof is complete, so suppose $\lambda_0 = 0$ and without loss of generality $\lambda_m > 0$.

Define a subspace of $\mathbf{E}$ by $\mathbf{Y} = \{y \mid \langle a^m, y\rangle = 0\}$, so by assumption the system

$$\langle a^i, y\rangle \le 0 \;\; \text{for } i = 1, 2, \ldots, m-1, \quad \langle c, y\rangle > 0, \quad y \in \mathbf{Y},$$

or equivalently

$$\langle P_{\mathbf{Y}}a^i, y\rangle \le 0 \;\; \text{for } i = 1, 2, \ldots, m-1, \quad \langle P_{\mathbf{Y}}c, y\rangle > 0, \quad y \in \mathbf{Y},$$

has no solution.

By the induction hypothesis applied to the subspace $\mathbf{Y}$, there are nonnegative reals $\mu_1, \mu_2, \ldots, \mu_{m-1}$ satisfying $\sum_{i=1}^{m-1} \mu_i P_{\mathbf{Y}} a^i = P_{\mathbf{Y}} c$, so the vector $c - \sum_1^{m-1} \mu_i a^i$ is orthogonal to the subspace $\mathbf{Y} = (\mathrm{span}\,(a^m))^\perp$. Thus some real μ_m satisfies

$$\mu_m a^m = c - \sum_1^{m-1} \mu_i a^i. \tag{2.2.10}$$

If μ_m is nonnegative we immediately obtain a solution of (2.2.8), and if not then we can substitute $a^m = -\lambda_m^{-1} \sum_1^{m-1} \lambda_i a^i$ in equation (2.2.10) to obtain a solution. □

Just like Gordan's theorem, the Farkas lemma has an important geometric interpretation which gives an alternative approach to its proof (Exercise 6): any point c not lying in the *finitely generated cone*

$$C = \Big\{ \sum_1^m \mu_i a^i \,\Big|\, 0 \le \mu_1, \mu_2, \ldots, \mu_m \in \mathbf{R} \Big\} \tag{2.2.11}$$

can be separated from C by a hyperplane. If x solves system (2.2.9) then C is contained in the closed halfspace $\{a \mid \langle a, x \rangle \le 0\}$, whereas c is contained in the complementary open halfspace. In particular, it follows that any finitely generated cone is closed.

Exercises and Commentary

Gordan's theorem appeared in [84], and the Farkas lemma appeared in [75]. The standard modern approach to theorems of the alternative (Exercises 7 and 8, for example) is via linear programming duality (see, for example, [53]). The approach we take to Gordan's theorem was suggested by Hiriart–Urruty [95]. Schur-convexity (Exercise 9) is discussed extensively in [134].

1. Prove the implications (ii) ⇒ (iii) ⇒ (i) in Theorem 2.2.6.
2. (a) Prove the orthogonal projection $P_{\mathbf{Y}} : \mathbf{E} \to \mathbf{Y}$ is a linear map.
   (b) Give a direct proof of the Farkas lemma for the case $m = 1$.
3. Use the Basic separation theorem (2.1.6) to give another proof of Gordan's theorem.
4. * Deduce Gordan's theorem from the Farkas lemma. (Hint: Consider the elements $(a^i, 1)$ of the space $\mathbf{E} \times \mathbf{R}$.)
5. * **(Carathéodory's theorem [52])** Suppose $\{a^i \mid i \in I\}$ is a finite set of points in $\mathbf{E}$. For any subset J of I, define the cone

$$C_J = \Big\{ \sum_{i \in J} \mu_i a^i \,\Big|\, 0 \le \mu_i \in \mathbf{R},\ (i \in J) \Big\}.$$

(a) Prove the cone C_I is the union of those cones C_J for which the set $\{a^i \mid i \in J\}$ is linearly independent. Furthermore, prove directly that any such cone C_J is closed.

(b) Deduce that any finitely generated cone is closed.

(c) If the point x lies in $\operatorname{conv}\{a^i \mid i \in I\}$, prove that in fact there is a subset $J \subset I$ of size at most $1 + \dim \mathbf{E}$ such that x lies in $\operatorname{conv}\{a^i \mid i \in J\}$. (Hint: Apply part (a) to the vectors $(a^i, 1)$ in $\mathbf{E} \times \mathbf{R}$.)

(d) Use part (c) to prove that if a subset of $\mathbf{E}$ is compact then so is its convex hull.

6. * Give another proof of the Farkas lemma by applying the Basic separation theorem (2.1.6) to the set defined by (2.2.11) and using the fact that any finitely generated cone is closed.

7. ** **(Ville's theorem)** With the function f defined by (2.2.5) (with $\mathbf{E} = \mathbf{R}^n$), consider the optimization problem

$$\inf\{f(x) \mid x \geq 0\} \tag{2.2.12}$$

and its relationship with the two systems

$$\sum_{i=0}^{m} \lambda_i a^i \geq 0, \quad \sum_{i=0}^{m} \lambda_i = 1,$$
$$0 \leq \lambda_0, \lambda_1, \ldots, \lambda_m \in \mathbf{R} \tag{2.2.13}$$

and

$$\langle a^i, x \rangle < 0 \ \text{ for } i = 0, 1, \ldots, m, \quad x \in \mathbf{R}^n_+. \tag{2.2.14}$$

Imitate the proof of Gordan's theorem (using Section 2.1, Exercise 14) to prove the following are equivalent:

(i) Problem (2.2.12) is bounded below.

(ii) System (2.2.13) is solvable.

(iii) System (2.2.14) is unsolvable.

Generalize by considering the problem $\inf\{f(x) \mid x_j \geq 0 \ (j \in J)\}$.

8. ** **(Stiemke's theorem)** Consider the optimization problem (2.2.4) and its relationship with the two systems

$$\sum_{i=0}^{m} \lambda_i a^i = 0, \quad 0 < \lambda_0, \lambda_1, \ldots, \lambda_m \in \mathbf{R} \tag{2.2.15}$$

and

$$\langle a^i, x \rangle \leq 0 \ \text{ for } i = 0, 1, \ldots, m, \ \text{ not all } 0, \quad x \in \mathbf{E}. \tag{2.2.16}$$

Prove the following are equivalent:

(i) Problem (2.2.4) has an optimal solution.
(ii) System (2.2.15) is solvable.
(iii) System (2.2.16) is unsolvable.

Hint: Complete the following steps.

(a) Prove (i) implies (ii) by Proposition 2.1.1.
(b) Prove (ii) implies (iii).
(c) If problem (2.2.4) has no optimal solution, prove that neither does the problem

$$\inf\Big\{\sum_{i=0}^{m} \exp y_i \,\Big|\, y \in K\Big\}, \tag{2.2.17}$$

where K is the subspace $\{(\langle a^i, x\rangle)_{i=0}^m \mid x \in \mathbf{E}\} \subset \mathbf{R}^{m+1}$. Hence, by considering a minimizing sequence for (2.2.17), deduce system (2.2.16) is solvable.

Generalize by considering the problem $\inf\{f(x) \mid x_j \geq 0 \ (j \in J)\}$.

9. ** **(Schur-convexity)** The *dual cone* of the cone $\mathbf{R}^n_\geq$ is defined by

$$\big(\mathbf{R}^n_\geq\big)^+ = \{y \in \mathbf{R}^n \mid \langle x, y\rangle \geq 0 \text{ for all } x \text{ in } \mathbf{R}^n_\geq\}.$$

(a) Prove a vector y lies in $(\mathbf{R}^n_\geq)^+$ if and only if

$$\sum_1^j y_i \geq 0 \ \text{ for } j = 1, 2, \ldots, n-1, \ \ \sum_1^n y_i = 0.$$

(b) By writing $\sum_1^j [x]_i = \max_k \langle a^k, x\rangle$ for some suitable set of vectors a^k, prove that the function $x \mapsto \sum_1^j [x]_i$ is convex. (Hint: Use Section 1.1, Exercise 7.)
(c) Deduce that the function $x \mapsto [x]$ is $(\mathbf{R}^n_\geq)^+$*-convex*, that is:

$$\lambda[x] + (1-\lambda)[y] - [\lambda x + (1-\lambda)y] \in \big(\mathbf{R}^n_\geq\big)^+ \ \text{ for } 0 \leq \lambda \leq 1.$$

(d) Use Gordan's theorem and Proposition 1.2.4 to deduce that for any x and y in $\mathbf{R}^n_\geq$, if $y - x$ lies in $(\mathbf{R}^n_\geq)^+$ then x lies in $\operatorname{conv}(\mathbf{P}^n y)$.
(e) A function $f : \mathbf{R}^n_\geq \to \mathbf{R}$ is *Schur-convex* if

$$x, y \in \mathbf{R}^n_\geq, \ y - x \in \big(\mathbf{R}^n_\geq\big)^+ \ \Rightarrow \ f(x) \leq f(y).$$

Prove that if f is convex, then it is Schur-convex if and only if it is the restriction to $\mathbf{R}^n_\geq$ of a *symmetric* convex function $g : \mathbf{R}^n \to \mathbf{R}$ (where by symmetric we mean $g(x) = g(\Pi x)$ for any x in $\mathbf{R}^n$ and any permutation matrix Π).

2.3 Max-functions

This section is an elementary exposition of the first order necessary conditions for a local minimizer of a differentiable function subject to differentiable inequality constraints. Throughout this section we use the term "differentiable" in the Gâteaux sense, defined in Section 2.1. Our approach, which relies on considering the local minimizers of a *max-function*

$$g(x) = \max_{i=0,1,\ldots,m} \{g_i(x)\}, \tag{2.3.1}$$

illustrates a pervasive analytic idea in optimization: *nonsmoothness*. Even if the functions $g_0, g_1, \ldots, g_m$ are smooth, g may not be, and hence the gradient may no longer be a useful notion.

Proposition 2.3.2 (Directional derivatives of max-functions) *Let $\bar{x}$ be a point in the interior of a set $C \subset \mathbf{E}$. Suppose that continuous functions $g_0, g_1, \ldots, g_m : C \to \mathbf{R}$ are differentiable at $\bar{x}$, that g is the max-function (2.3.1), and define the index set $K = \{i \mid g_i(\bar{x}) = g(\bar{x})\}$. Then for all directions d in $\mathbf{E}$, the directional derivative of g is given by*

$$g'(\bar{x}; d) = \max_{i \in K} \{\langle \nabla g_i(\bar{x}), d \rangle\}. \tag{2.3.3}$$

Proof. By continuity we can assume, without loss of generality, $K = \{0, 1, \ldots, m\}$; those g_i not attaining the maximum in (2.3.1) will not affect $g'(\bar{x}; d)$. Now for each i, we have the inequality

$$\liminf_{t \downarrow 0} \frac{g(\bar{x} + td) - g(\bar{x})}{t} \geq \lim_{t \downarrow 0} \frac{g_i(\bar{x} + td) - g_i(\bar{x})}{t} = \langle \nabla g_i(\bar{x}), d \rangle.$$

Suppose

$$\limsup_{t \downarrow 0} \frac{g(\bar{x} + td) - g(\bar{x})}{t} > \max_i \{\langle \nabla g_i(\bar{x}), d \rangle\}.$$

Then some real sequence $t_k \downarrow 0$ and real $\epsilon > 0$ satisfy

$$\frac{g(\bar{x} + t_k d) - g(\bar{x})}{t_k} \geq \max_i \{\langle \nabla g_i(\bar{x}), d \rangle\} + \epsilon \text{ for all } k \in \mathbf{N}$$

(where $\mathbf{N}$ denotes the sequence of natural numbers). We can now choose a subsequence R of $\mathbf{N}$ and a fixed index j so that all integers k in R satisfy $g(\bar{x} + t_k d) = g_j(\bar{x} + t_k d)$. In the limit we obtain the contradiction

$$\langle \nabla g_j(\bar{x}), d \rangle \geq \max_i \{\langle \nabla g_i(\bar{x}), d \rangle\} + \epsilon.$$

Hence

$$\limsup_{t \downarrow 0} \frac{g(\bar{x} + td) - g(\bar{x})}{t} \leq \max_i \{\langle \nabla g_i(\bar{x}), d \rangle\},$$

and the result follows. □

For most of this book we consider optimization problems of the form

$$\left.\begin{array}{lrcl}\inf & f(x) & & \\ \text{subject to} & g_i(x) & \leq & 0 \text{ for } i \in I \\ & h_j(x) & = & 0 \text{ for } j \in J \\ & x & \in & C,\end{array}\right\} \quad (2.3.4)$$

where C is a subset of $\mathbf{E}$, I and J are finite index sets, and the *objective function* f and *inequality* and *equality constraint functions* g_i $(i \in I)$ and h_j $(j \in J)$, respectively, are continuous from C to $\mathbf{R}$. A point x in C is *feasible* if it satisfies the constraints, and the set of all feasible x is called the *feasible region.* If the problem has no feasible points, we call it *inconsistent.* We say a feasible point $\bar{x}$ is a *local minimizer* if $f(x) \geq f(\bar{x})$ for all feasible x close to $\bar{x}$. We aim to derive first order necessary conditions for local minimizers.

We begin in this section with the differentiable inequality constrained problem

$$\left.\begin{array}{lrcl}\inf & f(x) & & \\ \text{subject to} & g_i(x) & \leq & 0 \text{ for } i = 1, 2, \ldots, m \\ & x & \in & C.\end{array}\right\} \quad (2.3.5)$$

For a feasible point $\bar{x}$ we define the *active set* $I(\bar{x}) = \{i \mid g_i(\bar{x}) = 0\}$. For this problem, assuming $\bar{x} \in \operatorname{int} C$, we call a vector $\lambda \in \mathbf{R}^m_+$ a *Lagrange multiplier vector* for $\bar{x}$ if $\bar{x}$ is a critical point of the *Lagrangian*

$$L(x; \lambda) = f(x) + \sum_{i=1}^{m} \lambda_i g_i(x)$$

(in other words, $\nabla f(\bar{x}) + \sum \lambda_i \nabla g_i(\bar{x}) = 0$), and *complementary slackness* holds: $\lambda_i = 0$ for indices i not in $I(\bar{x})$.

Theorem 2.3.6 (Fritz John conditions) *Suppose problem (2.3.5) has a local minimizer $\bar{x} \in \operatorname{int} C$. If the functions f, g_i $(i \in I(\bar{x}))$ are differentiable at $\bar{x}$ then there exist $\lambda_0, \lambda_i \in \mathbf{R}_+$ $(i \in I(\bar{x}))$, not all zero, satisfying*

$$\lambda_0 \nabla f(\bar{x}) + \sum_{i \in I(\bar{x})} \lambda_i \nabla g_i(\bar{x}) = 0.$$

Proof. Consider the function

$$g(x) = \max\{f(x) - f(\bar{x}),\ g_i(x) \mid i \in I(\bar{x})\}.$$

Since $\bar{x}$ is a local minimizer for the problem (2.3.5), it is a local minimizer of the function g, so all directions $d \in \mathbf{E}$ satisfy the inequality

$$g'(\bar{x}; d) = \max\{\langle \nabla f(\bar{x}), d\rangle, \langle \nabla g_i(\bar{x}), d\rangle \mid i \in I(\bar{x})\} \geq 0,$$

by the First order necessary condition (2.1.1) and Proposition 2.3.2 (Directional derivatives of max-functions). Thus the system

$$\langle \nabla f(\bar{x}), d \rangle < 0, \quad \langle \nabla g_i(\bar{x}), d \rangle < 0 \quad \text{for } i \in I(\bar{x})$$

has no solution, and the result follows by Gordan's theorem (2.2.1). □

One obvious disadvantage remains with the Fritz John first order conditions above: if $\lambda_0 = 0$ then the conditions are independent of the objective function f. To rule out this possibility we need to impose a regularity condition or "constraint qualification", an approach which is another recurring theme. The easiest such condition in this context is simply the linear independence of the gradients of the active constraints $\{\nabla g_i(\bar{x}) \mid i \in I(\bar{x})\}$. The culminating result of this section uses the following weaker condition.

Assumption 2.3.7 (The Mangasarian–Fromovitz constraint qualification) *There is a direction d in $\mathbf{E}$ satisfying $\langle \nabla g_i(\bar{x}), d \rangle < 0$ for all indices i in the active set $I(\bar{x})$.*

Theorem 2.3.8 (Karush–Kuhn–Tucker conditions) *Suppose problem (2.3.5) has a local minimizer $\bar{x}$ in* $\operatorname{int} C$. *If the functions f, g_i (for $i \in I(\bar{x})$) are differentiable at $\bar{x}$, and if the Mangasarian–Fromovitz constraint qualification (2.3.7) holds, then there is a Lagrange multiplier vector for $\bar{x}$.*

Proof. By the trivial implication in Gordan's theorem (2.2.1), the constraint qualification ensures $\lambda_0 \neq 0$ in the Fritz John conditions (2.3.6). □

Exercises and Commentary

The approach to first order conditions of this section is due to [95]. The Fritz John conditions appeared in [107]. The Karush–Kuhn–Tucker conditions were first published (under a different regularity condition) in [117], although the conditions appear earlier in an unpublished master's thesis [111]. The Mangasarian–Fromovitz constraint qualification appeared in [133]. A nice collection of optimization problems involving the determinant, similar to Exercise 8 (Minimum volume ellipsoid), appears in [47] (see also [183]). The classic reference for inequalities is [91].

1. Prove by induction that if the functions $g_0, g_1, \ldots, g_m : \mathbf{E} \to \mathbf{R}$ are all continuous at the point $\bar{x}$ then so is the max-function $g(x) = \max_i\{g_i(x)\}$.

2. **(Failure of Karush–Kuhn–Tucker)** Consider the following problem:

$$\begin{array}{lrcl} \inf & (x_1+1)^2 + x_2^2 & & \\ \text{subject to} & -x_1^3 + x_2^2 & \leq & 0 \\ & x & \in & \mathbf{R}^2. \end{array}$$

(a) Sketch the feasible region and hence solve the problem.

(b) Find multipliers λ_0 and λ satisfying the Fritz John conditions (2.3.6).

(c) Prove there exists no Lagrange multiplier vector for the optimal solution. Explain why not.

3. **(Linear independence implies Mangasarian–Fromovitz)** If the set of vectors $\{a^1, a^2, \ldots, a^m\}$ in $\mathbf{E}$ is linearly independent, prove directly there exists a direction d in $\mathbf{E}$ satisfying $\langle a^i, d\rangle < 0$ for $i = 1, 2, \ldots, m$.

4. For each of the following problems, explain why there must exist an optimal solution, and find it by using the Karush–Kuhn–Tucker conditions.

(a)
$$\begin{array}{lrl} \inf & x_1^2 + x_2^2 & \\ \text{subject to} & -2x_1 - x_2 + 10 & \leq 0 \\ & -x_1 & \leq 0. \end{array}$$

(b)
$$\begin{array}{lrl} \inf & 5x_1^2 + 6x_2^2 & \\ \text{subject to} & x_1 - 4 & \leq 0 \\ & 25 - x_1^2 - x_2^2 & \leq 0. \end{array}$$

5. **(Cauchy–Schwarz and steepest descent)** For a nonzero vector y in $\mathbf{E}$, use the Karush–Kuhn–Tucker conditions to solve the problem

$$\inf\{\langle y, x\rangle \mid \|x\|^2 \leq 1\}.$$

Deduce the Cauchy–Schwarz inequality.

6. * **(Hölder's inequality)** For real $p > 1$, define q by $p^{-1} + q^{-1} = 1$, and for x in $\mathbf{R}^n$ define

$$\|x\|_p = \Big(\sum_1^n |x_i|^p\Big)^{1/p}.$$

For a nonzero vector y in $\mathbf{R}^n$, consider the optimization problem

$$\inf\{\langle y, x\rangle \mid \|x\|_p^p \leq 1\}. \tag{2.3.9}$$

(a) Prove $\frac{d}{du}|u|^p/p = u|u|^{p-2}$ for all real u.

(b) Prove reals u and v satisfy $v = u|u|^{p-2}$ if and only if $u = v|v|^{q-2}$.

(c) Prove problem (2.3.9) has a nonzero optimal solution.

(d) Use the Karush–Kuhn–Tucker conditions to find the unique optimal solution.

(e) Deduce that any vectors x and y in $\mathbf{R}^n$ satisfy $\langle y, x \rangle \leq \|y\|_q \|x\|_p$.

(We develop another approach to this theory in Section 4.1, Exercise 11.)

7. * Consider a matrix A in $\mathbf{S}^n_{++}$ and a real $b > 0$.

   (a) Assuming the problem
   $$\inf\{-\log\det X \mid \operatorname{tr} AX \leq b,\ X \in \mathbf{S}^n_{++}\}$$
   has a solution, find it.

   (b) Repeat using the objective function $\operatorname{tr} X^{-1}$.

   (c) Prove the problems in parts (a) and (b) have optimal solutions. (Hint: Section 1.2, Exercise 14.)

8. ** **(Minimum volume ellipsoid)**

   (a) For a point y in $\mathbf{R}^n$ and the function $g : \mathbf{S}^n \to \mathbf{R}$ defined by $g(X) = \|Xy\|^2$, prove $\nabla g(X) = Xyy^T + yy^T X$ for all matrices X in $\mathbf{S}^n$.

   (b) Consider a set $\{y^1, y^2, \ldots, y^m\} \subset \mathbf{R}^n$. Prove this set spans $\mathbf{R}^n$ if and only if the matrix $\sum_i y^i (y^i)^T$ is positive definite.

   Now suppose the vectors $y^1, y^2, \ldots, y^m$ span $\mathbf{R}^n$.

   (c) Prove the problem
   $$\begin{array}{lrcl} \inf & -\log\det X & & \\ \text{subject to} & \|Xy^i\|^2 - 1 & \leq & 0 \quad \text{for } i = 1, 2, \ldots, m \\ & X & \in & \mathbf{S}^n_{++} \end{array}$$
   has an optimal solution. (Hint: Use part (b) and Section 1.2, Exercise 14.)

   Now suppose $\bar{X}$ is an optimal solution for the problem in part (c). (In this case the set $\{y \in \mathbf{R}^n \mid \|\bar{X}y\| \leq 1\}$ is a minimum volume ellipsoid (centered at the origin) containing the vectors $y^1, y^2, \ldots, y^m$.)

   (d) Show the Mangasarian–Fromovitz constraint qualification holds at $\bar{X}$ by considering the direction $d = -\bar{X}$.

   (e) Write down the Karush–Kuhn–Tucker conditions that $\bar{X}$ must satisfy.

   (f) When $\{y^1, y^2, \ldots, y^m\}$ is the standard basis of $\mathbf{R}^n$, the optimal solution of the problem in part (c) is $\bar{X} = I$. Find the corresponding Lagrange multiplier vector.

Chapter 3

Fenchel Duality

3.1 Subgradients and Convex Functions

We have already seen, in the First order sufficient condition (2.1.2), one benefit of convexity in optimization: critical points of convex functions are global minimizers. In this section we extend the types of functions we consider in two important ways:

(i) We do not require f to be differentiable.

(ii) We allow f to take the value $+\infty$.

Our derivation of first order conditions in Section 2.3 illustrates the utility of considering nonsmooth functions even in the context of smooth problems. Allowing the value $+\infty$ lets us rephrase a problem like

$$\inf\{g(x) \mid x \in C\}$$

as $\inf(g + \delta_C)$, where the *indicator function* $\delta_C(x)$ is 0 for x in C and $+\infty$ otherwise.

The *domain* of a function $f : \mathbf{E} \to (\infty, +\infty]$ is the set

$$\operatorname{dom} f = \{x \in \mathbf{E} \mid f(x) < +\infty\}.$$

We say f is *convex* if it is convex on its domain, and *proper* if its domain is nonempty. We call a function $g : \mathbf{E} \to [-\infty, +\infty)$ *concave* if $-g$ is convex, although for reasons of simplicity we will consider primarily convex functions. If a convex function f satisfies the stronger condition

$$f(\lambda x + \mu y) \leq \lambda f(x) + \mu f(y) \text{ for all } x, y \in \mathbf{E},\ \lambda, \mu \in \mathbf{R}_+$$

we say f is *sublinear*. If $f(\lambda x) = \lambda f(x)$ for all x in $\mathbf{E}$ and λ in $\mathbf{R}_+$ then f is *positively homogeneous*: in particular this implies $f(0) = 0$. (Recall the convention $0 \cdot (+\infty) = 0$.) If $f(x+y) \leq f(x) + f(y)$ for all x and y in $\mathbf{E}$ then we say f is *subadditive*. It is immediate that if the function f is sublinear then $-f(x) \leq f(-x)$ for all x in $\mathbf{E}$. The *lineality space* of a sublinear function f is the set

$$\operatorname{lin} f = \{x \in \mathbf{E} \mid -f(x) = f(-x)\}.$$

The following result (whose proof is left as an exercise) shows this set is a subspace.

Proposition 3.1.1 (Sublinearity) *A function $f : \mathbf{E} \to (\infty, +\infty]$ is sublinear if and only if it is positively homogeneous and subadditive. For a sublinear function f, the lineality space* $\operatorname{lin} f$ *is the largest subspace of* $\mathbf{E}$ *on which f is linear.*

As in the First order sufficient condition (2.1.2), it is easy to check that if the point $\bar{x}$ lies in the domain of the convex function f then the directional derivative $f'(\bar{x}; \cdot)$ is well-defined and positively homogeneous, taking values in $[-\infty, +\infty]$. The *core* of a set C (written $\operatorname{core}(C)$) is the set of points x in C such that for any direction d in $\mathbf{E}$, $x + td$ lies in C for all small real t. This set clearly contains the interior of C, although it may be larger (Exercise 2).

Proposition 3.1.2 (Sublinearity of the directional derivative) *If the function $f : \mathbf{E} \to (\infty, +\infty]$ is convex then, for any point $\bar{x}$ in* $\operatorname{core}(\operatorname{dom} f)$, *the directional derivative $f'(\bar{x}; \cdot)$ is everywhere finite and sublinear.*

Proof. For d in $\mathbf{E}$ and nonzero t in $\mathbf{R}$, define

$$g(d; t) = \frac{f(\bar{x} + td) - f(\bar{x})}{t}.$$

By convexity we deduce, for $0 < t \leq s \in \mathbf{R}$, the inequality

$$g(d; -s) \leq g(d; -t) \leq g(d; t) \leq g(d; s).$$

Since $\bar{x}$ lies in $\operatorname{core}(\operatorname{dom} f)$, for small $s > 0$ both $g(d; -s)$ and $g(d; s)$ are finite, so as $t \downarrow 0$ we have

$$+\infty > g(d; s) \geq g(d; t) \downarrow f'(\bar{x}; d) \geq g(d; -s) > -\infty. \tag{3.1.3}$$

Again by convexity we have, for any directions d and e in $\mathbf{E}$ and real $t > 0$,

$$g(d + e; t) \leq g(d; 2t) + g(e; 2t).$$

Now letting $t \downarrow 0$ gives subadditivity of $f'(\bar{x}; \cdot)$. The positive homogeneity is easy to check. $\square$

The idea of the derivative is fundamental in analysis because it allows us to approximate a wide class of functions using *linear functions.* In optimization we are concerned specifically with the minimization of functions, and hence often a *one-sided approximation* is sufficient. In place of the gradient we therefore consider *subgradients*, those elements ϕ of $\mathbf{E}$ satisfying

$$\langle \phi, x - \bar{x} \rangle \leq f(x) - f(\bar{x}) \quad \text{for all points } x \text{ in } \mathbf{E}. \tag{3.1.4}$$

We denote the set of subgradients (called the *subdifferential*) by $\partial f(\bar{x})$, defining $\partial f(\bar{x}) = \emptyset$ for $\bar{x}$ not in $\operatorname{dom} f$. The subdifferential is always a closed convex set. We can think of $\partial f(\bar{x})$ as the value at $\bar{x}$ of the "multifunction" or "set-valued map" $\partial f : \mathbf{E} \to \mathbf{E}$. The importance of such maps is another of our themes. We define its *domain*

$$\operatorname{dom} \partial f = \{x \in \mathbf{E} \mid \partial f(x) \neq \emptyset\}$$

(Exercise 19). We say f is *essentially strictly convex* if it is strictly convex on any convex subset of $\operatorname{dom} \partial f$.

The following very easy observation suggests the fundamental significance of subgradients in optimization.

Proposition 3.1.5 (Subgradients at optimality) *For any proper function $f : \mathbf{E} \to (\infty, +\infty]$, the point $\bar{x}$ is a (global) minimizer of f if and only if the condition $0 \in \partial f(\bar{x})$ holds.*

Alternatively put, minimizers of f correspond exactly to "zeroes" of ∂f.

The derivative is a local property whereas the subgradient definition (3.1.4) describes a global property. The main result of this section shows that the set of subgradients of a convex function is usually *nonempty*, and that we can describe it locally in terms of the directional derivative. We begin with another simple exercise.

Proposition 3.1.6 (Subgradients and directional derivatives) *If the function $f : \mathbf{E} \to (\infty, +\infty]$ is convex and the point $\bar{x}$ lies in $\operatorname{dom} f$, then an element ϕ of $\mathbf{E}$ is a subgradient of f at $\bar{x}$ if and only if it satisfies $\langle \phi, \cdot \rangle \leq f'(\bar{x}; \cdot)$.*

The idea behind the construction of a subgradient for a function f that we present here is rather simple. We recursively construct a decreasing sequence of sublinear functions which, after translation, minorize f. At each step we guarantee one extra direction of linearity. The basic step is summarized in the following exercise.

Lemma 3.1.7 *Suppose that the function $p : \mathbf{E} \to (\infty, +\infty]$ is sublinear and that the point $\bar{x}$ lies in $\operatorname{core}(\operatorname{dom} p)$. Then the function $q(\cdot) = p'(\bar{x}; \cdot)$ satisfies the conditions* (*i*) $q(\lambda\bar{x}) = \lambda p(\bar{x})$ *for all real* λ,

(*ii*) $q \le p$, *and*

(*iii*) $\operatorname{lin} q \supset \operatorname{lin} p + \operatorname{span}\{\bar{x}\}$.

With this tool we are now ready for the main result, which gives conditions guaranteeing the existence of a subgradient. Proposition 3.1.6 showed how to identify subgradients from directional derivatives; this next result shows how to move in the reverse direction.

Theorem 3.1.8 (Max formula) *If the function* $f : \mathbf{E} \to (\infty, +\infty]$ *is convex then any point* $\bar{x}$ *in* core (dom f) *and any direction* d *in* $\mathbf{E}$ *satisfy*

$$f'(\bar{x}; d) = \max\{\langle \phi, d\rangle \mid \phi \in \partial f(\bar{x})\}. \tag{3.1.9}$$

In particular, the subdifferential $\partial f(\bar{x})$ *is nonempty.*

Proof. In view of Proposition 3.1.6, we simply have to show that for any fixed d in $\mathbf{E}$ there is a subgradient ϕ satisfying $\langle \phi, d\rangle = f'(\bar{x}; d)$. Choose a basis $\{e_1, e_2, \ldots, e_n\}$ for $\mathbf{E}$ with $e_1 = d$ if d is nonzero. Now define a sequence of functions $p_0, p_1, \ldots, p_n$ recursively by $p_0(\cdot) = f'(\bar{x}; \cdot)$, and $p_k(\cdot) = p'_{k-1}(e_k; \cdot)$ for $k = 1, 2, \ldots, n$. We essentially show that $p_n(\cdot)$ is the required subgradient.

First note that, by Proposition 3.1.2, each p_k is everywhere finite and sublinear. By part (iii) of Lemma 3.1.7 we know

$$\operatorname{lin} p_k \supset \operatorname{lin} p_{k-1} + \operatorname{span}\{e_k\} \quad \text{for } k = 1, 2, \ldots, n,$$

so p_n is linear. Thus there is an element ϕ of $\mathbf{E}$ satisfying $\langle \phi, \cdot\rangle = p_n(\cdot)$.

Part (ii) of Lemma 3.1.7 implies $p_n \le p_{n-1} \le \ldots \le p_0$, so certainly, by Proposition 3.1.6, any point x in $\mathbf{E}$ satisfies

$$p_n(x - \bar{x}) \le p_0(x - \bar{x}) = f'(\bar{x}; x - \bar{x}) \le f(x) - f(\bar{x}).$$

Thus ϕ is a subgradient. If d is zero then we have $p_n(0) = 0 = f'(\bar{x}; 0)$. Finally, if d is nonzero then by part (i) of Lemma 3.1.7 we see

$$\begin{aligned} p_n(d) \le p_0(d) = p_0(e_1) = -p'_0(e_1; -e_1) = \\ -p_1(-e_1) = -p_1(-d) \le -p_n(-d) = p_n(d), \end{aligned}$$

whence $p_n(d) = p_0(d) = f'(\bar{x}; d)$. □

Corollary 3.1.10 (Differentiability of convex functions) *Suppose the function* $f : \mathbf{E} \to (\infty, +\infty]$ *is convex and the point* $\bar{x}$ *lies in* core (dom f). *Then* f *is Gâteaux differentiable at* $\bar{x}$ *exactly when* f *has a unique subgradient at* $\bar{x}$ *(in which case this subgradient is the derivative).*

We say the convex function f is *essentially smooth* if it is Gâteaux differentiable on $\operatorname{dom} \partial f$. (In this definition, we also require f to be "lower semicontinuous"; we defer discussion of lower semicontinuity until we need it, in Section 4.2.) We see later (Section 4.1, Exercise 21) that a function is essentially smooth if and only if its subdifferential is always singleton or empty.

The Max formula (Theorem 3.1.8) shows that convex functions typically have subgradients. In fact this property characterizes convexity (Exercise 12). This leads to a number of important ways of recognizing convex functions, one of which is the following example. Notice how a locally defined analytic condition results in a global geometric conclusion. The proof is outlined in the exercises.

Theorem 3.1.11 (Hessian characterization of convexity) *Given an open convex set $S \subset \mathbf{R}^n$, suppose the continuous function $f : \operatorname{cl} S \to \mathbf{R}$ is twice continuously differentiable on S. Then f is convex if and only if its Hessian matrix is positive semidefinite everywhere on S.*

Exercises and Commentary

The algebraic proof of the Max formula we follow here is due to [22]. The exercises below develop several standard characterizations of convexity—see for example [167]. The convexity of $-\log\det$ (Exercise 21) may be found in [99], for example. We shall see that the core and interior of a convex set in fact coincide (Theorem 4.1.4).

1. Prove Proposition 3.1.1 (Sublinearity).

2. **(Core versus interior)** Consider the set in $\mathbf{R}^2$

$$D = \{(x, y) \mid y = 0 \text{ or } |y| \geq x^2\}.$$

   Prove $0 \in \operatorname{core}(D) \setminus \operatorname{int}(D)$.

3. Prove the subdifferential is a closed convex set.

4. **(Subgradients and normal cones)** If a point $\bar{x}$ lies in a set $C \subset \mathbf{E}$, prove $\partial \delta_C(\bar{x}) = N_C(\bar{x})$.

5. Prove the following functions $x \in \mathbf{R} \mapsto f(x)$ are convex and calculate ∂f:

   (a) $$|x|$$

   (b) $$\delta_{\mathbf{R}_+}$$

   (c) $$\begin{cases} -\sqrt{x} & \text{if } x \geq 0 \\ +\infty & \text{otherwise} \end{cases}$$

(d)
$$\begin{cases} 0 & \text{if } x < 0 \\ 1 & \text{if } x = 0 \\ +\infty & \text{otherwise.} \end{cases}$$

6. Prove Proposition 3.1.6 (Subgradients and directional derivatives).

7. Prove Lemma 3.1.7.

8. **(Subgradients of norm)** Calculate $\partial\|\cdot\|$. Generalize your result to an arbitrary sublinear function.

9. **(Subgradients of maximum eigenvalue)** Prove
$$\partial\lambda_1(0) = \{Y \in \mathbf{S}^n_+ \mid \operatorname{tr} Y = 1\}.$$

10. ** For any vector μ in the cone $\mathbf{R}^n_{\geq}$, prove
$$\partial\langle\mu, [\cdot]\rangle(0) = \operatorname{conv}(\mathbf{P}^n\mu)$$
(see Section 2.2, Exercise 9 (Schur-convexity)).

11. * Define a function $f : \mathbf{R}^n \to \mathbf{R}$ by $f(x_1, x_2, \ldots, x_n) = \max_j\{x_j\}$, let $\bar{x} = 0$ and $d = (1, 1, \ldots, 1)^T$, and let $e_k = (1, 1, \ldots, 1, 0, \ldots, 0)^T$ (ending in $(k-1)$ zeroes). Calculate the functions p_k defined in the proof of Theorem 3.1.8 (Max formula), using Proposition 2.3.2 (Directional derivatives of max functions).

12. * **(Recognizing convex functions)** Suppose the set $S \subset \mathbf{R}^n$ is open and convex, and consider a function $f : S \to \mathbf{R}$. For points $x \notin S$, define $f(x) = +\infty$.

    (a) Prove $\partial f(x)$ is nonempty for all x in S if and only if f is convex. (Hint: For points u and v in S and real λ in $[0, 1]$, use the subgradient inequality (3.1.4) at the points $\bar{x} = \lambda u + (1 - \lambda)v$ and $x = u, v$ to check the definition of convexity.)

    (b) Prove that if $I \subset \mathbf{R}$ is an open interval and $g : I \to \mathbf{R}$ is differentiable then g is convex if and only if g' is nondecreasing on I, and g is strictly convex if and only if g' is strictly increasing on I. Deduce that if g is twice differentiable then g is convex if and only if g'' is nonnegative on I, and g is strictly convex if g'' is strictly positive on I.

    (c) Deduce that if f is twice continuously differentiable on S then f is convex if and only if its Hessian matrix is positive semidefinite everywhere on S, and f is strictly convex if its Hessian matrix is positive definite everywhere on S. (Hint: Apply part (b) to the function g defined by $g(t) = f(x + td)$ for small real t, points x in S, and directions d in $\mathbf{E}$.)

(d) Find a strictly convex function $f : (-1, 1) \to \mathbf{R}$ with $f''(0) = 0$.

(e) Prove that a continuous function $h : \operatorname{cl} S \to \mathbf{R}$ is convex if and only if its restriction to S is convex. What about strictly convex functions?

13. **(Local convexity)** Suppose the function $f : \mathbf{R}^n \to \mathbf{R}$ is twice continuously differentiable near 0 and $\nabla^2 f(0)$ is positive definite. Prove $f|_{\delta B}$ is convex for some real $\delta > 0$.

14. **(Examples of convex functions)** As we shall see in Section 4.2, most natural convex functions occur in pairs. The table in Section 3.3 lists many examples on $\mathbf{R}$. Use Exercise 12 to prove each function f and f^* in the table is convex.

15. **(Examples of convex functions)** Prove the following functions of $x \in \mathbf{R}$ are convex:

(a) $\log\left(\dfrac{\sinh ax}{\sinh x}\right)$ for $a \geq 1$.

(b) $\log\left(\dfrac{e^{ax}-1}{e^x-1}\right)$ for $a \geq 1$.

16. * **(Bregman distances [48])** For a function $\phi : \mathbf{E} \to (\infty, +\infty]$ that is strictly convex and differentiable on $\operatorname{int}(\operatorname{dom}\phi)$, define the *Bregman distance* $d_\phi : \operatorname{dom}\phi \times \operatorname{int}(\operatorname{dom}\phi) \to \mathbf{R}$ by

$$d_\phi(x, y) = \phi(x) - \phi(y) - \phi'(y)(x - y).$$

(a) Prove $d_\phi(x, y) \geq 0$, with equality if and only if $x = y$.

(b) Compute d_ϕ when $\phi(t) = t^2/2$ and when ϕ is the function p defined in Exercise 27.

(c) Suppose ϕ is three times differentiable. Prove d_ϕ is convex if and only if $-1/\phi''$ is convex on $\operatorname{int}(\operatorname{dom}\phi)$.

(d) Extend the results above to the function

$$D_\phi : (\operatorname{dom}\phi)^n \times (\operatorname{int}(\operatorname{dom}\phi))^n \to \mathbf{R}$$

defined by $D_\phi(x, y) = \sum_i d_\phi(x_i, y_i)$.

17. * **(Convex functions on $\mathbf{R}^2$)** Prove the following functions of $x \in \mathbf{R}^2$ are convex:

(a)
$$\begin{cases} (x_1 - x_2)(\log x_1 - \log x_2) & \text{if } x \in \mathbf{R}^2_{++} \\ 0 & \text{if } x = 0 \\ +\infty & \text{otherwise.} \end{cases}$$

(Hint: See Exercise 16.)

(b)

$$\begin{cases} \dfrac{x_1^2}{x_2} & \text{if } x_2 > 0 \\ 0 & \text{if } x = 0 \\ +\infty & \text{otherwise.} \end{cases}$$

18. * Prove the function

$$f(x) = \begin{cases} -(x_1 x_2 \ldots x_n)^{1/n} & \text{if } x \in \mathbf{R}^n_+ \\ +\infty & \text{otherwise} \end{cases}$$

is convex.

19. **(Domain of subdifferential)** If the function $f : \mathbf{R}^2 \to (\infty, +\infty]$ is defined by

$$f(x_1, x_2) = \begin{cases} \max\{1 - \sqrt{x_1}, |x_2|\} & \text{if } x_1 \geq 0 \\ +\infty & \text{otherwise,} \end{cases}$$

prove that f is convex but that $\operatorname{dom} \partial f$ is not convex.

20. * **(Monotonicity of gradients)** Suppose that the set $S \subset \mathbf{R}^n$ is open and convex and that the function $f : S \to \mathbf{R}$ is differentiable. Prove f is convex if and only if

$$\langle \nabla f(x) - \nabla f(y), x - y \rangle \geq 0 \text{ for all } x, y \in S,$$

and f is strictly convex if and only if the above inequality holds strictly whenever $x \neq y$. (You may use Exercise 12.)

21. ** **(The log barrier)** Use Exercise 20 (Monotonicity of gradients), Exercise 10 in Section 2.1 and Exercise 8 in Section 1.2 to prove that the function $f : \mathbf{S}^n_{++} \to \mathbf{R}$ defined by $f(X) = -\log \det X$ is strictly convex. Deduce the uniqueness of the minimum volume ellipsoid in Section 2.3, Exercise 8, and the matrix completion in Section 2.1, Exercise 12.

22. Prove the function (2.2.5) is convex on $\mathbf{R}^n$ by calculating its Hessian.

23. * If the function $f : \mathbf{E} \to (\infty, +\infty]$ is essentially strictly convex, prove all distinct points x and y in $\mathbf{E}$ satisfy $\partial f(x) \cap \partial f(y) = \emptyset$. Deduce that f has at most one minimizer.

24. **(Minimizers of essentially smooth functions)** Prove that any minimizer of an essentially smooth function f must lie in core $(\operatorname{dom} f)$.

25. ** **(Convex matrix functions)** Consider a matrix C in $\mathbf{S}^n_+$.

(a) For matrices X in $\mathbf{S}^n_{++}$ and D in $\mathbf{S}^n$, use a power series expansion to prove

$$\frac{d^2}{dt^2}\mathrm{tr}\,(C(X+tD)^{-1})\Big|_{t=0} \geq 0.$$

(b) Deduce $X \in \mathbf{S}^n_{++} \mapsto \mathrm{tr}\,(CX^{-1})$ is convex.

(c) Prove similarly the function $X \in \mathbf{S}^n \mapsto \mathrm{tr}\,(CX^2)$ and the function $X \in \mathbf{S}^n_+ \mapsto -\mathrm{tr}\,(CX^{1/2})$ are convex.

26. ** **(Log-convexity)** Given a convex set $C \subset \mathbf{E}$, we say that a function $f : C \to \mathbf{R}_{++}$ is *log-convex* if $\log f(\cdot)$ is convex.

(a) Prove any log-convex function is convex, using Section 1.1, Exercise 9 (Composing convex functions).

(b) If a polynomial $p : \mathbf{R} \to \mathbf{R}$ has all real roots, prove $1/p$ is log-convex on any interval on which p is strictly positive.

(c) One version of *Hölder's inequality* states, for real $p, q > 1$ satisfying $p^{-1} + q^{-1} = 1$ and functions $u, v : \mathbf{R}_+ \to \mathbf{R}$,

$$\int uv \leq \Big(\int |u|^p\Big)^{1/p}\Big(\int |v|^q\Big)^{1/q}$$

when the right hand side is well-defined. Use this to prove the *gamma function* $\Gamma : \mathbf{R} \to \mathbf{R}$ given by

$$\Gamma(x) = \int_0^\infty t^{x-1}e^{-t}\,dt$$

is log-convex.

27. ** **(Maximum entropy [36])** Define a convex function $p : \mathbf{R} \to (-\infty, +\infty]$ by

$$p(u) = \begin{cases} u\log u - u & \text{if } u > 0 \\ 0 & \text{if } u = 0 \\ +\infty & \text{if } u < 0 \end{cases}$$

and a convex function $f : \mathbf{R}^n \to (-\infty, +\infty]$ by

$$f(x) = \sum_{i=1}^n p(x_i).$$

Suppose $\hat{x}$ lies in the interior of $\mathbf{R}^n_+$.

(a) Prove f is strictly convex on $\mathbf{R}^n_+$ with compact level sets.

(b) Prove $f'(x; \hat{x} - x) = -\infty$ for any point x on the boundary of $\mathbf{R}^n_+$.

(c) Suppose the map $G : \mathbf{R}^n \to \mathbf{R}^m$ is linear with $G\hat{x} = b$. Prove for any vector c in $\mathbf{R}^n$ that the problem

$$\inf\{f(x) + \langle c, x\rangle \mid Gx = b,\ x \in \mathbf{R}^n\}$$

has a unique optimal solution $\bar{x}$, lying in $\mathbf{R}^n_{++}$.

(d) Use Corollary 2.1.3 (First order conditions for linear constraints) to prove that some vector λ in $\mathbf{R}^m$ satisfies $\nabla f(\bar{x}) = G^*\lambda - c$, and deduce $\bar{x}_i = \exp(G^*\lambda - c)_i$.

28. ** (*DAD* **problems [36])** Consider the following example of Exercise 27 (Maximum entropy). Suppose the $k \times k$ matrix A has each entry a_{ij} nonnegative. We say *A has doubly stochastic pattern* if there is a doubly stochastic matrix with exactly the same zero entries as A. Define a set $Z = \{(i,j) | a_{ij} > 0\}$, and let $\mathbf{R}^Z$ denote the set of vectors with components indexed by Z and $\mathbf{R}^Z_+$ denote those vectors in $\mathbf{R}^Z$ with all nonnegative components. Consider the problem

$$\begin{array}{lrcl} \inf & \sum_{(i,j)\in Z}(p(x_{ij}) - x_{ij}\log a_{ij}) & & \\ \text{subject to} & \sum_{i:(i,j)\in Z} x_{ij} & = & 1 \ \text{ for } j = 1,2,\ldots,k \\ & \sum_{j:(i,j)\in Z} x_{ij} & = & 1 \ \text{ for } i = 1,2,\ldots,k \\ & x & \in & \mathbf{R}^Z. \end{array}$$

(a) Suppose A has doubly stochastic pattern. Prove there is a point $\hat{x}$ in the interior of $\mathbf{R}^Z_+$ which is feasible for the problem above. Deduce that the problem has a unique optimal solution $\bar{x}$, and, for some vectors λ and μ in $\mathbf{R}^k$, $\bar{x}$ satisfies

$$\bar{x}_{ij} = a_{ij}\exp(\lambda_i + \mu_j) \ \text{ for } (i,j) \in Z.$$

(b) Deduce that A has doubly stochastic pattern if and only if there are diagonal matrices D_1 and D_2 with strictly positive diagonal entries and D_1AD_2 doubly stochastic.

29. ** **(Relativizing the Max formula)** If $f : \mathbf{E} \to (\infty, +\infty]$ is a convex function then for points $\bar{x}$ in $\mathrm{ri}\,(\mathrm{dom}\, f)$ and directions d in $\mathbf{E}$, prove the subdifferential $\partial f(\bar{x})$ is nonempty and

$$f'(\bar{x}; d) = \sup\{\langle \phi, d\rangle \mid \phi \in \partial f(\bar{x})\},$$

with attainment when finite.

3.2 The Value Function

In this section we describe another approach to the Karush–Kuhn–Tucker conditions (2.3.8) in the convex case using the existence of subgradients we established in the previous section. We consider an *(inequality-constrained) convex program*

$$\inf\{f(x) \mid g_i(x) \leq 0 \text{ for } i = 1, 2, \ldots, m, \ x \in \mathbf{E}\}, \tag{3.2.1}$$

where the functions $f, g_1, g_2, \ldots, g_m : \mathbf{E} \to (\infty, +\infty]$ are convex and satisfy $\emptyset \neq \operatorname{dom} f \subset \cap_i \operatorname{dom} g_i$. Denoting the vector with components $g_i(x)$ by $g(x)$, the function $L : \mathbf{E} \times \mathbf{R}^m_+ \to (\infty, +\infty]$ defined by

$$L(x; \lambda) = f(x) + \lambda^T g(x), \tag{3.2.2}$$

is called the *Lagrangian.* A *feasible solution* is a point x in $\operatorname{dom} f$ satisfying the constraints.

We should emphasize that the term "Lagrange multiplier" has different meanings in different contexts. In the present context we say a vector $\bar{\lambda} \in \mathbf{R}^m_+$ is a *Lagrange multiplier vector* for a feasible solution $\bar{x}$ if $\bar{x}$ minimizes the function $L(\,\cdot\,; \bar{\lambda})$ over $\mathbf{E}$ and $\bar{\lambda}$ satisfies the complementary slackness conditions: $\bar{\lambda}_i = 0$ whenever $g_i(\bar{x}) < 0$.

We can often use the following principle to solve simple optimization problems.

Proposition 3.2.3 (Lagrangian sufficient conditions) *If the point $\bar{x}$ is feasible for the convex program (3.2.1) and there is a Lagrange multiplier vector, then $\bar{x}$ is optimal.*

The proof is immediate, and in fact does not rely on convexity.

The Karush–Kuhn–Tucker conditions (2.3.8) are a converse to the above result when the functions $f, g_1, g_2, \ldots, g_m$ are convex and differentiable. We next follow a very different, and surprising, route to this result, circumventing differentiability. We *perturb* the problem (3.2.1), and analyze the resulting *(optimal) value function* $v : \mathbf{R}^m \to [-\infty, +\infty]$, defined by the equation

$$v(b) = \inf\{f(x) \mid g(x) \leq b\}. \tag{3.2.4}$$

We show that Lagrange multiplier vectors $\bar{\lambda}$ correspond to subgradients of v (Exercise 9).

Our old definition of convexity for functions does not naturally extend to functions $h : \mathbf{E} \to [-\infty, +\infty]$ (due to the possible occurrence of $\infty - \infty$). To generalize the definition we introduce the idea of the *epigraph* of h:

$$\operatorname{epi}(h) = \{(y, r) \in \mathbf{E} \times \mathbf{R} \mid h(y) \leq r\}, \tag{3.2.5}$$

and we say h is a *convex function* if $\text{epi}\,(h)$ is a convex set. An exercise shows in this case that the domain

$$\text{dom}\,(h) = \{y \mid h(y) < +\infty\}$$

is convex, and further that the value function v defined by equation (3.2.4) is convex. We say h is *proper* if $\text{dom}\, h$ is nonempty and h never takes the value $-\infty$: if we wish to demonstrate the existence of subgradients for v using the results in the previous section then we need to exclude $-\infty$.

Lemma 3.2.6 *If the function $h : \mathbf{E} \to [-\infty, +\infty]$ is convex and some point $\hat{y}$ in $\text{core}\,(\text{dom}\, h)$ satisfies $h(\hat{y}) > -\infty$, then h never takes the value $-\infty$.*

Proof. Suppose some point y in $\mathbf{E}$ satisfies $h(y) = -\infty$. Since $\hat{y}$ lies in $\text{core}\,(\text{dom}\, h)$, there is a real $t > 0$ with $\hat{y} + t(\hat{y} - y)$ in $\text{dom}\,(h)$, and hence a real r with $(\hat{y} + t(\hat{y} - y), r)$ in $\text{epi}\,(h)$. Now for any real s, (y, s) lies in $\text{epi}\,(h)$, so we know

$$\left(\hat{y}, \frac{r + ts}{1 + t}\right) = \frac{1}{1 + t}(\hat{y} + t(\hat{y} - y), r) + \frac{t}{1 + t}(y, s) \in \text{epi}\,(h),$$

Letting $s \to -\infty$ gives a contradiction. □

In Section 2.3 we saw that the Karush–Kuhn–Tucker conditions needed a regularity condition. In this approach we will apply a different condition, known as the *Slater constraint qualification*, for the problem (3.2.1):

$$\text{There exists } \hat{x} \text{ in } \text{dom}\,(f) \text{ with } g_i(\hat{x}) < 0 \text{ for } i = 1, 2, \ldots, m. \qquad (3.2.7)$$

Theorem 3.2.8 (Lagrangian necessary conditions) *Suppose that the point $\bar{x}$ in $\text{dom}\,(f)$ is optimal for the convex program (3.2.1) and that the Slater condition (3.2.7) holds. Then there is a Lagrange multiplier vector for $\bar{x}$.*

Proof. Defining the value function v by equation (3.2.4), certainly $v(0) > -\infty$, and the Slater condition shows $0 \in \text{core}\,(\text{dom}\, v)$, so in particular Lemma 3.2.6 shows that v never takes the value $-\infty$. (An incidental consequence, from Section 4.1, is the continuity of v at 0.) We now deduce the existence of a subgradient $-\bar{\lambda}$ of v at 0, by the Max formula (3.1.8).

Any vector b in $\mathbf{R}^m_+$ obviously satisfies $g(\bar{x}) \le b$, whence the inequality

$$f(\bar{x}) = v(0) \le v(b) + \bar{\lambda}^T b \le f(\bar{x}) + \bar{\lambda}^T b.$$

Hence, $\bar{\lambda}$ lies in $\mathbf{R}^m_+$. Furthermore, any point x in $\text{dom}\, f$ clearly satisfies

$$f(x) \ge v(g(x)) \ge v(0) - \bar{\lambda}^T g(x) = f(\bar{x}) - \bar{\lambda}^T g(x).$$

The case $x = \bar{x}$, using the inequalities $\bar{\lambda} \geq 0$ and $g(\bar{x}) \leq 0$, shows $\bar{\lambda}^T g(\bar{x}) = 0$, which yields the complementary slackness conditions. Finally, all points x in $\mathrm{dom}\, f$ must satisfy $f(x) + \bar{\lambda}^T g(x) \geq f(\bar{x}) = f(\bar{x}) + \bar{\lambda}^T g(\bar{x})$. □

In particular, if in the above result $\bar{x}$ lies in core (dom f) and the functions $f, g_1, g_2, \ldots, g_m$ are differentiable at $\bar{x}$ then

$$\nabla f(\bar{x}) + \sum_{i=1}^{m} \bar{\lambda}_i \nabla g_i(\bar{x}) = 0,$$

so we recapture the Karush–Kuhn–Tucker conditions (2.3.8). In fact, in this case it is easy to see that the Slater condition is equivalent to the Mangasarian–Fromovitz constraint qualification (Assumption 2.3.7).

Exercises and Commentary

Versions of the Lagrangian necessary conditions above appeared in [182] and [110]; for a survey see [158]. The approach here is analogous to [81]. The Slater condition first appeared in [173].

1. Prove the Lagrangian sufficient conditions (3.2.3).

2. Use the Lagrangian sufficient conditions (3.2.3) to solve the following problems.

(a)
$$\begin{array}{ll} \inf & x_1^2 + x_2^2 - 6x_1 - 2x_2 + 10 \\ \text{subject to} & 2x_1 + x_2 - 2 \leq 0 \\ & x_2 - 1 \leq 0 \\ & x \in \mathbf{R}^2. \end{array}$$

(b)
$$\begin{array}{ll} \inf & -2x_1 + x_2 \\ \text{subject to} & x_1^2 - x_2 \leq 0 \\ & x_2 - 4 \leq 0 \\ & x \in \mathbf{R}^2. \end{array}$$

(c)
$$\begin{array}{ll} \inf & x_1 + \dfrac{2}{x_2} \\ \text{subject to} & -x_2 + \dfrac{1}{2} \leq 0 \\ & -x_1 + x_2^2 \leq 0 \\ & x \in \{(x_1, x_2) \mid x_2 > 0\}. \end{array}$$

3. Given strictly positive reals $a_1, a_2, \ldots, a_n, c_1, c_2, \ldots, c_n$ and b, use the Lagrangian sufficient conditions to solve the problem

$$\inf \Big\{ \sum_{i=1}^{n} \frac{c_i}{x_i} \;\Big|\; \sum_{i=1}^{n} a_i x_i \leq b,\ x \in \mathbf{R}^n_{++} \Big\}.$$

4. For a matrix A in $\mathbf{S}^n_{++}$ and a real $b > 0$, use the Lagrangian sufficient conditions to solve the problem

$$\inf\{-\log\det X \mid \operatorname{tr} AX \leq b,\ X \in \mathbf{S}^n_{++}\}.$$

You may use the fact that the objective function is convex with derivative $-X^{-1}$ (see Section 3.1, Exercise 21 (The log barrier)).

5. * **(Mixed constraints)** Consider the convex program (3.2.1) with some additional linear constraints $\langle a^j, x\rangle = d_j$ for vectors a^j in $\mathbf{E}$ and reals d_j. By rewriting each equality as two inequalities (or otherwise), prove a version of the Lagrangian sufficient conditions for this problem.

6. **(Extended convex functions)**

   (a) Give an example of a convex function that takes the values 0 and $-\infty$.

   (b) Prove the value function v defined by equation (3.2.4) is convex.

   (c) Prove that a function $h : \mathbf{E} \to [-\infty, +\infty]$ is convex if and only if it satisfies the inequality

   $$h(\lambda x + (1-\lambda)y) \leq \lambda h(x) + (1-\lambda)h(y)$$

   for any points x and y in $\operatorname{dom} h$ (or $\mathbf{E}$ if h is proper) and any real λ in $(0, 1)$.

   (d) Prove that if the function $h : \mathbf{E} \to [-\infty, +\infty]$ is convex then $\operatorname{dom}(h)$ is convex.

7. **(Nonexistence of multiplier)** For the function $f : \mathbf{R} \to (\infty, +\infty]$ defined by $f(x) = -\sqrt{x}$ for x in $\mathbf{R}_+$ and $+\infty$ otherwise, show there is no Lagrange multiplier at the optimal solution of $\inf\{f(x) \,|\, x \leq 0\}$.

8. **(Duffin's duality gap)** Consider the following problem (for real b):

$$\inf\{e^{x_2} \mid \|x\| - x_1 \leq b,\ x \in \mathbf{R}^2\}. \tag{3.2.9}$$

   (a) Sketch the feasible region for $b > 0$ and for $b = 0$.

   (b) Plot the value function v.

   (c) Show that when $b = 0$ there is no Lagrange multiplier for any feasible solution. Explain why the Lagrangian necessary conditions (3.2.8) do not apply.

   (d) Repeat the above exercises with the objective function e^{x_2} replaced by x_2.

9. ** **(Karush–Kuhn–Tucker vectors [167])** Consider the convex program (3.2.1). Suppose the value function v given by equation (3.2.4) is finite at 0. We say the vector $\bar{\lambda}$ in $\mathbf{R}^m_+$ is a *Karush–Kuhn–Tucker vector* if it satisfies $v(0) = \inf\{L(x; \bar{\lambda}) \mid x \in \mathbf{E}\}$.

   (a) Prove that the set of Karush–Kuhn–Tucker vectors is $-\partial v(0)$.

   (b) Suppose the point $\bar{x}$ is an optimal solution of problem (3.2.1). Prove that the set of Karush–Kuhn–Tucker vectors coincides with the set of Lagrange multiplier vectors for $\bar{x}$.

   (c) Prove the Slater condition ensures the existence of a Karush–Kuhn–Tucker vector.

   (d) Suppose $\bar{\lambda}$ is a Karush–Kuhn–Tucker vector. Prove a feasible point $\bar{x}$ is optimal for problem (3.2.1) if and only if $\bar{\lambda}$ is a Lagrange multiplier vector for $\bar{x}$.

10. Prove the equivalence of the Slater and Mangasarian–Fromovitz conditions asserted at the end of the section.

11. **(Normals to epigraphs)** For a function $f : \mathbf{E} \to (\infty, +\infty]$ and a point $\bar{x}$ in core (dom f), calculate the normal cone $N_{\text{epi } f}(\bar{x}, f(\bar{x}))$.

12. * **(Normals to level sets)** Suppose the function $f : \mathbf{E} \to (\infty, +\infty]$ is convex. If the point $\bar{x}$ lies in core (dom f) and is not a minimizer for f, prove that the normal cone at $\bar{x}$ to the level set

$$C = \{x \in \mathbf{E} \mid f(x) \leq f(\bar{x})\}$$

   is given by $N_C(\bar{x}) = \mathbf{R}_+ \partial f(\bar{x})$. Is the assumption $\bar{x} \in$ core (dom f) and $f(\bar{x}) > \inf f$ necessary?

13. * **(Subdifferential of max-function)** Consider convex functions

$$g_1, g_2, \ldots, g_m : \mathbf{E} \to (\infty, +\infty],$$

   and define a function $g(x) = \max_i g_i(x)$ for all points x in $\mathbf{E}$. For a fixed point $\bar{x}$ in $\mathbf{E}$, define the index set $I = \{i \mid g_i(\bar{x}) = g(\bar{x})\}$ and let

$$C = \bigcup \Big\{ \partial\Big(\sum_{i \in I} \lambda_i g_i\Big)(\bar{x}) \,\Big|\, \lambda \in \mathbf{R}^I_+,\ \sum_{i \in I} \lambda_i = 1 \Big\}.$$

   (a) Prove $C \subset \partial g(\bar{x})$.

   (b) Suppose $0 \in \partial g(\bar{x})$. By considering the convex program

$$\inf_{t \in \mathbf{R},\ x \in \mathbf{E}} \{t \mid g_i(x) - t \leq 0 \ \text{ for } i = 1, 2, \ldots, m\},$$

   prove $0 \in C$.

(c) Deduce $\partial g(\bar{x}) = C$.

14. ** **(Minimum volume ellipsoid)** Denote the standard basis of $\mathbf{R}^n$ by $\{e^1, e^2, \ldots, e^n\}$ and consider the minimum volume ellipsoid problem (see Section 2.3, Exercise 8)

$$\begin{array}{lrcl} \inf & -\log\det X & & \\ \text{subject to} & \|Xe^i\|^2 - 1 & \leq & 0 \ \text{ for } i = 1, 2, \ldots, n \\ & X & \in & \mathbf{S}^n_{++}. \end{array}$$

Use the Lagrangian sufficient conditions (3.2.3) to prove $X = I$ is the unique optimal solution. (Hint: Use Section 3.1, Exercise 21 (The log barrier).) Deduce the following special case of *Hadamard's inequality*: Any matrix $(x^1 \ x^2 \ \ldots \ x^n)$ in $\mathbf{S}^n_{++}$ satisfies

$$\det(x^1 \ x^2 \ \ldots \ x^n) \leq \|x^1\| \|x^2\| \ldots \|x^n\|.$$

3.3 The Fenchel Conjugate

In the next few sections we sketch a little of the elegant and concise theory of Fenchel conjugation, and we use it to gain a deeper understanding of the Lagrangian necessary conditions for convex programs (3.2.8). The *Fenchel conjugate* of a function $h : \mathbf{E} \to [-\infty, +\infty]$ is the function $h^* : \mathbf{E} \to [-\infty, +\infty]$ defined by

$$h^*(\phi) = \sup_{x \in \mathbf{E}} \{\langle \phi, x \rangle - h(x)\}.$$

The function h^* is convex and if the domain of h is nonempty then h^* never takes the value $-\infty$. Clearly the conjugacy operation is *order-reversing*: for functions $f, g : \mathbf{E} \to [-\infty, +\infty]$, the inequality $f \geq g$ implies $f^* \leq g^*$.

Conjugate functions are ubiquitous in optimization. For example, we have already seen the conjugate of the exponential, defined by

$$\exp^*(t) = \begin{cases} t \log t - t & \text{if } t > 0 \\ 0 & \text{if } t = 0 \\ +\infty & \text{if } t < 0 \end{cases}$$

(see Section 3.1, Exercise 27). A rather more subtle example is the function $g : \mathbf{E} \to (\infty, +\infty]$ defined, for points $a^0, a^1, \ldots, a^m$ in $\mathbf{E}$, by

$$g(z) = \inf_{x \in \mathbf{R}^{m+1}} \Big\{ \sum_i \exp^*(x_i) \ \Big| \ \sum_i x_i = 1, \ \sum_i x_i a^i = z \Big\}. \tag{3.3.1}$$

The conjugate is the function we used in Section 2.2 to prove various theorems of the alternative:

$$g^*(y) = 1 + \log \Big(\sum_i \exp \langle a^i, y \rangle \Big) \tag{3.3.2}$$

(see Exercise 7).

As we shall see later (Section 4.2), many important convex functions h equal their *biconjugates* h^{**}. Such functions thus occur as natural pairs, h and h^*. Table 3.1 shows some elegant examples on $\mathbf{R}$, and Table 3.2 describes some simple transformations of these examples.

The following result summarizes the properties of two particularly important convex functions.

Proposition 3.3.3 (Log barriers) *The functions* $\mathrm{lb} : \mathbf{R}^n \to (\infty, +\infty]$ *and* $\mathrm{ld} : \mathbf{S}^n \to (\infty, +\infty]$ *defined by*

$$\mathrm{lb}(x) = \begin{cases} -\sum_{i=1}^n \log x_i & \text{if } x \in \mathbf{R}^n_{++} \\ +\infty & \text{otherwise} \end{cases}$$

$f(x) = g^*(x)$	$\operatorname{dom} f$	$g(y) = f^*(y)$	$\operatorname{dom} g$
0	$\mathbf{R}$	0	$\{0\}$
0	$\mathbf{R}_+$	0	$-\mathbf{R}_+$
0	$[-1,1]$	$\lvert y \rvert$	$\mathbf{R}$
0	$[0,1]$	y^+	$\mathbf{R}$
$\lvert x \rvert^p/p, \quad p>1$	$\mathbf{R}$	$\lvert y \rvert^q/q \ \ (\frac{1}{p}+\frac{1}{q}=1)$	$\mathbf{R}$
$\lvert x \rvert^p/p, \quad p>1$	$\mathbf{R}_+$	$\lvert y^+ \rvert^q/q \ \ (\frac{1}{p}+\frac{1}{q}=1)$	$\mathbf{R}$
$-x^p/p, \quad 0<p<1$	$\mathbf{R}_+$	$-(-y)^q/q \ \ (\frac{1}{p}+\frac{1}{q}=1)$	$-\mathbf{R}_{++}$
$\sqrt{1+x^2}$	$\mathbf{R}$	$-\sqrt{1-y^2}$	$[-1,1]$
$-\log x$	$\mathbf{R}_{++}$	$-1-\log(-y)$	$-\mathbf{R}_{++}$
$\cosh x$	$\mathbf{R}$	$y \sinh^{-1}(y) - \sqrt{1+y^2}$	$\mathbf{R}$
$-\log(\cos x)$	$(-\frac{\pi}{2},\frac{\pi}{2})$	$y \tan^{-1}(y) - \frac{1}{2}\log(1+y^2)$	$\mathbf{R}$
$\log(\cosh x)$	$\mathbf{R}$	$y \tanh^{-1}(y) + \frac{1}{2}\log(1-y^2)$	$(-1,1)$
e^x	$\mathbf{R}$	$\begin{cases} y\log y - y & (y>0) \\ 0 & (y=0) \end{cases}$	$\mathbf{R}_+$
$\log(1+e^x)$	$\mathbf{R}$	$\begin{cases} y\log y + (1-y)\log(1-y) & (y \in (0,1)) \\ 0 & (y=0,1) \end{cases}$	$[0,1]$
$-\log(1-e^x)$	$\mathbf{R}$	$\begin{cases} y\log y - (1+y)\log(1+y) & (y>0) \\ 0 & (y=0) \end{cases}$	$\mathbf{R}_+$

Table 3.1: Conjugate pairs of convex functions on $\mathbf{R}$.

$f = g^*$	$g = f^*$
$f(x)$	$g(y)$
$h(ax)$ $(a \neq 0)$	$h^*(y/a)$
$h(x+b)$	$h^*(y) - by$
$ah(x)$ $(a > 0)$	$ah^*(y/a)$

Table 3.2: Transformed conjugates.

and

$$\mathrm{ld}\,(X) = \begin{cases} -\log\det X & \text{if } X \in \mathbf{S}^n_{++} \\ +\infty & \text{otherwise} \end{cases}$$

are essentially smooth, and strictly convex on their domains. They satisfy the conjugacy relations

$$\begin{aligned} \mathrm{lb}^*(x) &= \mathrm{lb}\,(-x) - n \quad \text{for all } x \in \mathbf{R}^n, \text{ and} \\ \mathrm{ld}^*(X) &= \mathrm{ld}\,(-X) - n \quad \text{for all } X \in \mathbf{S}^n. \end{aligned}$$

The perturbed functions $\mathrm{lb} + \langle c, \cdot \rangle$ *and* $\mathrm{ld} + \langle C, \cdot \rangle$ *have compact level sets for any vector* $c \in \mathbf{R}^n_{++}$ *and matrix* $C \in \mathbf{S}^n_{++}$*, respectively.*

(See Section 3.1, Exercise 21 (The log barrier), and Section 1.2, Exercise 14 (Level sets of perturbed log barriers); the conjugacy formulas are simple calculations.) Notice the simple relationships $\mathrm{lb} = \mathrm{ld} \circ \mathrm{Diag}$ and $\mathrm{ld} = \mathrm{lb} \circ \lambda$ between these two functions.

The next elementary but important result relates conjugation with the subgradient. The proof is an exercise.

Proposition 3.3.4 (Fenchel–Young inequality) *Any points* ϕ *in* $\mathbf{E}$ *and* x *in the domain of a function* $h : \mathbf{E} \to (\infty, +\infty]$ *satisfy the inequality*

$$h(x) + h^*(\phi) \geq \langle \phi, x \rangle.$$

Equality holds if and only if $\phi \in \partial h(x)$.

In Section 3.2 we analyzed the standard inequality-constrained convex program by studying its optimal value under perturbations. A similar approach works for another model for convex programming, particularly suited to problems with linear constraints. An interesting byproduct is a convex analogue of the chain rule for differentiable functions,

$$\nabla(f + g \circ A)(x) = \nabla f(x) + A^* \nabla g(Ax)$$

(for a linear map A). When A is the identity map we obtain a *sum rule.*

In this section we fix a Euclidean space $\mathbf{Y}$. We denote the set of points where a function $g : \mathbf{Y} \to [-\infty, +\infty]$ is finite and continuous by $\operatorname{cont} g$.

Theorem 3.3.5 (Fenchel duality and convex calculus) *For given functions $f : \mathbf{E} \to (\infty, +\infty]$ and $g : \mathbf{Y} \to (\infty, +\infty]$ and a linear map $A : \mathbf{E} \to \mathbf{Y}$, let $p, d \in [-\infty, +\infty]$ be primal and dual values defined, respectively, by the* **Fenchel problems**

$$p = \inf_{x \in \mathbf{E}} \{f(x) + g(Ax)\} \tag{3.3.6}$$

$$d = \sup_{\phi \in \mathbf{Y}} \{-f^*(A^*\phi) - g^*(-\phi)\}. \tag{3.3.7}$$

These values satisfy the **weak duality** *inequality $p \geq d$. If, furthermore, f and g are convex and satisfy the condition*

$$0 \in \operatorname{core}(\operatorname{dom} g - A\operatorname{dom} f) \tag{3.3.8}$$

or the stronger condition

$$A\operatorname{dom} f \cap \operatorname{cont} g \neq \emptyset \tag{3.3.9}$$

then the values are equal ($p = d$), and the supremum in the dual problem (3.3.7) is attained if finite.

At any point x in $\mathbf{E}$, the calculus rule

$$\partial(f + g \circ A)(x) \supset \partial f(x) + A^* \partial g(Ax) \tag{3.3.10}$$

holds, with equality if f and g are convex and either condition (3.3.8) or (3.3.9) holds.

Proof. The weak duality inequality follows immediately from the Fenchel–Young inequality (3.3.4). To prove equality we define an optimal value function $h : \mathbf{Y} \to [-\infty, +\infty]$ by

$$h(u) = \inf_{x \in \mathbf{E}} \{f(x) + g(Ax + u)\}.$$

It is easy to check h is convex and $\operatorname{dom} h = \operatorname{dom} g - A\operatorname{dom} f$. If p is $-\infty$ there is nothing to prove, while if condition (3.3.8) holds and p is finite then Lemma 3.2.6 and the Max formula (3.1.8) show there is a subgradient $-\phi \in \partial h(0)$. Hence we deduce, for all u in $\mathbf{Y}$ and x in $\mathbf{E}$, the inequalities

$$\begin{aligned} h(0) &\leq h(u) + \langle \phi, u \rangle \\ &\leq f(x) + g(Ax+u) + \langle \phi, u \rangle \\ &= \{f(x) - \langle A^*\phi, x \rangle\} + \{g(Ax+u) - \langle -\phi, Ax+u \rangle\}. \end{aligned}$$

Taking the infimum over all points u, and then over all points x, gives the inequalities

$$h(0) \leq -f^*(A^*\phi) - g^*(-\phi) \leq d \leq p = h(0).$$

Thus ϕ attains the supremum in problem (3.3.7), and $p = d$. An easy exercise shows that condition (3.3.9) implies condition (3.3.8). The proof of the calculus rule in the second part of the theorem is a simple consequence of the first part (Exercise 9). □

The case of the Fenchel theorem above, when the function g is simply the indicator function of a point, gives the following particularly elegant and useful corollary.

Corollary 3.3.11 (Fenchel duality for linear constraints) *Given any function $f : \mathbf{E} \to (\infty, +\infty]$, any linear map $A : \mathbf{E} \to \mathbf{Y}$, and any element b of $\mathbf{Y}$, the weak duality inequality*

$$\inf_{x \in \mathbf{E}} \{f(x) \mid Ax = b\} \geq \sup_{\phi \in \mathbf{Y}} \{\langle b, \phi \rangle - f^*(A^*\phi)\}$$

holds. If f is convex and b belongs to $\mathrm{core}\,(A\,\mathrm{dom}\, f)$ *then equality holds, and the supremum is attained when finite.*

A pretty application of the Fenchel duality circle of ideas is the calculation of polar cones. The *(negative) polar cone* of the set $K \subset \mathbf{E}$ is the convex cone

$$K^- = \{\phi \in \mathbf{E} \mid \langle \phi, x \rangle \leq 0 \ \text{ for all } x \in K\},$$

and the cone K^{--} is called the *bipolar*. A particularly important example of the polar cone is the normal cone to a convex set $C \subset \mathbf{E}$ at a point x in C, since $N_C(x) = (C - x)^-$.

We use the following two examples extensively; the proofs are simple exercises.

Proposition 3.3.12 (Self-dual cones)

$$(\mathbf{R}^n_+)^- = -\mathbf{R}^n_+ \ \ \textit{and} \ \ (\mathbf{S}^n_+)^- = -\mathbf{S}^n_+.$$

The next result shows how the calculus rules above can be used to derive geometric consequences.

Corollary 3.3.13 (Krein–Rutman polar cone calculus) *Any cones $H \subset \mathbf{Y}$ and $K \subset \mathbf{E}$ and linear map $A : \mathbf{E} \to \mathbf{Y}$ satisfy*

$$(K \cap A^{-1}H)^- \supset A^*H^- + K^-.$$

Equality holds if H and K are convex and satisfy $H - AK = Y$ (or in particular $AK \cap \operatorname{int} H \neq \emptyset$).

Proof. Rephrasing the definition of the polar cone shows that for any cone $K \subset \mathbf{E}$, the polar cone K^- is just $\partial \delta_K(0)$. The result now follows by the Fenchel theorem above. □

The polarity operation arises naturally from Fenchel conjugation, since for any cone $K \subset \mathbf{E}$ we have $\delta_{K^-} = \delta_K^*$, whence $\delta_{K^{--}} = \delta_K^{**}$. The next result, which is an elementary application of the Basic separation theorem (2.1.6), leads naturally into the development of the next chapter by identifying K^{--} as the closed convex cone generated by K.

Theorem 3.3.14 (Bipolar cone) *The bipolar cone of any nonempty set $K \subset \mathbf{E}$ is given by $K^{--} = \operatorname{cl}(\operatorname{conv}(\mathbf{R}_+ K))$.*

For example, we deduce immediately that the normal cone $N_C(x)$ to a convex set C at a point x in C, and the *(convex) tangent cone* to C at x defined by $T_C(x) = \operatorname{cl} \mathbf{R}_+(C - x)$, are polars of each other.

Exercise 20 outlines how to use these two results about cones to characterize *pointed* cones (those closed convex cones K satisfying $K \cap -K = \{0\}$).

Theorem 3.3.15 (Pointed cones) *If $K \subset \mathbf{E}$ is a closed convex cone, then K is pointed if and only if there is an element y of $\mathbf{E}$ for which the set*

$$C = \{x \in K \mid \langle x, y \rangle = 1\}$$

is compact and generates K (that is, $K = \mathbf{R}_+ C$).

Exercises and Commentary

The conjugation operation has been closely associated with the names of Legendre, Moreau, and Rockafellar, as well as Fenchel; see [167, 70]. Fenchel's original work is [76]. A good reference for properties of convex cones is [151]; see also [20]. The log barriers of Proposition 3.3.3 play a key role in interior point methods for linear and semidefinite programming—see, for example, [148]. The self-duality of the positive semidefinite cone is due to Fejer [99]. Hahn–Banach extension (Exercise 13(e)) is a key technique in functional analysis; see, for example, [98]. Exercise 21 (Order subgradients) is aimed at multicriteria optimization; a good reference is [176]. Our approach may be found, for example, in [20]. The last three functions g in Table 3.1 are respectively known as the *Boltzmann–Shannon*, *Fermi–Dirac*, and *Bose–Einstein* entropies.

1. For each of the functions f in Table 3.1, check the calculation of f^* and check $f = f^{**}$.

2. **(Quadratics)** For all matrices A in $\mathbf{S}^n_{++}$, prove the function $x \in \mathbf{R}^n \mapsto x^T Ax/2$ is convex and calculate its conjugate. Use the order-reversing property of the conjugacy operation to prove
$$A \succeq B \Leftrightarrow B^{-1} \succeq A^{-1} \text{ for } A \text{ and } B \text{ in } \mathbf{S}^n_{++}.$$

3. Verify the conjugates of the log barriers lb and ld claimed in Proposition 3.3.3.

4. * **(Self-conjugacy)** Consider functions $f : \mathbf{E} \to (\infty, +\infty]$.
   (a) Prove $f = f^*$ if and only if $f(x) = \|x\|^2/2$ for all points x in $\mathbf{E}$.
   (b) Find two distinct functions f satisfying $f(-x) = f^*(x)$ for all points x in $\mathbf{E}$.

5. * **(Support functions)** The conjugate of the indicator function of a nonempty set $C \subset \mathbf{E}$, namely $\delta_C^* : \mathbf{E} \to (\infty, +\infty]$, is called the *support function* of C. Calculate it for the following sets:
   (a) the halfspace $\{x \mid \langle a, x\rangle \le b\}$ for $0 \ne a \in \mathbf{E}$ and $b \in \mathbf{R}$
   (b) the unit ball B
   (c) $\{x \in \mathbf{R}^n_+ \mid \|x\| \le 1\}$
   (d) the *polytope* conv $\{a^1, a^2, \ldots, a^m\}$ for given points $a^1, a^2, \ldots, a^m$ in $\mathbf{E}$
   (e) a cone K
   (f) the epigraph of a convex function $f : \mathbf{E} \to (\infty, +\infty]$
   (g) the subdifferential $\partial f(\bar{x})$, where the function $f : \mathbf{E} \to (\infty, +\infty]$ is convex and the point $\bar{x}$ lies in core (dom f)
   (h) $\{Y \in \mathbf{S}^n_+ \mid \operatorname{tr} Y = 1\}$

6. Calculate the conjugate and biconjugate of the function
$$f(x_1, x_2) = \begin{cases} \dfrac{x_1^2}{2x_2} + x_2 \log x_2 - x_2 & \text{if } x_2 > 0 \\ 0 & \text{if } x_1 = x_2 = 0 \\ +\infty & \text{otherwise.} \end{cases}$$

7. ** **(Maximum entropy example)**

   (a) Prove the function g defined by (3.3.1) is convex.

   (b) For any point y in $\mathbf{R}^{m+1}$, prove

$$g^*(y) = \sup_{x \in \mathbf{R}^{m+1}} \Big\{ \sum_i (x_i \langle a^i, y \rangle - \exp^*(x_i)) \Bigm| \sum_i x_i = 1 \Big\}.$$

   (c) Apply Exercise 27 in Section 3.1 to deduce the conjugacy formula (3.3.2).

   (d) Compute the conjugate of the function of $x \in \mathbf{R}^{m+1}$,

$$\begin{cases} \sum_i \exp^*(x_i) & \text{if } \sum_i x_i = 1 \\ +\infty & \text{otherwise.} \end{cases}$$

8. Prove the Fenchel–Young inequality.

9. * **(Fenchel duality and convex calculus)** Fill in the details for the proof of Theorem 3.3.5 as follows.

   (a) Prove the weak duality inequality.

   (b) Prove the inclusion (3.3.10).

   Now assume f and g are convex.

   (c) Prove the function h defined in the proof is convex with domain $\operatorname{dom} g - A \operatorname{dom} f$.

   (d) Prove the implication (3.3.9) $\Rightarrow$ (3.3.8).

   Finally, assume in addition that condition (3.3.8) holds.

   (e) Suppose $\phi \in \partial(f + g \circ A)(\bar{x})$. Use the first part of the theorem and the fact that $\bar{x}$ is an optimal solution of the problem

$$\inf_{x \in \mathbf{E}} \{(f(x) - \langle \phi, x \rangle) + g(Ax)\}$$

   to deduce equality in part (b).

   (f) Prove points $\bar{x} \in \mathbf{E}$ and $\bar{\phi} \in \mathbf{Y}$ are optimal for problems (3.3.6) and (3.3.7), respectively, if and only if they satisfy the conditions $A^*\bar{\phi} \in \partial f(\bar{x})$ and $-\bar{\phi} \in \partial g(A\bar{x})$.

10. **(Normals to an intersection)** If the point x lies in two convex subsets C and D of $\mathbf{E}$ satisfying $0 \in \operatorname{core}(C - D)$ (or in particular $C \cap \operatorname{int} D \neq \emptyset$), use Section 3.1, Exercise 4 (Subgradients and normal cones) to prove

$$N_{C \cap D}(x) = N_C(x) + N_D(x).$$

11. * **(Failure of convex calculus)**

(a) Find convex functions $f, g : \mathbf{R} \to (\infty, +\infty]$ with

$$\partial f(0) + \partial g(0) \neq \partial (f + g)(0).$$

(Hint: Section 3.1, Exercise 5.)

(b) Find a convex function $g : \mathbf{R}^2 \to (\infty, +\infty]$ and a linear map $A : \mathbf{R} \to \mathbf{R}^2$ with $A^* \partial g(0) \neq \partial (g \circ A)(0)$.

12. * **(Infimal convolution)** If the functions $f, g : \mathbf{E} \to (-\infty, +\infty]$ are convex, we define the *infimal convolution* $f \odot g : \mathbf{E} \to [-\infty, +\infty]$ by

$$(f \odot g)(y) = \inf_x \{f(x) + g(y - x)\}.$$

(a) Prove $f \odot g$ is convex. (On the other hand, if g is concave prove so is $f \odot g$.)

(b) Prove $(f \odot g)^* = f^* + g^*$.

(c) If $\operatorname{dom} f \cap \operatorname{cont} g \neq \emptyset$, prove $(f + g)^* = f^* \odot g^*$.

(d) Given a nonempty set $C \subset \mathbf{E}$, define the *distance function* by

$$d_C(x) = \inf_{y \in C} \|x - y\|.$$

(i) Prove d_C^2 is a difference of convex functions, by observing

$$(d_C(x))^2 = \frac{\|x\|^2}{2} - \Big(\frac{\|\cdot\|^2}{2} + \delta_C\Big)^*(x).$$

Now suppose C is convex.

(ii) Prove d_C is convex and $d_C^* = \delta_B + \delta_C^*$.

(iii) For x in C prove $\partial d_C(x) = B \cap N_C(x)$.

(iv) If C is closed and $x \notin C$, prove

$$\nabla d_C(x) = d_C(x)^{-1}(x - P_C(x)),$$

where $P_C(x)$ is the nearest point to x in C.

(v) If C is closed, prove

$$\nabla \frac{d_C^2}{2}(x) = x - P_C(x)$$

for all points x.

(e) Define the *Lambert W-function* $W : \mathbf{R}_+ \to \mathbf{R}_+$ as the inverse of $y \in \mathbf{R}_+ \mapsto ye^y$. Prove the conjugate of the function

$$x \in \mathbf{R} \mapsto \exp^*(x) + \frac{x^2}{2}$$

is the function

$$y \in \mathbf{R} \mapsto W(e^y) + \frac{(W(e^y))^2}{2}.$$

13. * **(Applications of Fenchel duality)**

(a) **(Sandwich theorem)** Let the functions $f : \mathbf{E} \to (\infty, +\infty]$ and $g : \mathbf{Y} \to (\infty, +\infty]$ be convex and the map $A : \mathbf{E} \to \mathbf{Y}$ be linear. Suppose $f \geq -g \circ A$ and $0 \in \operatorname{core}(\operatorname{dom} g - A \operatorname{dom} f)$ (or $A \operatorname{dom} f \cap \operatorname{cont} g \neq \emptyset$). Prove there is an affine function $\alpha : \mathbf{E} \to \mathbf{R}$ satisfying $f \geq \alpha \geq -g \circ A$.

(b) Interpret the Sandwich theorem geometrically in the case when A is the identity.

(c) **(Pshenichnii–Rockafellar conditions [159])** If the convex set C in $\mathbf{E}$ satisfies the condition $C \cap \operatorname{cont} f \neq \emptyset$ (or the condition $\operatorname{int} C \cap \operatorname{dom} f \neq \emptyset$), and if f is bounded below on C, use part (a) to prove there is an affine function $\alpha \leq f$ with $\inf_C f = \inf_C \alpha$. Deduce that a point $\bar{x}$ minimizes f on C if and only if it satisfies $0 \in \partial f(\bar{x}) + N_C(\bar{x})$.

(d) Apply part (c) to the following two cases:

(i) C a single point $\{x^0\} \subset \mathbf{E}$

(ii) C a polyhedron $\{x \mid Ax \leq b\}$, where $b \in \mathbf{R}^n = \mathbf{Y}$

(e) **(Hahn–Banach extension)** If the function $f : \mathbf{E} \to \mathbf{R}$ is everywhere finite and sublinear, and for some linear subspace L of $\mathbf{E}$ the function $h : L \to \mathbf{R}$ is linear and *dominated* by f (in other words $f \geq h$ on L), prove there is a linear function $\alpha : \mathbf{E} \to \mathbf{R}$, dominated by f, which agrees with h on L.

14. Fill in the details of the proof of the Krein–Rutman calculus (3.3.13).

15. * **(Bipolar theorem)** For any nonempty set $K \subset \mathbf{E}$, prove the set $\operatorname{cl}(\operatorname{conv}(\mathbf{R}_+ K))$ is the smallest closed convex cone containing K. Deduce Theorem 3.3.14 (Bipolar cones).

16. * **(Sums of closed cones)**

(a) Prove that any cones $H, K \subset \mathbf{E}$ satisfy $(H + K)^- = H^- \cap K^-$.

(b) Deduce that if H and K are closed convex cones then they satisfy $(H \cap K)^- = \mathrm{cl}\,(H^- + K^-)$, and prove that the closure can be omitted under the condition $K \cap \mathrm{int}\, H \neq \emptyset$.

In $\mathbf{R}^3$, define sets

$$\begin{aligned} H &= \{x \mid x_1^2 + x_2^2 \leq x_3^2,\ x_3 \leq 0\} \text{ and} \\ K &= \{x \mid x_2 = -x_3\}. \end{aligned}$$

(c) Prove H and K are closed convex cones.

(d) Calculate the polar cones H^-, K^-, and $(H \cap K)^-$.

(e) Prove $(1,1,1) \in (H \cap K)^- \setminus (H^- + K^-)$, and deduce that the sum of two closed convex cones is not necessarily closed.

17. * **(Subdifferential of a max-function)** With the notation of Section 3.2, Exercise 13, suppose

$$\mathrm{dom}\, g_j \cap \bigcap_{i \in I \setminus \{j\}} \mathrm{cont}\, g_i \neq \emptyset$$

for some index j in I. Prove

$$\partial(\max_i g_i)(\bar{x}) = \mathrm{conv} \bigcup_{i \in I} \partial g_i(\bar{x}).$$

18. * **(Order convexity)** Given a Euclidean space $\mathbf{Y}$ and a closed convex cone $S \subset \mathbf{Y}$, we write $u \leq_S v$ for points u and v in $\mathbf{Y}$ if $v - u$ lies in S.

(a) Identify the partial order $\leq_S$ in the following cases:

(i) $S = \{0\}$

(ii) $S = \mathbf{Y}$

(iii) $\mathbf{Y} = \mathbf{R}^n$ and $S = \mathbf{R}^n_+$

Given a convex set $C \subset \mathbf{E}$, we say a function $F : C \to \mathbf{Y}$ is *S-convex* if it satisfies

$$F(\lambda x + \mu z) \leq_S \lambda F(x) + \mu F(z)$$

for all points x and z in $\mathbf{E}$ and nonnegative reals λ and μ satisfying $\lambda + \mu = 1$. If, furthermore, C is a cone and this inequality holds for *all* λ and μ in $\mathbf{R}_+$ then we say F is *S-sublinear*.

(b) Identify S-convexity in the cases listed in part (a).

(c) Prove F is S-convex if and only if the function $\langle \phi, F(\cdot) \rangle$ is convex for all elements ϕ of $-S^-$.

(d) Prove the following functions are $\mathbf{S}^n_+$-convex:

(i) $X \in \mathbf{S}^n \mapsto X^2$

(ii) $X \in \mathbf{S}^n_{++} \mapsto X^{-1}$

(iii) $X \in \mathbf{S}^n_+ \mapsto -X^{1/2}$

Hint: Use Exercise 25 in Section 3.1.

(e) Prove the function $X \in \mathbf{S}^2 \mapsto X^4$ is not $\mathbf{S}^2_+$-convex. Hint: Consider the matrices

$$\begin{bmatrix} 4 & 2 \\ 2 & 1 \end{bmatrix} \text{ and } \begin{bmatrix} 4 & 0 \\ 0 & 8 \end{bmatrix}.$$

19. **(Order convexity of inversion)** For any matrix A in $\mathbf{S}^n_{++}$, define a function $q_A : \mathbf{R}^n \to \mathbf{R}$ by $q_A(x) = x^T Ax/2$.

(a) Prove $q_A^* = q_{A^{-1}}$.

(b) For any other matrix B in $\mathbf{S}^n_{++}$, prove $2(q_A \odot q_B) \le q_{(A+B)/2}$. (See Exercise 12.)

(c) Deduce $(A^{-1} + B^{-1})/2 \succeq ((A + B)/2)^{-1}$.

20. ** **(Pointed cones and bases)** Consider a closed convex cone K in $\mathbf{E}$. A *base* for K is a convex set C with $0 \notin \operatorname{cl} C$ and $K = \mathbf{R}_+ C$. Using Exercise 16, prove the following properties are equivalent by showing the implications

$$(a) \Rightarrow (b) \Rightarrow (c) \Rightarrow (d) \Rightarrow (e) \Rightarrow (f) \Rightarrow (a).$$

(a) K is pointed.

(b) $\operatorname{cl}(K^- - K^-) = \mathbf{E}$.

(c) $K^- - K^- = \mathbf{E}$.

(d) K^- has nonempty interior. (Here you may use the fact that K^- has nonempty relative interior—see Section 1.1, Exercise 13.)

(e) There exists a vector y in $\mathbf{E}$ and real $\epsilon > 0$ with $\langle y, x \rangle \ge \epsilon \|x\|$ for all points x in K.

(f) K has a bounded base.

21. ** **(Order-subgradients)** This exercise uses the terminology of Exercise 18, and we assume the cone $S \subset \mathbf{Y}$ is pointed: $S \cap -S = \{0\}$. An element y of $\mathbf{Y}$ is the *S-infimum* of a set $D \subset \mathbf{Y}$ (written $y = \inf_S D$) if the conditions

(i) $D \subset y + S$ and

(ii) $D \subset z + S$ for some z in Y implies $y \in z + S$

both hold.

(a) Verify that this notion corresponds to the usual infimum when $\mathbf{Y} = \mathbf{R}$ and $S = \mathbf{R}_+$.

(b) Prove every subset of $\mathbf{Y}$ has at most one S-infimum.

(c) Prove *decreasing* sequences in S converge:

$$x_0 \geq_S x_1 \geq_S x_2 \ldots \geq_S 0$$

implies $\lim_n x_n$ exists and equals $\inf_S(x_n)$. (Hint: Prove first that $S \cap (x_0 - S)$ is compact using Section 1.1, Exercise 6 (Recession cones).)

An *S-subgradient* of F at a point x in C is a linear map $T : \mathbf{E} \to \mathbf{Y}$ satisfying

$$T(z - x) \leq_S F(z) - F(x) \quad \text{for all } z \text{ in } C.$$

The set of S-subgradients is denoted $\partial_S F(x)$. Suppose now $x \in \operatorname{core} C$. Generalize the arguments of Section 3.1 in the following steps.

(d) For any direction h in $\mathbf{E}$, prove

$$\nabla_S F(x; h) = \inf_S \{t^{-1}(F(x + th) - F(x)) \mid t > 0, \; x + th \in C\}$$

exists and, as a function of h, is S-sublinear.

(e) For any S-subgradient $T \in \partial_S F(x)$ and direction $h \in \mathbf{E}$, prove $Th \leq_S \nabla_S F(x; h)$.

(f) Given h in $\mathbf{E}$, prove there exists T in $\partial_S F(x)$ satisfying $Th = \nabla_S F(x; h)$. Deduce the max formula

$$\nabla_S F(x; h) = \max\{Th \mid T \in \partial_S F(x)\}$$

and, in particular, that $\partial_S F(x)$ is nonempty. (You should interpret the "max" in the formula.)

(g) The function F is *Gâteaux differentiable* at x (with derivative the linear map $\nabla F(x) : \mathbf{E} \to \mathbf{Y}$) if

$$\lim_{t \to 0} t^{-1}(F(x + th) - F(x)) = (\nabla F(x))h$$

holds for all h in $\mathbf{E}$. Prove this is the case if and only if $\partial_S F(x)$ is a singleton.

Now fix an element ϕ of $-\operatorname{int}(S^-)$.

(h) Prove $\langle \phi, F(\cdot) \rangle'(x; h) = \langle \phi, \nabla_S F(x; h) \rangle$.

(i) Prove F is Gâteaux differentiable at x if and only if $\langle \phi, F(\cdot)\rangle$ is likewise.

22. ** **(Linearly constrained examples)** Prove Corollary 3.3.11 (Fenchel duality for linear constraints). Deduce duality theorems for the following problems.

(a) **Separable problems**

$$\inf\Big\{\sum_{i=1}^{n} p(x_i)\ \Big|\ Ax = b\Big\},$$

where the map $A : \mathbf{R}^n \to \mathbf{R}^m$ is linear, $b \in \mathbf{R}^m$, and the function $p : \mathbf{R} \to (\infty, +\infty]$ is convex, defined as follows:

(i) **(Nearest points in polyhedrons)** $p(t) = t^2/2$ with domain $\mathbf{R}_+$.

(ii) **(Analytic center)** $p(t) = -\log t$ with domain $\mathbf{R}_{++}$.

(iii) **(Maximum entropy)** $p = \exp^*$.

What happens if the objective function is replaced by $\sum_i p_i(x_i)$?

(b) The **BFGS update** problem in Section 2.1, Exercise 13.

(c) The **DAD problem** in Section 3.1, Exercise 28.

(d) Example (3.3.1).

23. * **(Linear inequalities)** What does Corollary 3.3.11 (Fenchel duality for linear constraints) become if we replace the constraint $Ax = b$ by $Ax \in b + K$ where $K \subset \mathbf{Y}$ is a convex cone? Write down the dual problem for Section 3.2, Exercise 2, part (a), solve it, and verify the duality theorem.

24. **(Symmetric Fenchel duality)** For functions $f, g : \mathbf{E} \to [-\infty, +\infty]$, define the *concave conjugate* $g_* : \mathbf{E} \to [-\infty, +\infty]$ by

$$g_*(\phi) = \inf_{x\in\mathbf{E}} \{\langle \phi, x\rangle - g(x)\}.$$

Prove

$$\inf(f - g) \geq \sup(g_* - f^*),$$

with equality if f is convex, g is concave, and

$$0 \in \operatorname{core}(\operatorname{dom} f - \operatorname{dom}(-g)).$$

25. ** **(Divergence bounds [135])**

(a) Prove the function

$$t \in \mathbf{R} \mapsto 2(2+t)(\exp^* t + 1) - 3(t-1)^2$$

is convex and is minimized when $t = 1$.

(b) For v in $\mathbf{R}_{++}$ and u in $\mathbf{R}_+$, deduce the inequality

$$3(u-v)^2 \leq 2(u+2v)\Big(u \log\left(\frac{u}{v}\right) - u + v\Big).$$

Now suppose the vector p in $\mathbf{R}^n_{++}$ satisfies $\sum_1^n p_i = 1$.

(c) If the vector $q \in \mathbf{R}^n_{++}$ satisfies $\sum_1^n q_i = 1$, use the Cauchy–Schwarz inequality to prove the inequality

$$\Big(\sum_1^n |p_i - q_i|\Big)^2 \leq 3 \sum_1^n \frac{(p_i - q_i)^2}{p_i + 2q_i},$$

and deduce the inequality

$$\sum_1^n p_i \log\left(\frac{p_i}{q_i}\right) \geq \frac{1}{2}\Big(\sum_1^n |p_i - q_i|\Big)^2.$$

(d) Hence show the inequality

$$\log n + \sum_1^n p_i \log p_i \geq \frac{1}{2}\Big(\sum_1^n \Big|p_i - \frac{1}{n}\Big|\Big)^2.$$

(e) Use convexity to prove the inequality

$$\sum_1^n p_i \log p_i \leq \log \sum_1^n p_i^2.$$

(f) Deduce the bound

$$\log n + \sum_1^n p_i \log p_i \leq \frac{\max p_i}{\min p_i} - 1.$$

Chapter 4

Convex Analysis

4.1 Continuity of Convex Functions

We have already seen that linear functions are always continuous. More generally, a remarkable feature of convex functions on $\mathbf{E}$ is that they must be continuous on the interior of their domains. Part of the surprise is that an *algebraic/geometric* assumption (convexity) leads to a *topological* conclusion (continuity). It is this powerful fact that guarantees the usefulness of regularity conditions like $A\operatorname{dom} f \cap \operatorname{cont} g \neq \emptyset$ (3.3.9), which we studied in the previous section.

Clearly an arbitrary function f is bounded above on some neighbourhood of any point in $\operatorname{cont} f$. For convex functions the converse is also true, and in a rather strong sense, needing the following definition. For a real $L \geq 0$, we say that a function $f : \mathbf{E} \to (\infty, +\infty]$ is *Lipschitz (with constant L)* on a subset C of $\operatorname{dom} f$ if $|f(x) - f(y)| \leq L\|x - y\|$ for any points x and y in C. If f is Lipschitz on a neighbourhood of a point z then we say that f is *locally Lipschitz around* z. If $\mathbf{Y}$ is another Euclidean space we make analogous definitions for functions $F : \mathbf{E} \to \mathbf{Y}$, with $\|F(x) - F(y)\|$ replacing $|f(x) - f(y)|$.

Theorem 4.1.1 (Local boundedness) *Let $f : \mathbf{E} \to (\infty, +\infty]$ be a convex function. Then f is locally Lipschitz around a point z in its domain if and only if it is bounded above on a neighbourhood of z.*

Proof. One direction is clear, so let us without loss of generality take $z = 0$, $f(0) = 0$, and suppose $f \leq 1$ on $2B$; we shall deduce f is Lipschitz on B.

Notice first the bound $f \geq -1$ on $2B$, since convexity implies $f(-x) \geq -f(x)$ on $2B$. Now for any distinct points x and y in B, define $\alpha = \|y - x\|$ and fix a point $w = y + \alpha^{-1}(y - x)$, which lies in $2B$. By convexity we obtain

$$f(y) - f(x) \leq \frac{1}{1+\alpha} f(x) + \frac{\alpha}{1+\alpha} f(w) - f(x) \leq \frac{2\alpha}{1+\alpha} \leq 2\|y - x\|,$$

and the result now follows, since x and y may be interchanged. □

This result makes it easy to identify the set of points at which a convex function on $\mathbf{E}$ is continuous. First we prove a key lemma.

Lemma 4.1.2 *Let* Δ *be the* **simplex** $\{x \in \mathbf{R}^n_+ \mid \sum x_i \leq 1\}$. *If the function* $g : \Delta \to \mathbf{R}$ *is convex then it is continuous on* $\operatorname{int} \Delta$.

Proof. By the above result, we just need to show g is bounded above on Δ. But any point x in Δ satisfies

$$\begin{aligned} g(x) &= g\Big(\sum_1^n x_i e^i + (1 - \textstyle\sum x_i) 0\Big) \leq \sum_1^n x_i g(e^i) + (1 - \textstyle\sum x_i) g(0) \\ &\leq \max\{g(e^1), g(e^2), \ldots, g(e^n), g(0)\} \end{aligned}$$

(where $\{e^1, e^2, \ldots, e^n\}$ is the standard basis in $\mathbf{R}^n$). □

Theorem 4.1.3 (Convexity and continuity) *Let* $f : \mathbf{E} \to (\infty, +\infty]$ *be a convex function. Then* f *is continuous (in fact locally Lipschitz) on the interior of its domain.*

Proof. We lose no generality if we restrict ourselves to the case $\mathbf{E} = \mathbf{R}^n$. For any point x in $\operatorname{int}(\operatorname{dom} f)$ we can choose a neighbourhood of x in $\operatorname{dom} f$ that is a scaled down, translated copy of the simplex (since the simplex is bounded with nonempty interior). The proof of the preceding lemma now shows f is bounded above on a neighbourhood of x, and the result follows by Theorem 4.1.1 (Local boundedness). □

Since it is easy to see that if the convex function f is locally Lipschitz around a point $\bar{x}$ in $\operatorname{int}(\operatorname{dom} f)$ with constant L then $\partial f(\bar{x}) \subset LB$, we can also conclude that $\partial f(\bar{x})$ is a nonempty compact convex set. Furthermore, this result allows us to conclude quickly that "all norms on $\mathbf{E}$ are equivalent" (see Exercise 2).

We have seen that for a convex function f, the two sets $\operatorname{cont} f$ and $\operatorname{int}(\operatorname{dom} f)$ are identical. By contrast, our algebraic approach to the existence of subgradients involved $\operatorname{core}(\operatorname{dom} f)$. It transpires that this is the same set. To see this we introduce the idea of the *gauge function* $\gamma_C : \mathbf{E} \to (\infty, +\infty]$ associated with a nonempty set C in $\mathbf{E}$:

$$\gamma_C(x) = \inf\{\lambda \in \mathbf{R}_+ \mid x \in \lambda C\}.$$

It is easy to check γ_C is sublinear (and in particular convex) when C is convex. Notice $\gamma_B = \|\cdot\|$.

Theorem 4.1.4 (Core and interior) *The core and the interior of any convex set in* $\mathbf{E}$ *are identical and convex.*

Proof. Any convex set $C \subset \mathbf{E}$ clearly satisfies $\operatorname{int} C \subset \operatorname{core} C$. If we suppose, without loss of generality, $0 \in \operatorname{core} C$, then γ_C is everywhere finite, and hence continuous by the previous result. We claim

$$\operatorname{int} C = \{x \mid \gamma_C(x) < 1\}.$$

To see this, observe that the right hand side is contained in C, and is open by continuity, and hence is contained in $\operatorname{int} C$. The reverse inclusion is easy, and we deduce $\operatorname{int} C$ is convex. Finally, since $\gamma_C(0) = 0$, we see $0 \in \operatorname{int} C$, which completes the proof. □

The conjugate of the gauge function γ_C is the indicator function of a set $C^\circ \subset \mathbf{E}$ defined by

$$C^\circ = \{\phi \in \mathbf{E} \mid \langle \phi, x \rangle \leq 1 \text{ for all } x \in C\}.$$

We call C° the *polar set* for C. Clearly it is a closed convex set containing 0, and when C is a cone it coincides with the polar cone C^-. The following result therefore generalizes the Bipolar cone theorem (3.3.14).

Theorem 4.1.5 (Bipolar set) *The bipolar set of any subset* C *of* $\mathbf{E}$ *is given by*

$$C^{\circ\circ} = \operatorname{cl}(\operatorname{conv}(C \cup \{0\})).$$

The ideas of polarity and separating hyperplanes are intimately related. The separation-based proof of the above result (Exercise 5) is a good example, as is the next theorem, whose proof is outlined in Exercise 6.

Theorem 4.1.6 (Supporting hyperplane) *Suppose that the convex set* $C \subset \mathbf{E}$ *has nonempty interior and that the point* $\bar{x}$ *lies on the boundary of* C*. Then there is a* **supporting hyperplane** *to* C *at* $\bar{x}$*: there is a nonzero element* a *of* $\mathbf{E}$ *satisfying* $\langle a, x \rangle \geq \langle a, \bar{x} \rangle$ *for all points* x *in* C.

(The set $\{x \in \mathbf{E} \mid \langle a, x - \bar{x} \rangle = 0\}$ is the supporting hyperplane.)

To end this section we use this result to prove a remarkable theorem of Minkowski describing an extremal representation of finite-dimensional compact convex sets. An *extreme point* of a convex set $C \subset \mathbf{E}$ is a point x in C whose complement $C \setminus \{x\}$ is convex. We denote the set of extreme points by $\operatorname{ext} C$. We start with another exercise.

Lemma 4.1.7 *Given a supporting hyperplane* H *of a convex set* $C \subset \mathbf{E}$*, any extreme point of* $C \cap H$ *is also an extreme point of* C.

Our proof of Minkowski's theorem depends on two facts: first, any convex set that spans $\mathbf{E}$ and contains the origin has nonempty interior (see Section 1.1, Exercise 13(b)); second, we can define the *dimension* of a set $C \subset \mathbf{E}$ (written $\dim C$) as the dimension of span $(C - x)$ for *any* point x in C (see Section 1.1, Exercise 12 (Affine sets)).

Theorem 4.1.8 (Minkowski) *Any compact convex set $C \subset \mathbf{E}$ is the convex hull of its extreme points.*

Proof. Our proof is by induction on $\dim C$; clearly the result holds when $\dim C = 0$. Assume the result holds for all sets of dimension less than $\dim C$. We will deduce it for the set C.

By translating C and redefining $\mathbf{E}$, we can assume $0 \in C$ and $\operatorname{span} C = \mathbf{E}$. Thus C has nonempty interior.

Given any point x in $\operatorname{bd} C$, the Supporting hyperplane theorem (4.1.6) shows C has a supporting hyperplane H at x. By the induction hypothesis applied to the set $C \cap H$ we deduce, using Lemma 4.1.7,

$$x \in \operatorname{conv}(\operatorname{ext}(C \cap H)) \subset \operatorname{conv}(\operatorname{ext} C).$$

Thus we have proved $\operatorname{bd} C \subset \operatorname{conv}(\operatorname{ext} C)$, so $\operatorname{conv}(\operatorname{bd} C) \subset \operatorname{conv}(\operatorname{ext} C)$. But since C is compact it is easy to see $\operatorname{conv}(\operatorname{bd} C) = C$, and the result now follows. □

Exercises and Commentary

An easy introduction to convex analysis in finite dimensions is [181]. The approach we adopt here (and in the exercises) extends easily to infinite dimensions; see [98, 131, 153]. The Lipschitz condition was introduced in [129]. Minkowski's theorem first appeared in [141, 142]. The Open mapping theorem (Exercise 9) is another fundamental tool of functional analysis [98]. For recent references on Pareto minimization (Exercise 12), see [44].

1. * **(Points of continuity)** Suppose the function $f : \mathbf{E} \to (\infty, +\infty]$ is convex.

   (a) Use the Local boundedness theorem (4.1.1) to prove that f is continuous and finite at x if and only if it minorizes a function $g : \mathbf{E} \to (\infty, +\infty]$ which is continuous and finite at x.

   (b) Suppose f is continuous at some point y in $\operatorname{dom} f$. Use part (a) to prove directly that f is continuous at any point z in $\operatorname{core}(\operatorname{dom} f)$. (Hint: Pick a point u in $\operatorname{dom} f$ such that $z = \delta y + (1 - \delta)u$ for some real $\delta \in (0, 1)$; now observe that the function

   $$x \in \mathbf{E} \mapsto \delta^{-1}(f(\delta x + (1 - \delta)u) - (1 - \delta)f(u))$$

minorizes f.)

(c) Prove that f is continuous at a point x in $\mathrm{dom}\, f$ if and only if

$$(x, f(x) + \epsilon) \in \mathrm{int}\,(\mathrm{epi}\, f)$$

for some (all) real $\epsilon > 0$.

(d) Assuming $0 \in \mathrm{cont}\, f$, prove f^* has bounded level sets. Deduce that the function $X \in \mathbf{S}^n \mapsto \langle C, X \rangle + \mathrm{ld}\,(X)$ has compact level sets for any matrix C in $\mathbf{S}^n_{++}$.

(e) Assuming $x \in \mathrm{cont}\, f$, prove $\partial f(x)$ is a nonempty compact convex set.

2. **(Equivalent norms)** A *norm* is a sublinear function $|||\cdot||| : \mathbf{E} \to \mathbf{R}_+$ that satisfies $|||x||| = |||-x||| > 0$ for all nonzero points x in $\mathbf{E}$. By considering the function $|||\cdot|||$ on the standard unit ball B, prove any norm $|||\cdot|||$ is *equivalent* to the Euclidean norm $\|\cdot\|$: that is, there are constants $K \geq k > 0$ with $k\|x\| \leq |||x||| \leq K\|x\|$ for all x.

3. **(Examples of polars)** Calculate the polars of the following sets:

   (a) $\mathrm{conv}\,\big(B \cup \{(1,1), (-1,-1)\}\big) \subset \mathbf{R}^2$.

   (b) $\left\{(x,y) \in \mathbf{R}^2 \,\middle|\, y \geq b + \dfrac{x^2}{2}\right\}$ $(b \in \mathbf{R})$.

4. **(Polar sets and cones)** Suppose the set $C \subset \mathbf{E}$ is closed, convex, and contains 0. Prove the convex cones in $\mathbf{E} \times \mathbf{R}$

$$\mathrm{cl}\, \mathbf{R}_+(C \times \{1\}) \quad \text{and} \quad \mathrm{cl}\, \mathbf{R}_+(C^\circ \times \{-1\})$$

are mutually polar.

5. * **(Polar sets)** Suppose C is a nonempty subset of $\mathbf{E}$.

   (a) Prove $\gamma_C^* = \delta_{C^\circ}$.

   (b) Prove C° is a closed convex set containing 0.

   (c) Prove $C \subset C^{\circ\circ}$.

   (d) If C is a cone, prove $C^\circ = C^-$.

   (e) For a subset D of $\mathbf{E}$, prove $C \subset D$ implies $D^\circ \subset C^\circ$.

   (f) Prove C is bounded if and only if $0 \in \mathrm{int}\, C^\circ$.

   (g) For any closed halfspace $H \subset \mathbf{E}$ containing 0, prove $H^{\circ\circ} = H$.

   (h) Prove Theorem 4.1.5 (Bipolar set).

6. * **(Polar sets and strict separation)** Fix a nonempty set C in $\mathbf{E}$.

   (a) For points x in $\operatorname{int} C$ and ϕ in C°, prove $\langle \phi, x \rangle < 1$.

   (b) Assume further that C is a convex set. Prove γ_C is sublinear.

   (c) Assume in addition $0 \in \operatorname{core} C$. Deduce

   $$\operatorname{cl} C = \{x \mid \gamma_C(x) \leq 1\}.$$

   (d) Finally, suppose in addition that $D \subset \mathbf{E}$ is a convex set disjoint from the interior of C. By considering the Fenchel problem $\inf\{\delta_D + \gamma_C\}$, prove there is a closed halfspace containing D but disjoint from the interior of C.

7. * **(Polar calculus [23])** Suppose C and D are subsets of $\mathbf{E}$.

   (a) Prove $(C \cup D)^\circ = C^\circ \cap D^\circ$.

   (b) If C and D are convex, prove

   $$\operatorname{conv}(C \cup D) = \bigcup_{\lambda \in [0,1]} (\lambda C + (1 - \lambda) D).$$

   (c) If C is a convex cone and the convex set D contains 0, prove

   $$C + D \subset \operatorname{cl} \operatorname{conv}(C \cup D).$$

   Now suppose the closed convex sets K and H of $\mathbf{E}$ both contain 0.

   (d) Prove $(K \cap H)^\circ = \operatorname{cl} \operatorname{conv}(K^\circ \cup H^\circ)$.

   (e) If furthermore K is a cone, prove $(K \cap H)^\circ = \operatorname{cl}(K^\circ + H^\circ)$.

8. ** **(Polar calculus [23])** Suppose P is a cone in $\mathbf{E}$ and C is a nonempty subset of a Euclidean space $\mathbf{Y}$.

   (a) Prove $(P \times C)^\circ = P^\circ \times C^\circ$.

   (b) If furthermore C is compact and convex (possibly not containing 0), and K is a cone in $\mathbf{E} \times \mathbf{Y}$, prove

   $$(K \cap (P \times C))^\circ = (K \cap (P \times C^{\circ\circ}))^\circ.$$

   (c) If furthermore K and P are closed and convex, use Exercise 7 to prove

   $$(K \cap (P \times C))^\circ = \operatorname{cl}(K^\circ + (P^\circ \times C^\circ)).$$

   (d) Find a counterexample to part (c) when C is unbounded.

9. * **(Open mapping theorem)** Suppose the linear map $A : \mathbf{E} \to \mathbf{Y}$ is surjective.

   (a) Prove any set $C \subset \mathbf{E}$ satisfies $A \operatorname{core} C \subset \operatorname{core} AC$.

   (b) Deduce A is an *open map*: that is, the image of any open set is open.

   (c) Prove another condition ensuring condition (3.3.8) in the Fenchel theorem is that there is a point $\hat{x}$ in $\operatorname{int}(\operatorname{dom} f)$ with $A\hat{x}$ in $\operatorname{dom} g$ and A is surjective. Prove similarly that a sufficient condition for Fenchel duality with linear constraints (Corollary 3.3.11) to hold is A surjective and $b \in A(\operatorname{int}(\operatorname{dom} f))$.

   (d) Deduce that any cones $H \subset \mathbf{Y}$ and $K \subset \mathbf{E}$, and any surjective linear map $A : \mathbf{E} \to \mathbf{Y}$ satisfy $(K \cap A^{-1}H)^- = A^*H^- + K^-$, providing $H \cap A(\operatorname{int} K) \neq \emptyset$.

10. * **(Conical absorption)**

    (a) If the set $A \subset \mathbf{E}$ is convex, the set $C \subset \mathbf{E}$ is bounded, and $\mathbf{R}_+ A = \mathbf{E}$, prove there exists a real $\delta > 0$ such that $\delta C \subset A$.

    Now define two sets in $\mathbf{S}^2_+$ by

$$A = \left\{ \begin{bmatrix} y & x \\ x & z \end{bmatrix} \in \mathbf{S}^2_+ \;\middle|\; |x| \leq y^{2/3} \right\}, \quad \text{and}$$
$$C = \{X \in \mathbf{S}^2_+ \mid \operatorname{tr} X \leq 1\}.$$

    (b) Prove that both A and C are closed, convex, and contain 0, and that C is bounded.

    (c) Prove $\mathbf{R}_+ A = \mathbf{S}^2_+ = \mathbf{R}_+ C$.

    (d) Prove there is no real $\delta > 0$ such that $\delta C \subset A$.

11. **(Hölder's inequality)** This question develops an alternative approach to the theory of the p-norm $\|\cdot\|_p$ defined in Section 2.3, Exercise 6.

    (a) Prove $p^{-1}\|x\|_p^p$ is a convex function, and deduce the set

$$B_p = \{x \mid \|x\|_p \leq 1\}$$

    is convex.

    (b) Prove the gauge function $\gamma_{B_p}(\cdot)$ is exactly $\|\cdot\|_p$, and deduce $\|\cdot\|_p$ is convex.

    (c) Use the Fenchel–Young inequality (3.3.4) to prove that any vectors x and ϕ in $\mathbf{R}^n$ satisfy the inequality

$$p^{-1}\|x\|_p^p + q^{-1}\|\phi\|_q^q \geq \langle \phi, x \rangle.$$

(d) Assuming $\|u\|_p = \|v\|_q = 1$, deduce $\langle u, v \rangle \leq 1$, and hence prove that any vectors x and ϕ in $\mathbf{R}^n$ satisfy the inequality

$$\langle \phi, x \rangle \leq \|\phi\|_q \|x\|_p.$$

(e) Calculate B_p°.

12. * **(Pareto minimization)** We use the notation of Section 3.3, Exercise 18 (Order convexity), and we assume the cone S is pointed and has nonempty interior. Given a set $D \subset \mathbf{Y}$, we say a point y in D is a *Pareto minimum of D (with respect to S)* if

$$(y - D) \cap S = \{0\},$$

and a *weak minimum* if

$$(y - D) \cap \operatorname{int} S = \emptyset.$$

(a) Prove y is a Pareto (respectively weak) minimum of D if and only if it is a Pareto (respectively weak) minimum of $D + S$.

(b) The map $X \in \mathbf{S}_+^n \mapsto X^{1/2}$ is $\mathbf{S}_+^n$-order-preserving (Section 1.2, Exercise 5). Use this fact to prove, for any matrix Z in $\mathbf{S}_+^n$, the unique Pareto minimum of the set

$$\{X \in \mathbf{S}^n \mid X^2 \succeq Z^2\}$$

with respect to $\mathbf{S}_+^n$ is Z.

For a convex set $C \subset \mathbf{E}$ and an S-convex function $F : C \to \mathbf{Y}$, we say a point $\bar{x}$ in C is a *Pareto* (respectively, *weak*) *minimum* of the *vector optimization problem*

$$\inf\{F(x) \mid x \in C\} \tag{4.1.9}$$

if $F(\bar{x})$ is a Pareto (respectively weak) minimum of $F(C)$.

(c) Prove $F(C) + S$ is convex.

(d) **(Scalarization)** Suppose $\bar{x}$ is a weak minimum of the problem (4.1.9). By separating $(F(\bar{x}) - F(C) - S)$ and $\operatorname{int} S$ (using Exercise 6), prove there is a nonzero element ϕ of $-S^-$ such that $\bar{x}$ solves the *scalarized* convex optimization problem

$$\inf\{\langle \phi, F(x) \rangle \mid x \in C\}.$$

Conversely, show any solution of this problem is a weak minimum of (4.1.9).

13. **(Existence of extreme points)** Prove any nonempty compact convex set $C \subset \mathbf{E}$ has an extreme point, without using Minkowski's theorem, by considering the furthest point in C from the origin.

14. Prove Lemma 4.1.7.

15. For any compact convex set $C \subset \mathbf{E}$, prove $C = \mathrm{conv}\,(\mathrm{bd}\, C)$.

16. * **(A converse of Minkowski's theorem)** Suppose D is a subset of a compact convex set $C \subset \mathbf{E}$ satisfying $\mathrm{cl}\,(\mathrm{conv}\, D) = C$. Prove $\mathrm{ext}\, C \subset \mathrm{cl}\, D$.

17. * **(Extreme points)** Consider a compact convex set $C \subset \mathbf{E}$.

    (a) If $\dim \mathbf{E} \leq 2$, prove the set $\mathrm{ext}\, C$ is closed.

    (b) If $\mathbf{E}$ is $\mathbf{R}^3$ and C is the convex hull of the set

    $$\{(x, y, 0) \mid x^2 + y^2 = 1\} \cup \{(1, 0, 1), (1, 0, -1)\},$$

    prove $\mathrm{ext}\, C$ is not closed.

18. * **(Exposed points)** A point x in a convex set $C \subset \mathbf{E}$ is called *exposed* if there is an element ϕ of $\mathbf{E}$ such that $\langle \phi, x \rangle > \langle \phi, z \rangle$ for all points $z \neq x$ in C.

    (a) Prove any exposed point is an extreme point.

    (b) Find a set in $\mathbf{R}^2$ with an extreme point which is not exposed.

19. ** **(Tangency conditions)** Let $\mathbf{Y}$ be a Euclidean space. Fix a convex set C in $\mathbf{E}$ and a point x in C.

    (a) Show $x \in \mathrm{core}\, C$ if and only if $T_C(x) = \mathbf{E}$. (You may use Exercise 20(a).)

    (b) For a linear map $A : \mathbf{E} \to \mathbf{Y}$, prove $AT_C(x) \subset T_{AC}(Ax)$.

    (c) For another convex set D in $\mathbf{Y}$ and a point y in D, prove

    $$\begin{aligned} N_{C \times D}(x, y) &= N_C(x) \times N_D(y) \quad \text{and} \\ T_{C \times D}(x, y) &= T_C(x) \times T_D(y). \end{aligned}$$

    (d) Suppose the point x also lies in the convex set $G \subset \mathbf{E}$. Prove $T_C(x) - T_G(x) \subset T_{C-G}(0)$, and deduce

    $$0 \in \mathrm{core}\,(C - G) \Leftrightarrow T_C(x) - T_G(x) = \mathbf{E}.$$

    (e) Show that the condition (3.3.8) in the Fenchel theorem can be replaced by the condition

    $$T_{\mathrm{dom}\, g}(Ax) - AT_{\mathrm{dom}\, f}(x) = \mathbf{Y}$$

    for an arbitrary point x in $\mathrm{dom}\, f \cap A^{-1} \mathrm{dom}\, g$.

20. ** **(Properties of the relative interior)** (We use Exercise 9 (Open mapping theorem), as well as Section 1.1, Exercise 13.)

    (a) Let D be a nonempty convex set in $\mathbf{E}$. Prove D is a linear subspace if and only if $\mathrm{cl}\, D$ is a linear subspace. (Hint: $\mathrm{ri}\, D \neq \emptyset$.)

    (b) For a point x in a convex set $C \subset \mathbf{E}$, prove the following properties are equivalent:

    (i) $x \in \mathrm{ri}\, C$.
    (ii) The tangent cone $\mathrm{cl}\, \mathbf{R}_+(C - x)$ is a linear subspace.
    (iii) The normal cone $N_C(x)$ is a linear subspace.
    (iv) $y \in N_C(x) \Rightarrow -y \in N_C(x)$.

    (c) For a convex set $C \subset \mathbf{E}$ and a linear map $A : \mathbf{E} \to \mathbf{Y}$, prove $A \mathrm{ri}\, C \supset \mathrm{ri}\, AC$, and deduce

$$A \mathrm{ri}\, C = \mathrm{ri}\, AC.$$

    (d) Suppose U and V are convex sets in $\mathbf{E}$. Deduce

$$\mathrm{ri}\,(U - V) = \mathrm{ri}\, U - \mathrm{ri}\, V.$$

    (e) Apply Section 3.1, Exercise 29 (Relativizing the Max formula) to conclude that the condition (3.3.8) in the Fenchel theorem (3.3.5) can be replaced by

$$\mathrm{ri}\,(\mathrm{dom}\, g) \cap A \mathrm{ri}\,(\mathrm{dom}\, f) \neq \emptyset.$$

    (f) Suppose the function $f : \mathbf{E} \to (\infty, +\infty]$ is bounded below on the convex set $C \subset \mathbf{E}$, and $\mathrm{ri}\, C \cap \mathrm{ri}\,(\mathrm{dom}\, f) \neq \emptyset$. Prove there is an affine function $\alpha \leq f$ with $\inf_C f = \inf_C \alpha$.

21. ** **(Essential smoothness)** For any convex function f and any point $x \in \mathrm{bd}(\mathrm{dom}\, f)$, prove $\partial f(x)$ is either empty or unbounded. Deduce that a function is essentially smooth if and only if its subdifferential is always singleton or empty.

22. ** **(Birkhoff's theorem [15])** We use the notation of Section 1.2.

    (a) Prove $\mathbf{P}^n = \{(z_{ij}) \in \mathbf{\Gamma}^n \mid z_{ij} = 0 \text{ or } 1 \text{ for all } i, j\}$.

    (b) Prove $\mathbf{P}^n \subset \mathrm{ext}\,(\mathbf{\Gamma}^n)$.

    (c) Suppose $(z_{ij}) \in \mathbf{\Gamma}^n \setminus \mathbf{P}^n$. Prove there exist sequences of distinct indices $i_1, i_2, \ldots, i_m$ and $j_1, j_2, \ldots, j_m$ such that

$$0 < z_{i_r j_r}, z_{i_{r+1} j_r} < 1 \quad (r = 1, 2, \ldots, m)$$

(where $i_{m+1} = i_1$). For these sequences, show the matrix (z'_{ij}) defined by

$$z'_{ij} - z_{ij} = \begin{cases} \epsilon & \text{if } (i,j) = (i_r, j_r) \text{ for some } r \\ -\epsilon & \text{if } (i,j) = (i_{r+1}, j_r) \text{ for some } r \\ 0 & \text{otherwise} \end{cases}$$

is doubly stochastic for all small real ϵ. Deduce $(z_{ij}) \notin \text{ext}\,(\mathbf{\Gamma}^n)$.

(d) Deduce $\text{ext}\,(\mathbf{\Gamma}^n) = \mathbf{P}^n$. Hence prove Birkhoff's theorem (1.2.5).

(e) Use Carathéodory's theorem (Section 2.2, Exercise 5) to bound the number of permutation matrices needed to represent a doubly stochastic matrix in Birkhoff's theorem.

4.2 Fenchel Biconjugation

We have seen that many important convex functions $h : \mathbf{E} \to (\infty, +\infty]$ agree identically with their biconjugates h^{**}. Table 3.1 in Section 3.3 lists many one-dimensional examples, and the Bipolar cone theorem (3.3.14) shows $\delta_K = \delta_K^{**}$ for any closed convex cone K. In this section we isolate exactly the circumstances when $h = h^{**}$.

We can easily check that h^{**} is a minorant of h (that is, $h^{**} \leq h$ pointwise). Our specific aim in this section is to find conditions on a point x in $\mathbf{E}$ guaranteeing $h^{**}(x) = h(x)$. This becomes the key relationship for the study of duality in optimization. As we see in this section, the conditions we need are both geometric and topological. This is neither particularly surprising or stringent. Since any conjugate function must have a closed convex epigraph, we cannot expect a function to agree with its biconjugate unless the function itself has a closed convex epigraph. On the other hand, this restriction is not particularly strong since, as we saw in the previous section, convex functions automatically have strong continuity properties.

We say the function $h : \mathbf{E} \to [-\infty, +\infty]$ is *closed* if its epigraph is a closed set. We say h is *lower semicontinuous* at a point x in $\mathbf{E}$ if

$$\liminf h(x^r) \ \Big(= \lim_{s\to\infty} \inf_{r\geq s} h(x^r)\Big) \geq h(x)$$

for any sequence $x^r \to x$. A function $h : \mathbf{E} \to [-\infty, +\infty]$ is *lower semicontinuous* if it is lower semicontinuous at every point in $\mathbf{E}$; this is in fact equivalent to h being closed, which in turn holds if and only if h has closed level sets. Any two functions h and g satisfying $h \leq g$ (in which case we call h a *minorant* of g) must satisfy $h^* \geq g^*$, and hence $h^{**} \leq g^{**}$.

Theorem 4.2.1 (Fenchel biconjugation) *The three properties below are equivalent for any function $h : \mathbf{E} \to (-\infty, +\infty]$:*

(i) h is closed and convex.

(ii) $h = h^{**}$.

(iii) For all points x in $\mathbf{E}$,

$$h(x) = \sup\{\alpha(x) \mid \alpha \text{ an affine minorant of } h\}.$$

Hence the conjugacy operation induces a bijection between proper closed convex functions.

Proof. We can assume h is proper. Since conjugate functions are always closed and convex we know property (ii) implies property (i). Also, any affine minorant α of h satisfies $\alpha = \alpha^{**} \leq h^{**} \leq h$, and hence property (iii) implies (ii). It remains to show (i) implies (iii).

Fix a point x^0 in $\mathbf{E}$. Assume first $x^0 \in \mathrm{cl}\,(\mathrm{dom}\, h)$, and fix any real $r < h(x^0)$. Since h is closed, the set $\{x \mid h(x) > r\}$ is open, so there is an open convex neighbourhood U of x^0 with $h(x) > r$ on U. Now note that the set $\mathrm{dom}\, h \cap \mathrm{cont}\, \delta_U$ is nonempty, so we can apply the Fenchel theorem (3.3.5) to deduce that some element ϕ of $\mathbf{E}$ satisfies

$$r \leq \inf_x \{h(x) + \delta_U(x)\} = \{-h^*(\phi) - \delta_U^*(-\phi)\}. \tag{4.2.2}$$

Now define an affine function $\alpha(\cdot) = \langle \phi, \cdot \rangle + \delta_U^*(-\phi) + r$. Inequality (4.2.2) shows that α minorizes h, and by definition we know $\alpha(x^0) \geq r$. Since r was arbitrary, (iii) follows at the point $x = x^0$.

Suppose on the other hand x^0 does not lie in $\mathrm{cl}\,(\mathrm{dom}\, h)$. By the Basic separation theorem (2.1.6) there is a real b and a nonzero element a of $\mathbf{E}$ satisfying

$$\langle a, x^0 \rangle > b \geq \langle a, x \rangle \text{ for all points } x \text{ in } \mathrm{dom}\, h.$$

The argument in the preceding paragraph shows there is an affine minorant α of h. But now the affine function $\alpha(\cdot) + k(\langle a, \cdot \rangle - b)$ is a minorant of h for all $k = 1, 2, \ldots$. Evaluating these functions at $x = x^0$ proves property (iii) at x^0. The final remark follows easily. □

We immediately deduce that a closed convex function $h : \mathbf{E} \to [-\infty, +\infty]$ equals its biconjugate if and only if it is proper or identically $+\infty$ or $-\infty$.

Restricting the conjugacy bijection to finite sublinear functions gives the following result.

Corollary 4.2.3 (Support functions) *Fenchel conjugacy induces a bijection between everywhere-finite sublinear functions and nonempty compact convex sets in* $\mathbf{E}$:

(a) *If the set $C \subset \mathbf{E}$ is compact, convex and nonempty then the* **support function** δ_C^* *is everywhere finite and sublinear.*

(b) *If the function $h : \mathbf{E} \to \mathbf{R}$ is sublinear then $h^* = \delta_C$, where the set*

$$C = \{\phi \in \mathbf{E} \mid \langle \phi, d \rangle \leq h(d) \text{ for all } d \in \mathbf{E}\}$$

is nonempty, compact, and convex.

Proof. See Exercise 9. □

Conjugacy offers a convenient way to recognize when a convex function has bounded level sets.

Theorem 4.2.4 (Moreau–Rockafellar) *A closed convex proper function on* $\mathbf{E}$ *has bounded level sets if and only if its conjugate is continuous at 0.*

Proof. By Proposition 1.1.5, a convex function $f : \mathbf{E} \to (\infty, +\infty]$ has bounded level sets if and only if it satisfies the growth condition

$$\liminf_{\|x\| \to \infty} \frac{f(x)}{\|x\|} > 0.$$

Since f is closed we can check that this is equivalent to the existence of a minorant of the form $\epsilon\|\cdot\| + k \leq f(\cdot)$ for some constants $\epsilon > 0$ and k. Taking conjugates, this is in turn equivalent to f^* being bounded above near 0, and the result then follows by Theorem 4.1.1 (Local boundedness). □

Strict convexity is also easy to recognize via conjugacy, using the following result (see Exercise 19 for the proof).

Theorem 4.2.5 (Strict-smooth duality) *A proper closed convex function on* $\mathbf{E}$ *is essentially strictly convex if and only if its conjugate is essentially smooth.*

What can we say about h^{**} when the function $h : \mathbf{E} \to [-\infty, +\infty]$ is not necessarily closed? To answer this question we introduce the idea of the *closure* of h, denoted $\mathrm{cl}\, h$, defined by

$$\mathrm{epi}\,(\mathrm{cl}\, h) = \mathrm{cl}\,(\mathrm{epi}\, h). \tag{4.2.6}$$

It is easy to verify that $\mathrm{cl}\, h$ is then well-defined. The definition immediately implies $\mathrm{cl}\, h$ is the largest closed function minorizing h. Clearly if h is convex, so is $\mathrm{cl}\, h$. We leave the proof of the next simple result as an exercise.

Proposition 4.2.7 (Lower semicontinuity and closure) *If a function* $f : \mathbf{E} \to [-\infty, +\infty]$ *is convex then it is lower semicontinuous at a point* x *where it is finite if and only if* $f(x) = (\mathrm{cl}\, f)(x)$*. In this case* f *is proper.*

We can now answer the question we posed at the beginning of the section.

Theorem 4.2.8 *Suppose the function* $h : \mathbf{E} \to [-\infty, +\infty]$ *is convex.*

(*a*) *If* h^{**} *is somewhere finite then* $h^{**} = \mathrm{cl}\, h$.

(*b*) *For any point* x *where* h *is finite,* $h(x) = h^{**}(x)$ *if and only if* h *is lower semicontinuous at* x.

Proof. Observe first that since h^{**} is closed and minorizes h, we know $h^{**} \leq \mathrm{cl}\, h \leq h$. If h^{**} is somewhere finite then h^{**} (and hence $\mathrm{cl}\, h$) is never $-\infty$ by applying Proposition 4.2.7 (Lower semicontinuity and closure) to h^{**}. On the other hand, if h is finite and lower semicontinuous at x then Proposition 4.2.7 shows $\mathrm{cl}\, h(x)$ is finite, and applying the proposition again to $\mathrm{cl}\, h$ shows once more that $\mathrm{cl}\, h$ is never $-\infty$. In either case, the Fenchel biconjugation theorem implies $\mathrm{cl}\, h = (\mathrm{cl}\, h)^{**} \leq h^{**} \leq \mathrm{cl}\, h$, so $\mathrm{cl}\, h = h^{**}$. Part (a) is now immediate, while part (b) follows by using Proposition 4.2.7 once more. □

Any proper convex function h with an affine minorant has its biconjugate h^{**} somewhere finite. (In fact, because $\mathbf{E}$ is finite-dimensional, h^{**} is somewhere finite if and only if h is proper—see Exercise 25.)

Exercises and Commentary

Our approach in this section again extends easily to infinite dimensions; see for example [70]. Our definition of a closed function is a little different to that in [167], although they coincide for proper functions. The original version of von Neumann's minimax theorem (Exercise 16) had both the sets C and D simplices. The proof was by Brouwer's fixed point theorem (8.1.3). The Fisher information function introduced in Exercise 24 is useful in signal reconstruction [35]. The inequality in Exercise 20 (Logarithmic homogeneity) is important for interior point methods [148, Prop. 2.4.1].

1. Prove that any function $h : \mathbf{E} \to [-\infty, +\infty]$ satisfies $h^{**} \leq h$.

2. **(Lower semicontinuity and closedness)** For any given function $h : \mathbf{E} \to [-\infty, +\infty]$, prove the following properties are equivalent:

   (a) h is lower semicontinuous.

   (b) h has closed level sets.

   (c) h is closed.

   Prove that such a function has a global minimizer on any nonempty, compact set.

3. **(Pointwise maxima)** If the functions $f_\gamma : \mathbf{E} \to [-\infty, +\infty]$ are all convex (respectively closed) then prove the function defined by $f(x) = \sup_\gamma f_\gamma(x)$ is convex (respectively closed). Deduce that for any function $h : \mathbf{E} \to [-\infty, +\infty]$, the conjugate function h^* is closed and convex.

4. Verify directly that any affine function equals its biconjugate.

5. * **(Midpoint convexity)**

   (a) A function $f : \mathbf{E} \to (\infty, +\infty]$ is *midpoint convex* if it satisfies

   $$f\Big(\frac{x+y}{2}\Big) \leq \frac{f(x)+f(y)}{2} \quad \text{for all } x \text{ and } y \text{ in } \mathbf{E}.$$

   Prove a closed function is convex if and only if it is midpoint convex.

   (b) Use the inequality

   $$2(X^2+Y^2) \succeq (X+Y)^2 \quad \text{for all } X \text{ and } Y \text{ in } \mathbf{S}^n$$

   to prove the function $Z \in \mathbf{S}^n_+ \mapsto -Z^{1/2}$ is $\mathbf{S}^n_+$-convex (see Section 3.3, Exercise 18 (Order convexity)).

6. Is the Fenchel biconjugation theorem (4.2.1) valid for arbitrary functions $h : \mathbf{E} \to [-\infty, +\infty]$?

7. **(Inverse of subdifferential)** For a function $h : \mathbf{E} \to (\infty, +\infty]$, if points x and ϕ in $\mathbf{E}$ satisfy $\phi \in \partial h(x)$, prove $x \in \partial h^*(\phi)$. Prove the converse if h is closed and convex.

8. * **(Closed subdifferential)** If a function $h : \mathbf{E} \to (\infty, +\infty]$ is closed, prove the multifunction ∂h is *closed*: that is,

   $$\phi_r \in \partial h(x_r),\ x_r \to x,\ \phi_r \to \phi \ \Rightarrow \phi \in \partial h(x).$$

   Deduce that if h is essentially smooth and a sequence of points x_r in $\mathrm{int}\,(\mathrm{dom}\,h)$ approaches a point in $\mathrm{bd}\,(\mathrm{dom}\,h)$ then $\|\nabla h(x_r)\| \to \infty$.

9. * **(Support functions)**

   (a) Prove that if the set $C \subset \mathbf{E}$ is nonempty then δ^*_C is a closed sublinear function and $\delta^{**}_C = \delta_{\mathrm{cl\ conv}C}$. Prove that if C is also bounded then δ^*_C is everywhere finite.

   (b) Prove that any sets $C, D \subset \mathbf{E}$ satisfy

   $$\begin{aligned} \delta^*_{C+D} &= \delta^*_C + \delta^*_D \quad \text{and} \\ \delta^*_{\mathrm{conv}(C \cup D)} &= \max(\delta^*_C, \delta^*_D). \end{aligned}$$

   (c) Suppose the function $h : \mathbf{E} \to (-\infty, +\infty]$ is positively homogeneous, and define a closed convex set

   $$C = \{\phi \in \mathbf{E} \mid \langle \phi, d \rangle \leq h(d)\ \forall d\}.$$

   Prove $h^* = \delta_C$. Prove that if h is in fact sublinear and everywhere finite then C is nonempty and compact.

(d) Deduce Corollary 4.2.3 (Support functions).

10. * **(Almost homogeneous functions [19])** Prove that a function $f : \mathbf{E} \to \mathbf{R}$ has a representation

$$f(x) = \max_{i \in I}\{\langle a^i, x\rangle - b_i\} \quad (x \in \mathbf{E})$$

for a compact set $\{(a^i, b_i) \mid i \in I\} \subset \mathbf{E} \times \mathbf{R}$ if and only if f is convex and satisfies $\sup_{\mathbf{E}} |f - g| < \infty$ for some sublinear function g.

11. * Complete the details of the proof of the Moreau–Rockafellar theorem (4.2.4).

12. **(Compact bases for cones)** Consider a closed convex cone K. Using the Moreau–Rockafellar theorem (4.2.4), show that a point x lies in $\operatorname{int} K$ if and only if the set $\{\phi \in K^- \mid \langle \phi, x\rangle \geq -1\}$ is bounded. If the set $\{\phi \in K^- \mid \langle \phi, x\rangle = -1\}$ is nonempty and bounded, prove $x \in \operatorname{int} K$.

13. For any function $h : \mathbf{E} \to [-\infty, +\infty]$, prove the set $\operatorname{cl}(\operatorname{epi} h)$ is the epigraph of some function.

14. * **(Lower semicontinuity and closure)** For any convex function $h : \mathbf{E} \to [-\infty, +\infty]$ and any point x^0 in $\mathbf{E}$, prove

$$(\operatorname{cl} h)(x^0) = \lim_{\delta \downarrow 0} \inf_{\|x - x^0\| \leq \delta} h(x).$$

Deduce Proposition 4.2.7.

15. For any point x in $\mathbf{E}$ and any function $h : \mathbf{E} \to (-\infty, +\infty]$ with a subgradient at x, prove h is lower semicontinuous at x.

16. * **(Von Neumann's minimax theorem [185])** Suppose $\mathbf{Y}$ is a Euclidean space. Suppose that the sets $C \subset \mathbf{E}$ and $D \subset \mathbf{Y}$ are nonempty and convex with D closed and that the map $A : \mathbf{E} \to \mathbf{Y}$ is linear.

    (a) By considering the Fenchel problem

$$\inf_{x \in \mathbf{E}}\{\delta_C(x) + \delta_D^*(Ax)\}$$

prove

$$\inf_{x \in C} \sup_{y \in D} \langle y, Ax\rangle = \max_{y \in D} \inf_{x \in C} \langle y, Ax\rangle$$

(where the max is attained if finite), under the assumption

$$0 \in \operatorname{core}(\operatorname{dom} \delta_D^* - AC). \tag{4.2.9}$$

(b) Prove property (4.2.9) holds in either of the two cases

(i) D is bounded, or

(ii) A is surjective and 0 lies in $\operatorname{int} C$. (Hint: Use the Open mapping theorem, Section 4.1, Exercise 9.)

(c) Suppose both C and D are compact. Prove

$$\min_{x \in C} \max_{y \in D} \langle y, Ax \rangle = \max_{y \in D} \min_{x \in C} \langle y, Ax \rangle.$$

17. **(Recovering primal solutions)** Assume all the conditions for the Fenchel theorem (3.3.5) hold, and that in addition the functions f and g are closed.

(a) Prove that if the point $\bar{\phi} \in \mathbf{Y}$ is an optimal dual solution then the point $\bar{x} \in \mathbf{E}$ is optimal for the primal problem if and only if it satisfies the two conditions $\bar{x} \in \partial f^*(A^*\bar{\phi})$ and $A\bar{x} \in \partial g^*(-\bar{\phi})$.

(b) Deduce that if f^* is differentiable at the point $A^*\bar{\phi}$ then the only possible primal optimal solution is $\bar{x} = \nabla f^*(A^*\bar{\phi})$.

(c) ** Apply this result to the problems in Section 3.3, Exercise 22.

18. Calculate the support function δ_C^* of the set $C = \{x \in \mathbf{R}^2 \mid x_2 \geq x_1^2\}$. Prove the "contour" $\{y \mid \delta_C^*(y) = 1\}$ is not closed.

19. * **(Strict-smooth duality)** Consider a proper closed convex function $f : \mathbf{E} \to (\infty, +\infty]$.

(a) If f has Gâteaux derivative y at a point x in $\mathbf{E}$, prove the inequality

$$f^*(z) > f^*(y) + \langle x, z - y \rangle$$

for elements z of $\mathbf{E}$ distinct from y.

(b) If f is essentially smooth, prove that f^* is essentially strictly convex.

(c) Deduce the Strict-smooth duality theorem (4.2.5) using Exercise 23 in Section 3.1.

20. * **(Logarithmic homogeneity)** If the function $f : \mathbf{E} \to (\infty, +\infty]$ is closed, convex, and proper, then for any real $\nu > 0$ prove the inequality

$$f(x) + f^*(\phi) + \nu \log \langle x, -\phi \rangle \geq \nu \log \nu - \nu \quad \text{for all } x, \phi \in \mathbf{E}$$

holds (where we interpret $\log \alpha = -\infty$ when $\alpha \leq 0$) if and only f satisfies the condition

$$f(tx) = f(x) - \nu \log t \quad \text{for all } x \in \mathbf{E},\ t \in \mathbf{R}_{++}.$$

Hint: Consider first the case $\nu = 1$, and use the inequality

$$\alpha \leq -1 - \log(-\alpha).$$

21. * **(Cofiniteness)** Consider a function $h : \mathbf{E} \to (\infty, +\infty]$ and the following properties:

    (i) $h(\cdot) - \langle \phi, \cdot \rangle$ has bounded level sets for all ϕ in $\mathbf{E}$.

    (ii) $\lim_{\|x\| \to \infty} \|x\|^{-1} h(x) = +\infty$.

    (iii) h^* is everywhere finite.

    Complete the following steps.

    (a) Prove properties (i) and (ii) are equivalent.

    (b) If h is closed, convex and proper, use the Moreau–Rockafellar theorem (4.2.4) to prove properties (i) and (iii) are equivalent.

22. ** **(Computing closures)**

    (a) Prove any closed convex function $g : \mathbf{R} \to (\infty, +\infty]$ is continuous on its domain.

    (b) Consider a convex function $f : \mathbf{E} \to (\infty, +\infty]$. For any points x in $\mathbf{E}$ and y in $\mathrm{int}\,(\mathrm{dom}\, f)$, prove

    $$f^{**}(x) = \lim_{t \uparrow 1} f(y + t(x - y)).$$

    Hint: Use part (a) and the Accessibility lemma (Section 1.1, Exercise 11).

23. ** **(Recession functions)** This exercise uses Section 1.1, Exercise 6 (Recession cones). The *recession function* of a closed convex function $f : \mathbf{E} \to (\infty, +\infty]$ is defined by

    $$0^+ f(d) = \sup_{t \in \mathbf{R}_{++}} \frac{f(x + td) - f(x)}{t} \quad \text{for } d \text{ in } \mathbf{E},$$

    where x is any point in $\mathrm{dom}\, f$.

    (a) Prove $0^+ f$ is closed and sublinear.

    (b) Prove $\mathrm{epi}\,(0^+ f) = 0^+(\mathrm{epi}\, f)$, and deduce that $0^+ f$ is independent of the choice of the point x.

    (c) For any real $\alpha > \inf f$, prove

    $$0^+\{y \in \mathbf{E} \mid f(y) \leq \alpha\} = \{d \in \mathbf{E} \mid 0^+ f(d) \leq 0\}.$$

24. ** **(Fisher information function)** Let $f : \mathbf{R} \to (\infty, +\infty]$ be a given function, and define a function $g : \mathbf{R}^2 \to (\infty, +\infty]$ by

$$g(x, y) = \begin{cases} yf\left(\dfrac{x}{y}\right) & \text{if } y > 0 \\ +\infty & \text{otherwise.} \end{cases}$$

   (a) Prove g is convex if and only if f is convex.

   (b) Suppose f is essentially strictly convex. For y and v in $\mathbf{R}_{++}$ and x and u in $\mathbf{R}$, prove

$$g(x, y) + g(u, v) = g(x + y, u + v) \;\Leftrightarrow\; \frac{x}{y} = \frac{u}{v}.$$

   (c) Calculate g^*.

   (d) Suppose f is closed, convex, and finite at 0. Using Exercises 22 and 23, prove

$$g^{**}(x, y) = \begin{cases} yf\left(\dfrac{x}{y}\right) & \text{if } y > 0 \\ 0^+ f(x) & \text{if } y = 0 \\ +\infty & \text{otherwise.} \end{cases}$$

   (e) If $f(x) = x^2/2$ for all x in $\mathbf{R}$, calculate g.

   (f) Define a set $C = \{(x, y) \in \mathbf{R}^2 \mid x^2 \le y \le x\}$ and a function

$$h(x, y) = \begin{cases} \dfrac{x^3}{y^2} & \text{if } (x, y) \in C \setminus \{0\} \\ 0 & \text{if } (x, y) = 0 \\ +\infty & \text{otherwise.} \end{cases}$$

   Prove h is closed and convex but is not continuous relative to its (compact) domain C. Construct another such example with $\sup_C h$ finite.

25. ** **(Finiteness of biconjugate)** Consider a convex function $h : \mathbf{E} \to [-\infty, +\infty]$.

   (a) If h is proper and has an affine minorant, prove h^{**} is somewhere finite.

   (b) If h^{**} is somewhere finite, prove h is proper.

   (c) Use the fact that any proper convex function has a subgradient (Section 3.1, Exercise 29) to deduce that h^{**} is somewhere finite if and only if h is proper.

   (d) Deduce $h^{**} = \mathrm{cl}\, h$ for any convex function $h : E \to (\infty, +\infty]$.

26. ** **(Self-dual cones [8])** Consider a function $h : \mathbf{E} \to [-\infty, \infty)$ for which $-h$ is closed and sublinear, and suppose there is a point $\hat{x} \in \mathbf{E}$ satisfying $h(\hat{x}) > 0$. Define the *concave polar* of h as the function $h_\circ : \mathbf{E} \to [-\infty, \infty)$ given by

$$h_\circ(y) = \inf\{\langle x, y \rangle \mid h(x) \geq 1\}.$$

   (a) Prove $-h_\circ$ is closed and sublinear, and, for real $\lambda > 0$, we have $\lambda(\lambda h)_\circ = h_\circ$.

   (b) Prove the closed convex cone

$$K_h = \{(x, t) \in \mathbf{E} \times \mathbf{R} \mid |t| \leq h(x)\}$$

   has polar $(K_h)^- = -K_{h_\circ}$.

   (c) Suppose the vector $\alpha \in \mathbf{R}^n_{++}$ satisfies $\sum_i \alpha_i = 1$, and define a function $h^\alpha : \mathbf{R}^n \to [-\infty, +\infty)$ by

$$h^\alpha(x) = \begin{cases} \prod_i x_i^{\alpha_i} & \text{if } x \geq 0 \\ -\infty & \text{otherwise.} \end{cases}$$

   Prove $h^\alpha_\circ = h^\alpha / h^\alpha(\alpha)$, and deduce the cone

$$P_\alpha = K_{(h^\alpha(\alpha))^{-1/2} h^\alpha}$$

   is *self-dual*: $P_\alpha^- = -P_\alpha$.

   (d) Prove the cones

$$\begin{aligned} Q_2 &= \{(x, t, z) \in \mathbf{R}^3 \mid t^2 \leq 2xz, \ x, z \geq 0\} \ \text{ and} \\ Q_3 &= \{(x, t, z) \in \mathbf{R}^3 \mid 2|t|^3 \leq \sqrt{27} x z^2, \ x, z \geq 0\} \end{aligned}$$

   are self-dual.

   (e) Prove Q_2 is *isometric* to $\mathbf{S}^2_+$; in other words, there is a linear map $A : \mathbf{R}^3 \to \mathbf{S}^2_+$ preserving the norm and satisfying $AQ_2 = \mathbf{S}^2_+$.

27. ** **(Conical open mapping [8])** Define two closed convex cones in $\mathbf{R}^3$:

$$\begin{aligned} Q &= \{(x, y, z) \in \mathbf{R}^3 \mid y^2 \leq 2xz, \ x, z \geq 0\}. \ \text{ and} \\ S &= \{(w, x, y) \in \mathbf{R}^3 \mid 2|x|^3 \leq \sqrt{27} w y^2, \ w, y \geq 0\}. \end{aligned}$$

These cones are self-dual by Exercise 26. Now define convex cones in $\mathbf{R}^4$ by

$$C = (0 \times Q) + (S \times 0) \quad \text{and} \quad D = 0 \times \mathbf{R}^3.$$

   (a) Prove $C \cap D = \{0\} \times Q$.

(b) Prove $-C^- = (\mathbf{R} \times Q) \cap (S \times \mathbf{R})$.

(c) Define the projection $P : \mathbf{R}^4 \to \mathbf{R}^3$ by $P(w, x, y, z) = (x, y, z)$. Prove $P(C^-) = -Q$, or equivalently,

$$C^- + D^- = (C \cap D)^-.$$

(d) Deduce the normal cone formula

$$N_{C \cap D}(x) = N_C(x) + N_D(x) \quad \text{for all } x \text{ in } C \cap D$$

and, by taking polars, the tangent cone formula

$$T_{C \cap D}(x) = T_C(x) \cap T_D(x) \quad \text{for all } x \text{ in } C \cap D.$$

(e) Prove C^- is a closed convex pointed cone with nonempty interior and D^- is a line, and yet there is no constant $\epsilon > 0$ satisfying

$$(C^- + D^-) \cap \epsilon B \subset (C^- \cap B) + (D^- \cap B).$$

(Hint: Prove equivalently there is no $\epsilon > 0$ satisfying

$$P(C^-) \cap \epsilon B \subset P(C^- \cap B)$$

by considering the path $\{(t^2, t^3, t) \mid t \geq 0\}$ in Q.) Compare this with the situation when C and D are subspaces, using the Open mapping theorem (Section 4.1, Exercise 9).

(f) Consider the path

$$u(t) = \Big(\frac{2}{\sqrt{27}}, t^2, t^3, 0\Big) \quad \text{if } t \geq 0.$$

Prove $d_C(u(t)) = 0$ and $d_D(u(t)) = 2/\sqrt{27}$ for all $t \geq 0$, and yet

$$d_{C \cap D}(u(t)) \to +\infty \quad \text{as } t \to +\infty.$$

(Hint: Use the isometry in Exercise 26.)

28. ** **(Expected surprise [18])** An event occurs once every n days, with probability p_i on day i for $i = 1, 2, \ldots, n$. We seek a distribution maximizing the average surprise caused by the event. Define the "surprise" as minus the logarithm of the probability that the event occurs on day i given that it has not occurred so far. Using Bayes conditional probability rule, our problem is

$$\inf\Big\{S(p) \,\Big|\, \sum_1^n p_i = 1\Big\},$$

where we define the function $S : \mathbf{R}^n \to (\infty, +\infty]$ by

$$S(p) = \sum_{i=1}^{n} h\Big(p_i, \sum_{j=i}^{n} p_j\Big),$$

and the function $h : \mathbf{R}^2 \to (\infty, +\infty]$ by

$$h(x, y) = \begin{cases} x \log\left(\dfrac{x}{y}\right) & \text{if } x, y > 0 \\ 0 & \text{if } x \geq 0,\ y = 0 \\ +\infty & \text{otherwise.} \end{cases}$$

(a) Prove h is closed and convex using Exercise 24 (Fisher information function).

(b) Hence prove S is closed and convex.

(c) Prove the problem has an optimal solution.

(d) By imitating Section 3.1, Exercise 27 (Maximum entropy), show the solution $\bar{p}$ is unique and is expressed recursively by

$$\bar{p}_1 = \mu_1, \quad \bar{p}_k = \mu_k\Big(1 - \sum_{1}^{k-1} \bar{p}_j\Big) \text{ for } k = 2, 3, \ldots, n,$$

where the numbers μ_k are defined by the recursion

$$\mu_n = 1, \quad \mu_{k-1} = \mu_k e^{-\mu_k} \text{ for } k = 2, 3, \ldots, n.$$

(e) Deduce that the components of $\bar{p}$ form an increasing sequence and that $\bar{p}_{n-j}$ is independent of j.

(f) Prove $\bar{p}_1 \sim 1/n$ for large n.

4.3 Lagrangian Duality

The duality between a convex function h and its Fenchel conjugate h^* which we outlined earlier is an elegant piece of theory. The real significance, however, lies in its power to describe duality theory for convex programs, one of the most far-reaching ideas in the study of optimization.

We return to the convex program that we studied in Section 3.2:

$$\inf\{f(x) \mid g(x) \leq 0, \ x \in \mathbf{E}\}. \tag{4.3.1}$$

Here the function f and the components $g_1, g_2, \ldots, g_m : \mathbf{E} \to (\infty, +\infty]$ are convex, and satisfy $\emptyset \neq \operatorname{dom} f \subset \cap_1^m \operatorname{dom} g_i$. As before, the Lagrangian function $L : \mathbf{E} \times \mathbf{R}_+^m \to (\infty, +\infty]$ is defined by $L(x; \lambda) = f(x) + \lambda^T g(x)$.

Notice that the Lagrangian encapsulates all the information of the *primal problem* (4.3.1): clearly

$$\sup_{\lambda \in \mathbf{R}_+^m} L(x; \lambda) = \begin{cases} f(x) & \text{if } x \text{ is feasible} \\ +\infty & \text{otherwise,} \end{cases}$$

so if we denote the optimal value of (4.3.1) by $p \in [-\infty, +\infty]$, we could rewrite the problem in the following form:

$$p = \inf_{x \in \mathbf{E}} \sup_{\lambda \in \mathbf{R}_+^m} L(x; \lambda). \tag{4.3.2}$$

This makes it rather natural to consider an associated problem

$$d = \sup_{\lambda \in \mathbf{R}_+^m} \inf_{x \in \mathbf{E}} L(x; \lambda) \tag{4.3.3}$$

where $d \in [-\infty, +\infty]$ is called the *dual value*. Thus the *dual problem* consists of maximizing over vectors λ in $\mathbf{R}_+^m$ the *dual function* $\Phi(\lambda) = \inf_x L(x; \lambda)$. This dual problem is perfectly well-defined without any assumptions on the functions f and g. It is an easy exercise to show the "weak duality inequality" $p \geq d$. Notice Φ is concave.

It can happen that the primal value p is strictly larger than the dual value d (Exercise 5). In this case we say there is a *duality gap*. We next investigate conditions ensuring there is no duality gap. As in Section 3.2, the chief tool in our analysis is the primal value function $v : \mathbf{R}^m \to [-\infty, +\infty]$, defined by

$$v(b) = \inf\{f(x) \mid g(x) \leq b\}. \tag{4.3.4}$$

Below we summarize the relationships among these various ideas and pieces of notation.

Proposition 4.3.5 (Dual optimal value)

(a) *The primal optimal value* p *is* $v(0)$.

(b) *The conjugate of the value function satisfies*

$$v^*(-\lambda) = \begin{cases} -\Phi(\lambda) & \text{if } \lambda \geq 0 \\ +\infty & \text{otherwise.} \end{cases}$$

(c) *The dual optimal value* d *is* $v^{**}(0)$.

Proof. Part (a) is just the definition of p. Part (b) follows from the identities

$$\begin{aligned} v^*(-\lambda) &= \sup\{-\lambda^T b - v(b) \mid b \in \mathbf{R}^m\} \\ &= \sup\{-\lambda^T b - f(x) \mid g(x) + z = b,\ x \in \operatorname{dom} f,\ b \in \mathbf{R}^m,\ z \in \mathbf{R}^m_+\} \\ &= \sup\{-\lambda^T (g(x) + z) - f(x) \mid x \in \operatorname{dom} f,\ z \in \mathbf{R}^m_+\} \\ &= -\inf\{f(x) + \lambda^T g(x) \mid x \in \operatorname{dom} f\} + \sup\{-\lambda^T z \mid z \in \mathbf{R}^m_+\} \\ &= \begin{cases} -\Phi(\lambda) & \text{if } \lambda \geq 0 \\ +\infty & \text{otherwise.} \end{cases} \end{aligned}$$

Finally, we observe

$$d = \sup_{\lambda \in \mathbf{R}^m_+} \Phi(\lambda) = -\inf_{\lambda \in \mathbf{R}^m_+} -\Phi(\lambda) = -\inf_{\lambda \in \mathbf{R}^m_+} v^*(-\lambda) = v^{**}(0),$$

so part (c) follows. □

Notice the above result does not use convexity.

The reason for our interest in the relationship between a convex function and its biconjugate should now be clear, in light of parts (a) and (c) above.

Corollary 4.3.6 (Zero duality gap) *Suppose the value of the primal problem (4.3.1) is finite. Then the primal and dual values are equal if and only if the value function* v *is lower semicontinuous at 0. In this case the set of optimal dual solutions is* $-\partial v(0)$.

Proof. By the previous result, there is no duality gap exactly when the value function satisfies $v(0) = v^{**}(0)$, so Theorem 4.2.8 proves the first assertion. By part (b) of the previous result, dual optimal solutions λ are characterized by the property $0 \in \partial v^*(-\lambda)$ or equivalently $v^*(-\lambda) + v^{**}(0) = 0$. But we know $v(0) = v^{**}(0)$, so this property is equivalent to the condition $-\lambda \in \partial v(0)$. □

This result sheds new light on our proof of the Lagrangian necessary conditions (3.2.8); the proof in fact demonstrates the existence of a dual optimal solution. We consider below two distinct approaches to proving the absence of a duality gap. The first uses the Slater condition, as in Theorem 3.2.8, to force attainment in the dual problem. The second (dual) approach uses compactness to force attainment in the primal problem.

Theorem 4.3.7 (Dual attainment) *If the Slater condition holds for the primal problem (4.3.1) then the primal and dual values are equal, and the dual value is attained if finite.*

Proof. If p is $-\infty$ there is nothing to prove, since we know $p \geq d$. If on the other hand p is finite then, as in the proof of the Lagrangian necessary conditions (3.2.8), the Slater condition forces $\partial v(0) \neq \emptyset$. Hence v is finite and lower semicontinuous at 0 (Section 4.2, Exercise 15), and the result follows by Corollary 4.3.6 (Zero duality gap). $\square$

An indirect way of stating the Slater condition is that there is a point $\hat{x}$ in $\mathbf{E}$ for which the set $\{\lambda \in \mathbf{R}^m_+ \mid L(\hat{x};\lambda) \geq \alpha\}$ is compact for all real α. The second approach uses a "dual" condition to ensure the value function is closed.

Theorem 4.3.8 (Primal attainment) *Suppose that the functions*

$$f, g_1, g_2, \ldots, g_m : \mathbf{E} \to (\infty, +\infty]$$

are closed and that for some real $\hat{\lambda}_0 \geq 0$ and some vector $\hat{\lambda}$ in $\mathbf{R}^m_+$, the function $\hat{\lambda}_0 f + \hat{\lambda}^T g$ has compact level sets. Then the value function v defined by equation (4.3.4) is closed, and the infimum in this equation is attained when finite. Consequently, if the functions $f, g_1, g_2, \ldots, g_m$ are, in addition, convex and the dual value for the problem (4.3.1) is not $-\infty$, then the primal and dual values p and d are equal, and the primal value is attained when finite.

Proof. If the points (b^r, s_r) lie in $\operatorname{epi} v$ for $r = 1, 2, \ldots$ and approach the point (b, s) then for each integer r there is a point x^r in $\mathbf{E}$ satisfying $f(x^r) \leq s_r + r^{-1}$ and $g(x^r) \leq b^r$. Hence we deduce

$$(\hat{\lambda}_0 f + \hat{\lambda}^T g)(x^r) \leq \hat{\lambda}_0 (s_r + r^{-1}) + \hat{\lambda}^T b^r \to \hat{\lambda}_0 s + \hat{\lambda}^T b.$$

By the compact level set assumption, the sequence (x^r) has a subsequence converging to some point $\bar{x}$, and since all the functions are closed, we know $f(\bar{x}) \leq s$ and $g(\bar{x}) \leq b$. We deduce $v(b) \leq s$, so (b, s) lies in $\operatorname{epi} v$ as we required. When $v(b)$ is finite, the same argument with (b^r, s_r) replaced by $(b, v(b))$ for each r shows the infimum is attained.

If the functions $f, g_1, g_2, \ldots, g_m$ are convex then we know (from Section 3.2) v is convex. If d is $+\infty$ then again from the inequality $p \geq d$, there is nothing to prove. If d $(= v^{**}(0))$ is finite then Theorem 4.2.8 shows $v^{**} = \mathrm{cl}\, v$, and the above argument shows $\mathrm{cl}\, v = v$. Hence $p = v(0) = v^{**}(0) = d$, and the result follows. □

Notice that if either the objective function f or any one of the constraint functions $g_1, g_2, \ldots, g_m$ has compact level sets then the compact level set condition in the above result holds.

Exercises and Commentary

An attractive elementary account of finite-dimensional convex duality theory appears in [152]. A good reference for this kind of development in infinite dimensions is [98]. When the value function v is lower semicontinuous at 0 we say the problem (4.3.1) is *normal*; see [167]. If $\partial v(0) \neq \emptyset$ (or $v(0) = -\infty$) the problem is called *stable*; see, for example, [6]). For a straightforward account of interior point methods and the penalized linear program in Exercise 4 (Examples of duals) see [187, p. 40]. For more on the minimax theory in Exercise 14 see, for example, [60].

1. **(Weak duality)** Prove that the primal and dual values p and d defined by equations (4.3.2) and (4.3.3) satisfy $p \geq d$.

2. Calculate the Lagrangian dual of the problem in Section 3.2, Exercise 3.

3. **(Slater and compactness)** Prove the Slater condition holds for problem (4.3.1) if and only if there is a point $\hat{x}$ in $\mathbf{E}$ for which the level sets

$$\{\lambda \in \mathbf{R}^m_+ \mid -L(\hat{x}; \lambda) \leq \alpha\}$$

are compact for all real α.

4. **(Examples of duals)** Calculate the Lagrangian dual problem for the following problems (for given vectors $a^1, a^2, \ldots, a^m$, and c in $\mathbf{R}^n$).

   (a) The *linear program*

$$\inf_{x \in \mathbf{R}^n} \{\langle c, x\rangle \mid \langle a^i, x\rangle \leq b_i \;\text{ for } i = 1, 2, \ldots, m\}.$$

   (b) Another linear program

$$\inf_{x \in \mathbf{R}^n} \{\langle c, x\rangle + \delta_{\mathbf{R}^n_+}(x) \mid \langle a^i, x\rangle \leq b_i \;\text{ for } i = 1, 2, \ldots, m\}.$$

   (c) The *quadratic program* (for $C \in \mathbf{S}^n_{++}$)

$$\inf_{x \in \mathbf{R}^n} \Big\{\frac{x^T C x}{2} \;\Big|\; \langle a^i, x\rangle \leq b_i \;\text{ for } i = 1, 2, \ldots, m\Big\}.$$

(d) The separable problem

$$\inf_{x \in \mathbf{R}^n} \Big\{ \sum_{j=1}^{n} p(x_j) \,\Big|\, \langle a^i, x \rangle \le b_i \ \text{ for } i = 1, 2, \ldots, m \Big\}$$

for a given function $p : \mathbf{R} \to (\infty, +\infty]$.

(e) The *penalized linear program*

$$\inf_{x \in \mathbf{R}^n} \{ \langle c, x \rangle + \epsilon \mathrm{lb}\,(x) \mid \langle a^i, x \rangle \le b_i \ \text{ for } i = 1, 2, \ldots, m \}$$

for real $\epsilon > 0$.

For given matrices $A_1, A_2, \ldots, A_m$, and C in $\mathbf{S}^n$, calculate the dual of the *semidefinite program*

$$\inf_{X \in \mathbf{S}^n_+} \{ \mathrm{tr}\,(CX) + \delta_{\mathbf{S}^n_+}(X) \mid \mathrm{tr}\,(A_i X) \le b_i \ \text{ for } i = 1, 2, \ldots, m \},$$

and the *penalized semidefinite program*

$$\inf_{X \in \mathbf{S}^n_+} \{ \mathrm{tr}\,(CX) + \epsilon \mathrm{ld}\,X \mid \mathrm{tr}\,(A_i X) \le b_i \ \text{ for } i = 1, 2, \ldots, m \}$$

for real $\epsilon > 0$.

5. **(Duffin's duality gap, continued)**

(a) For the problem considered in Section 3.2, Exercise 8, namely

$$\inf_{x \in \mathbf{R}^2} \{ e^{x_2} \mid \|x\| - x_1 \le 0 \},$$

calculate the dual function, and hence find the dual value.

(b) Repeat part (a) with the objective function e^{x_2} replaced by x_2.

6. Consider the problem

$$\inf\{ \exp^*(x_1) + \exp^*(x_2) \mid x_1 + 2x_2 - 1 \le 0, \ x \in \mathbf{R}^2 \}.$$

Write down the Lagrangian dual problem, solve the primal and dual problems, and verify that the optimal values are equal.

7. Given a matrix C in $\mathbf{S}^n_{++}$, calculate

$$\inf_{X \in \mathbf{S}^n_{++}} \{ \mathrm{tr}\,(CX) \mid -\log(\det X) \le 0 \}$$

by Lagrangian duality.

8. * **(Mixed constraints)** Explain why an appropriate dual for the problem
$$\inf\{f(x) \mid g(x) \le 0,\ h(x) = 0\}$$
for a function $h : \mathrm{dom}\, f \to \mathbf{R}^k$ is
$$\sup_{\lambda \in \mathbf{R}^m_+,\ \mu \in \mathbf{R}^k} \ \inf_{x \in \mathrm{dom}\, f} \{f(x) + \lambda^T g(x) + \mu^T h(x)\}.$$

9. **(Fenchel and Lagrangian duality)** Let $\mathbf{Y}$ be a Euclidean space. By suitably rewriting the primal Fenchel problem
$$\inf_{x \in \mathbf{E}} \{f(x) + g(Ax)\}$$
for given functions $f : \mathbf{E} \to (\infty, +\infty]$, $g : \mathbf{Y} \to (\infty, +\infty]$, and linear $A : \mathbf{E} \to \mathbf{Y}$, interpret the dual Fenchel problem
$$\sup_{\phi \in \mathbf{Y}} \{-f^*(A^*\phi) - g^*(-\phi)\}$$
as a Lagrangian dual problem.

10. **(Trust region subproblem duality [175])** Given a matrix A in $\mathbf{S}^n$ and a vector b in $\mathbf{R}^n$, consider the *nonconvex* problem
$$\inf \left\{x^T Ax + b^T x \mid x^T x - 1 \le 0,\ x \in \mathbf{R}^n\right\}.$$
Complete the following steps to prove there is an optimal dual solution, with no duality gap.
    (i) Prove the result when A is positive semidefinite.
    (ii) If A is not positive definite, prove the primal optimal value does not change if we replace the inequality in the constraint by an equality.
    (iii) By observing for any real α the equality
$$\min \left\{x^T Ax + b^T x \mid x^T x = 1\right\} = \\ -\alpha + \min \left\{x^T (A + \alpha I)x + b^T x \mid x^T x = 1\right\},$$
    prove the general result.

11. ** If there is no duality gap, prove that dual optimal solutions are the same as Karush–Kuhn–Tucker vectors (Section 3.2, Exercise 9).

12. * **(Conjugates of compositions)** Consider the composition $g \circ f$ of a nondecreasing convex function $g : \mathbf{R} \to (\infty, +\infty]$ with a convex function $f : \mathbf{E} \to (\infty, +\infty]$. We interpret $g(+\infty) = +\infty$, and we

assume there is a point $\hat{x}$ in $\mathbf{E}$ satisfying $f(\hat{x}) \in \operatorname{int}(\operatorname{dom} g)$. Use Lagrangian duality to prove the formula, for ϕ in $\mathbf{E}$,

$$(g \circ f)^*(\phi) = \inf_{t \in \mathbf{R}_+} \left\{ g^*(t) + t f^*\left(\frac{\phi}{t}\right) \right\},$$

where we interpret

$$0 f^*\left(\frac{\phi}{0}\right) = \delta^*_{\operatorname{dom} f}(\phi).$$

13. ** **(A symmetric pair [28])**

(a) Given real $\gamma_1, \gamma_2, \ldots, \gamma_n > 0$, define $h : \mathbf{R}^n \to (\infty, +\infty]$ by

$$h(x) = \begin{cases} \prod_{i=1}^n x_i^{-\gamma_i} & \text{if } x \in \mathbf{R}^n_{++} \\ +\infty & \text{otherwise.} \end{cases}$$

By writing $g(x) = \exp(\log g(x))$ and using the composition formula in Exercise 12, prove

$$h^*(y) = \begin{cases} -(\gamma + 1) \displaystyle\prod_{i=1}^n \left(\frac{-y_i}{\gamma_i} \right)^{\gamma_i / (\gamma + 1)} & \text{if } -y \in \mathbf{R}^n_+ \\ +\infty & \text{otherwise,} \end{cases}$$

where $\gamma = \sum_i \gamma_i$.

(b) Given real $\alpha_1, \alpha_2, \ldots, \alpha_n > 0$, define $\alpha = \sum_i \alpha_i$ and suppose a real μ satisfies $\mu > \alpha + 1$. Now define a function $f : \mathbf{R}^n \times \mathbf{R} \to (\infty, +\infty]$ by

$$f(x, s) = \begin{cases} \mu^{-1} s^\mu \prod_i x_i^{-\alpha_i} & \text{if } x \in \mathbf{R}^n_{++},\ s \in \mathbf{R}_+ \\ +\infty & \text{otherwise.} \end{cases}$$

Use part (a) to prove

$$f^*(y, t) = \begin{cases} \rho \nu^{-1} t^\nu \prod_i (-y_i)^{-\beta_i} & \text{if } -y \in \mathbf{R}^n_{++},\ t \in \mathbf{R}_+ \\ +\infty & \text{otherwise} \end{cases}$$

for constants

$$\nu = \frac{\mu}{\mu - (\alpha + 1)}, \quad \beta_i = \frac{\alpha_i}{\mu - (\alpha + 1)}, \quad \rho = \prod_i \left(\frac{\alpha_i}{\mu} \right)^{\beta_i}.$$

(c) Deduce $f = f^{**}$, whence f is convex.

(d) Give an alternative proof of the convexity of f by using Section 4.2, Exercise 24(a) (Fisher information function) and induction.

(e) Prove f is strictly convex.

14. ** **(Convex minimax theory)** Suppose that $\mathbf{Y}$ is a Euclidean space, that the sets $C \subset \mathbf{Y}$ and $D \subset \mathbf{E}$ are nonempty, and consider a function $\psi : C \times D \to \mathbf{R}$.

(a) Prove the inequality

$$\sup_{y \in D} \inf_{x \in C} \psi(x, y) \le \inf_{x \in C} \sup_{y \in D} \psi(x, y).$$

(b) We call a point $(\bar{x}, \bar{y})$ in $C \times D$ a *saddlepoint* if it satisfies

$$\psi(\bar{x}, y) \le \psi(\bar{x}, \bar{y}) \le \psi(x, \bar{y}) \quad \text{for all } x \in C,\ y \in D.$$

In this case prove

$$\sup_{y \in D} \inf_{x \in C} \psi(x, y) = \psi(\bar{x}, \bar{y}) = \inf_{x \in C} \sup_{y \in D} \psi(x, y).$$

(c) Suppose the function $p_y : \mathbf{E} \to (\infty, +\infty]$ defined by

$$p_y(x) = \begin{cases} \psi(x, y) & \text{if } x \in C \\ +\infty & \text{otherwise} \end{cases}$$

is convex, for all y in D. Prove the function $h : \mathbf{Y} \to [-\infty, +\infty]$ defined by

$$h(z) = \inf_{x \in C} \sup_{y \in D} \{\psi(x, y) + \langle z, y \rangle\}$$

is convex.

(d) Suppose the function $q_x : \mathbf{Y} \to (\infty, +\infty]$ defined by

$$q_x(y) = \begin{cases} -\psi(x, y) & \text{if } y \in D \\ +\infty & \text{otherwise} \end{cases}$$

is closed and convex for all points x in C. Deduce

$$h^{**}(0) = \sup_{y \in D} \inf_{x \in C} \psi(x, y).$$

(e) Suppose that for all points y in D the function p_y defined in part (c) is closed and convex, and that for some point $\hat{y}$ in D, $p_{\hat{y}}$ has compact level sets. If h is finite at 0, prove it is lower semicontinuous there. If the assumption in part (d) also holds, deduce

$$\sup_{y \in D} \inf_{x \in C} \psi(x, y) = \min_{x \in C} \sup_{y \in D} \psi(x, y).$$

(f) Suppose the functions $f, g_1, g_2, \ldots, g_s : \mathbf{R}^t \to (\infty, +\infty]$ are closed and convex. Interpret the above results in the following two cases:

(i)

$$\begin{aligned} C &= (\operatorname{dom} f) \cap \Big(\bigcap_{i=1}^{s} \operatorname{dom} g_i \Big) \\ D &= \mathbf{R}^s_+ \\ \psi(u,w) &= f(u) + \textstyle\sum_{i=1}^{s} w_i g_i(u). \end{aligned}$$

(ii)

$$\begin{aligned} C &= \mathbf{R}^s_+ \\ D &= (\operatorname{dom} f) \cap \Big(\bigcap_{i=1}^{s} \operatorname{dom} g_i \Big) \\ \psi(u,w) &= -f(w) - \textstyle\sum_{i=1}^{s} u_i g_i(w). \end{aligned}$$

(g) **(Kakutani [109])** Suppose that the nonempty sets $C \subset \mathbf{Y}$ and $D \subset \mathbf{E}$ are compact and convex, that the function $\psi : C \times D \to \mathbf{R}$ is continuous, that $\psi(x, y)$ is convex in the variable x for all fixed y in D, and that $-\psi(x, y)$ is convex in the variable y for all points x in C. Deduce ψ has a saddlepoint.

Chapter 5

Special Cases

5.1 Polyhedral Convex Sets and Functions

In our earlier section on theorems of the alternative (Section 2.2), we observed that finitely generated cones are closed. Remarkably, a finite linear-algebraic assumption leads to a topological conclusion. In this section we pursue the consequences of this type of assumption in convex analysis.

There are two natural ways to impose a finite linear structure on the sets and functions we consider. The first we have already seen: a "polyhedron" (or *polyhedral* set) is a finite intersection of closed halfspaces in $\mathbf{E}$, and we say a function $f : \mathbf{E} \to [-\infty, +\infty]$ is *polyhedral* if its epigraph is polyhedral. On the other hand, a *polytope* is the convex hull of a finite subset of $\mathbf{E}$, and we call a subset of $\mathbf{E}$ *finitely generated* if it is the sum of a polytope and a finitely generated cone (in the sense of formula (2.2.11)). Notice we do not yet know if a cone that is a finitely generated set in this sense is finitely generated in the sense of (2.2.11); we return to this point later in the section. The function f is *finitely generated* if its epigraph is finitely generated. A central result of this section is that polyhedra and finitely generated sets in fact coincide.

We begin with some easy observations collected together in the following two results.

Proposition 5.1.1 (Polyhedral functions) *Suppose that the function $f : \mathbf{E} \to [-\infty, +\infty]$ is polyhedral. Then f is closed and convex and can be decomposed in the form*

$$f = \max_{i \in I} g_i + \delta_P, \tag{5.1.2}$$

where the index set I is finite (and possibly empty), the functions g_i are affine, and the set $P \subset \mathbf{E}$ is polyhedral (and possibly empty). Thus the domain of f is polyhedral and coincides with $\operatorname{dom} \partial f$ *if f is proper.*

Proof. Since any polyhedron is closed and convex, so is f, and the decomposition (5.1.2) follows directly from the definition. If f is proper then both the sets I and P are nonempty in this decomposition. At any point x in P ($= \operatorname{dom} f$) we know $0 \in \partial \delta_P(x)$, and the function $\max_i g_i$ certainly has a subgradient at x since it is everywhere finite. Hence we deduce the condition $\partial f(x) \neq \emptyset$. □

Proposition 5.1.3 (Finitely generated functions) *Suppose the function $f : \mathbf{E} \to [-\infty, +\infty]$ is finitely generated. Then f is closed and convex and* $\operatorname{dom} f$ *is finitely generated. Furthermore, f^* is polyhedral.*

Proof. Polytopes are compact and convex (by Carathéodory's theorem (Section 2.2, Exercise 5)), and finitely generated cones are closed and convex, so finitely generated sets (and therefore functions) are closed and convex (by Section 1.1, Exercise 5(a)). We leave the remainder of the proof as an exercise. □

An easy exercise shows that a set $P \subset \mathbf{E}$ is polyhedral (respectively, finitely generated) if and only if δ_P is polyhedral (respectively, finitely generated).

To prove that polyhedra and finitely generated sets in fact coincide, we consider the two extreme special cases: first, compact sets, and second, cones. Observe first that compact, finitely generated sets are just polytopes, directly from the definition.

Lemma 5.1.4 *Any polyhedron has at most finitely many extreme points.*

Proof. Fix a finite set of affine functions $\{g_i \mid i \in I\}$ on $\mathbf{E}$, and consider the polyhedron

$$P = \{x \in \mathbf{E} \mid g_i(x) \leq 0 \text{ for } i \in I\}.$$

For any point x in P, the "active set" is $\{i \in I \mid g_i(x) = 0\}$. Suppose two distinct extreme points x and y of P have the same active set. Then for any small real ϵ the points $x \pm \epsilon(y - x)$ both lie in P. But this contradicts the assumption that x is extreme. Hence different extreme points have different active sets, and the result follows. □

This lemma together with Minkowski's theorem (4.1.8) reveals the nature of compact polyhedra.

Theorem 5.1.5 *Any compact polyhedron is a polytope.*

We next turn to cones.

Lemma 5.1.6 *Any polyhedral cone is a finitely generated cone (in the sense of (2.2.11)).*

Proof. Given a polyhedral cone $P \subset \mathbf{E}$, define a subspace $L = P \cap -P$ and a pointed polyhedral cone $K = P \cap L^{\perp}$. Observe the decomposition $P = K \oplus L$. By the Pointed cone theorem (3.3.15), there is an element y of $\mathbf{E}$ for which the set

$$C = \{x \in K \mid \langle x, y \rangle = 1\}$$

is compact and satisfies $K = \mathbf{R}_+ C$. Since C is polyhedral, the previous result shows it is a polytope. Thus K is finitely generated, whence so is P. □

Theorem 5.1.7 (Polyhedrality) *A set or function is polyhedral if and only if it is finitely generated.*

Proof. For finite sets $\{a_i \mid i \in I\} \subset \mathbf{E}$ and $\{b_i \mid i \in I\} \subset \mathbf{R}$, consider the polyhedron in $\mathbf{E}$ defined by

$$P = \{x \in \mathbf{E} \mid \langle a_i, x \rangle \le b_i \text{ for } i \in I\}.$$

The polyhedral cone in $\mathbf{E} \times \mathbf{R}$ defined by

$$Q = \{(x, r) \in \mathbf{E} \times \mathbf{R} \mid \langle a_i, x \rangle - b_i r \le 0 \text{ for } i \in I\}$$

is finitely generated by the previous lemma, so there are finite subsets $\{x_j \mid j \in J\}$ and $\{y_t \mid t \in T\}$ of $\mathbf{E}$ with

$$Q = \Big\{ \sum_{j \in J} \lambda_j (x_j, 1) + \sum_{t \in T} \mu_t (y_t, 0) \,\Big|\, \lambda_j \in \mathbf{R}_+ \text{ for } j \in J,\ \mu_t \in \mathbf{R}_+ \text{ for } t \in T \Big\}.$$

We deduce

$$\begin{aligned} P &= \{x \mid (x, 1) \in Q\} \\ &= \operatorname{conv}\{x_j \mid j \in J\} + \Big\{ \sum_{t \in T} \mu_t y_y \,\Big|\, \mu_t \in \mathbf{R}_+ \text{ for } t \in T \Big\}, \end{aligned}$$

so P is finitely generated. We have thus shown that any polyhedral set (and hence function) is finitely generated.

Conversely, suppose the function $f : \mathbf{E} \to [-\infty, +\infty]$ is finitely generated. Consider first the case when f is proper. By Proposition 5.1.3, f^* is polyhedral, and hence (by the above argument) finitely generated. But f is closed and convex, by Proposition 5.1.3, so the Fenchel biconjugation theorem (4.2.1) implies $f = f^{**}$. By applying Proposition 5.1.3 once again we see f^{**} (and hence f) is polyhedral. We leave the improper case as an exercise. □

Notice these two results show our two notions of a finitely generated cone do indeed coincide.

The following collection of exercises shows that many linear-algebraic operations preserve polyhedrality.

Proposition 5.1.8 (Polyhedral algebra) *Consider a Euclidean space* $\mathbf{Y}$ *and a linear map* $A : \mathbf{E} \to \mathbf{Y}$.

(a) *If the set* $P \subset \mathbf{E}$ *is polyhedral then so is its image* AP.

(b) *If the set* $K \subset \mathbf{Y}$ *is polyhedral then so is its inverse image* $A^{-1}K$.

(c) *The sum and pointwise maximum of finitely many polyhedral functions are polyhedral.*

(d) *If the function* $g : \mathbf{Y} \to [-\infty, +\infty]$ *is polyhedral then so is the composite function* $g \circ A$.

(e) *If the function* $q : \mathbf{E} \times \mathbf{Y} \to [-\infty, +\infty]$ *is polyhedral then so is the function* $h : \mathbf{Y} \to [-\infty, +\infty]$ *defined by* $h(u) = \inf_{x \in \mathbf{E}} q(x, u)$.

Corollary 5.1.9 (Polyhedral Fenchel duality) *All the conclusions of the Fenchel duality theorem (3.3.5) remain valid if the regularity condition (3.3.8) is replaced by the assumption that the functions* f *and* g *are polyhedral with* $\operatorname{dom} g \cap A \operatorname{dom} f$ *nonempty.*

Proof. We follow the original proof, simply observing that the value function h defined in the proof is polyhedral by the Polyhedral algebra proposition above. Thus, when the optimal value is finite, h has a subgradient at 0. □

We conclude this section with a result emphasizing the power of Fenchel duality for convex problems with linear constraints.

Corollary 5.1.10 (Mixed Fenchel duality) *All the conclusions of the Fenchel duality theorem (3.3.5) remain valid if the regularity condition (3.3.8) is replaced by the assumption that* $\operatorname{dom} g \cap A \operatorname{cont} f$ *is nonempty and the function* g *is polyhedral.*

Proof. Assume without loss of generality the primal optimal value

$$p = \inf_{x \in \mathbf{E}} \{f(x) + g(Ax)\} = \inf_{x \in \mathbf{E},\ r \in \mathbf{R}} \{f(x) + r \mid g(Ax) \le r\}$$

is finite. By assumption there is a feasible point for the problem on the right at which the objective function is continuous, so there is an affine function $\alpha : \mathbf{E} \times \mathbf{R} \to \mathbf{R}$ minorizing the function $(x, r) \mapsto f(x) + r$ such that

$$p = \inf_{x \in \mathbf{E},\ r \in \mathbf{R}} \{\alpha(x, r) \mid g(Ax) \le r\}$$

(see Section 3.3, Exercise 13(c)). Clearly α has the form $\alpha(x, r) = \beta(x) + r$ for some affine minorant β of f, so

$$p = \inf_{x \in \mathbf{E}} \{\beta(x) + g(Ax)\}.$$

Now we apply polyhedral Fenchel duality (Corollary 5.1.9) to deduce the existence of an element ϕ of $\mathbf{Y}$ such that

$$p = -\beta^*(A^*\phi) - g^*(-\phi) \leq -f^*(A^*\phi) - g^*(-\phi) \leq p$$

(using the weak duality inequality), and the duality result follows. The calculus rules follow as before. □

It is interesting to compare this result with the version of Fenchel duality using the Open mapping theorem (Section 4.1, Exercise 9), where the assumption that g is polyhedral is replaced by surjectivity of A.

Exercises and Commentary

Our approach in this section is analogous to [181]. The key idea, Theorem 5.1.7 (Polyhedrality), is due to Minkowski [141] and Weyl [186]. A nice development of geometric programming (see Exercise 13) appears in [152].

1. Prove directly from the definition that any polyhedral function has a decomposition of the form (5.1.2).

2. Fill in the details for the proof of the Finitely generated functions proposition (5.1.3).

3. Use Proposition 4.2.7 (Lower semicontinuity and closure) to show that if a finitely generated function f is not proper then it has the form

$$f(x) = \begin{cases} +\infty & \text{if } x \notin K \\ -\infty & \text{if } x \in K \end{cases}$$

   for some finitely generated set K.

4. Prove a set $K \subset \mathbf{E}$ is polyhedral (respectively, finitely generated) if and only if δ_K is polyhedral (respectively, finitely generated). Do not use the Polyhedrality theorem (5.1.7).

5. Complete the proof of the Polyhedrality theorem (5.1.7) for improper functions using Exercise 3.

6. **(Tangents to polyhedra)** Prove the tangent cone to a polyhedron P at a point x in P is given by $T_P(x) = \mathbf{R}_+(P - x)$.

7. * **(Polyhedral algebra)** Prove Proposition 5.1.8 using the following steps.

   (i) Prove parts (a)–(d).

(ii) In the notation of part (e), consider the natural projection

$$P_{\mathbf{Y}\times\mathbf{R}} : \mathbf{E}\times\mathbf{Y}\times\mathbf{R} \to \mathbf{Y}\times\mathbf{R}.$$

Prove the inclusions

$$P_{\mathbf{Y}\times\mathbf{R}}(\operatorname{epi} q) \subset \operatorname{epi} h \subset \operatorname{cl}(P_{\mathbf{Y}\times\mathbf{R}}(\operatorname{epi} q)).$$

(iii) Deduce part (e).

8. If the function $f : \mathbf{E} \to (\infty, +\infty]$ is polyhedral, prove the subdifferential of f at a point x in $\operatorname{dom} f$ is a nonempty polyhedron and is bounded if and only if x lies in $\operatorname{int}(\operatorname{dom} f)$.

9. **(Polyhedral cones)** For any polyhedral cones $H \subset \mathbf{Y}$ and $K \subset \mathbf{E}$ and any linear map $A : \mathbf{E} \to \mathbf{Y}$, prove the relation

$$(K \cap A^{-1}H)^- = A^*H^- + K^-$$

using convex calculus.

10. Apply the Mixed Fenchel duality corollary (5.1.10) to the problem $\inf\{f(x) \mid Ax \le b\}$, for a linear map $A : \mathbf{E} \to \mathbf{R}^m$ and a point b in $\mathbf{R}^m$.

11. * **(Generalized Fenchel duality)** Consider convex functions

$$h_1, h_2, \ldots, h_m : \mathbf{E} \to (\infty, +\infty]$$

with $\cap_i \operatorname{cont} h_i$ nonempty. By applying the Mixed Fenchel duality corollary (5.1.10) to the problem

$$\inf_{x,x^1,x^2,\ldots,x^m \in \mathbf{E}} \Big\{ \sum_{i=1}^m h_i(x^i) \;\Big|\; x^i = x \text{ for } i = 1, 2, \ldots, m \Big\},$$

prove

$$\inf_{x\in\mathbf{E}} \sum_i h_i(x) = -\inf \Big\{ \sum_i h_i^*(\phi^i) \;\Big|\; \phi^1, \phi^2, \ldots, \phi^m \in \mathbf{E},\ \sum_i \phi^i = 0 \Big\}.$$

12. ** **(Relativizing Mixed Fenchel duality)** In the Mixed Fenchel duality corollary (5.1.10), prove the condition $\operatorname{dom} g \cap A \operatorname{cont} f \neq \emptyset$ can be replaced by $\operatorname{dom} g \cap A \operatorname{ri}(\operatorname{dom} f) \neq \emptyset$.

13. ** **(Geometric programming)** Consider the *constrained geometric program*

$$\inf_{x\in\mathbf{E}} \{h_0(x) \mid h_i(x) \le 1 \text{ for } i = 1, 2, \ldots, m\},$$

where each function h_i is a sum of functions of the form

$$x \in \mathbf{E} \mapsto c \log \left(\sum_{j=1}^{n} \exp \langle a^j, x \rangle \right)$$

for real $c > 0$ and elements $a^1, a^2, \ldots, a^n$ of $\mathbf{E}$. Write down the Lagrangian dual problem and simplify it using Exercise 11 and the form of the conjugate of each h_i given by (3.3.1). State a duality theorem.

5.2 Functions of Eigenvalues

Fenchel conjugacy gives a concise and beautiful avenue to many eigenvalue inequalities in classical matrix analysis. In this section we outline this approach.

The two cones $\mathbf{R}^n_+$ and $\mathbf{S}^n_+$ appear repeatedly in applications, as do their corresponding logarithmic barriers lb and ld, which we defined in Section 3.3. We can relate the vector and matrix examples, using the notation of Section 1.2, through the identities

$$\delta_{\mathbf{S}^n_+} = \delta_{\mathbf{R}^n_+} \circ \lambda \quad \text{and} \quad \mathrm{ld} = \mathrm{lb} \circ \lambda. \tag{5.2.1}$$

We see in this section that these identities fall into a broader pattern.

Recall the function $[\cdot] : \mathbf{R}^n \to \mathbf{R}^n$ rearranges components into nonincreasing order. We say a function f on $\mathbf{R}^n$ is *symmetric* if $f(x) = f([x])$ for all vectors x in $\mathbf{R}^n$; in other words, permuting components does not change the function value. We call a symmetric function of the eigenvalues of a symmetric matrix a *spectral function.* The following formula is crucial.

Theorem 5.2.2 (Spectral conjugacy) *If $f : \mathbf{R}^n \to [-\infty, +\infty]$ is a symmetric function, it satisfies the formula*

$$(f \circ \lambda)^* = f^* \circ \lambda.$$

Proof. By Fan's inequality (1.2.2) any matrix Y in $\mathbf{S}^n$ satisfies the inequalities

$$\begin{aligned}
(f \circ \lambda)^*(Y) &= \sup_{X \in \mathbf{S}^n} \{\mathrm{tr}\,(XY) - f(\lambda(X))\} \\
&\le \sup_X \{\lambda(X)^T \lambda(Y) - f(\lambda(X))\} \\
&\le \sup_{x \in \mathbf{R}^n} \{x^T \lambda(Y) - f(x)\} \\
&= f^*(\lambda(Y)).
\end{aligned}$$

On the other hand, fixing a spectral decomposition $Y = U^T(\mathrm{Diag}\,\lambda(Y))U$ for some matrix U in $\mathbf{O}^n$ leads to the reverse inequality

$$\begin{aligned}
f^*(\lambda(Y)) &= \sup_{x \in \mathbf{R}^n} \{x^T \lambda(Y) - f(x)\} \\
&= \sup_x \{\mathrm{tr}\,((\mathrm{Diag}\,x)UYU^T) - f(x)\} \\
&= \sup_x \{\mathrm{tr}\,(U^T(\mathrm{Diag}\,x)UY) - f(\lambda(U^T \mathrm{Diag}\,xU))\} \\
&\le \sup_{X \in \mathbf{S}^n} \{\mathrm{tr}\,(XY) - f(\lambda(X))\} \\
&= (f \circ \lambda)^*(Y),
\end{aligned}$$

which completes the proof. □

This formula, for example, makes it very easy to calculate ld * (see the Log barriers proposition (3.3.3)) and to check the self-duality of the cone $\mathbf{S}^n_+$.

Once we can compute conjugates easily, we can also recognize closed convex functions easily using the Fenchel biconjugation theorem (4.2.1).

Corollary 5.2.3 (Davis) *Suppose the function $f : \mathbf{R}^n \to (\infty, +\infty]$ is symmetric. Then the "spectral function" $f \circ \lambda$ is closed and convex if and only if f is closed and convex.*

We deduce immediately that the logarithmic barrier ld is closed and convex, as well as the function $X \mapsto \mathrm{tr}\,(X^{-1})$ on $\mathbf{S}^n_{++}$, for example.

Identifying subgradients is also easy using the conjugacy formula and the Fenchel–Young inequality (3.3.4).

Corollary 5.2.4 (Spectral subgradients) *If $f : \mathbf{R}^n \to (\infty, +\infty]$ is a symmetric function, then for any two matrices X and Y in $\mathbf{S}^n$, the following properties are equivalent:*

(*i*) $Y \in \partial(f \circ \lambda)(X)$.

(*ii*) *X and Y have a simultaneous ordered spectral decomposition and satisfy $\lambda(Y) \in \partial f(\lambda(X))$.*

(*iii*) *$X = U^T(\mathrm{Diag}\, x)U$ and $Y = U^T(\mathrm{Diag}\, y)U$ for some matrix U in $\mathbf{O}^n$ and vectors x and y in $\mathbf{R}^n$ satisfying $y \in \partial f(x)$.*

Proof. Notice the inequalities

$$(f \circ \lambda)(X) + (f \circ \lambda)^*(Y) = f(\lambda(X)) + f^*(\lambda(Y)) \geq \lambda(X)^T\lambda(Y) \geq \mathrm{tr}\,(XY).$$

The condition $Y \in \partial(f \circ \lambda)(X)$ is equivalent to equality between the left and right hand sides (and hence throughout), and the equivalence of properties (i) and (ii) follows using Fan's inequality (1.2.1). For the remainder of the proof, see Exercise 9. □

Corollary 5.2.5 (Spectral differentiability) *Suppose that the function $f : \mathbf{R}^n \to (\infty, +\infty]$ is symmetric, closed, and convex. Then $f \circ \lambda$ is differentiable at a matrix X in $\mathbf{S}^n$ if and only if f is differentiable at $\lambda(X)$.*

Proof. If $\partial(f \circ \lambda)(X)$ is a singleton, so is $\partial f(\lambda(X))$, by the Spectral subgradients corollary above. Conversely, suppose $\partial f(\lambda(X))$ consists only of the vector $y \in \mathbf{R}^n$. Using Exercise 9(b), we see the components of y are nonincreasing, so by the same corollary, $\partial(f \circ \lambda)(X)$ is the nonempty convex set

$$\{U^T(\mathrm{Diag}\, y)U \mid U \in \mathbf{O}^n,\ U^T\mathrm{Diag}\,(\lambda(X))U = X\}.$$

But every element of this set has the same norm (namely $\|y\|$), so the set must be a singleton. $\square$

Notice that the proof in fact shows that when f is differentiable at $\lambda(X)$ we have the formula

$$\nabla(f \circ \lambda)(X) = U^T(\mathrm{Diag}\,\nabla f(\lambda(X)))U \tag{5.2.6}$$

for any matrix U in $\mathbf{O}^n$ satisfying $U^T(\mathrm{Diag}\,\lambda(X))U = X$.

The pattern of these results is clear: many analytic and geometric properties of the matrix function $f \circ \lambda$ parallel the corresponding properties of the underlying function f. The following exercise is another example.

Corollary 5.2.7 *Suppose the function $f : \mathbf{R}^n \to (\infty, +\infty]$ is symmetric, closed, and convex. Then $f \circ \lambda$ is essentially strictly convex (respectively, essentially smooth) if and only if f is essentially strictly convex (respectively, essentially smooth).*

For example, the logarithmic barrier ld is both essentially smooth and essentially strictly convex.

Exercises and Commentary

Our approach in this section follows [120]. The Davis theorem (5.2.3) appeared in [58] (without the closure assumption). Many convexity properties of eigenvalues like Exercise 4 (Examples of convex spectral functions) can be found in [99] or [10], for example. Surveys of eigenvalue optimization appear in [128, 127].

1. Prove the identities (5.2.1).

2. Use the Spectral conjugacy theorem (5.2.2) to calculate ld^* and $\delta^*_{\mathbf{S}^n_+}$.

3. Prove the Davis characterization (Corollary 5.2.3) using the Fenchel biconjugation theorem (4.2.1).

4. **(Examples of convex spectral functions)** Use the Davis characterization (Corollary 5.2.3) to prove the following functions of a matrix $X \in \mathbf{S}^n$ are closed and convex:

   (a) $\mathrm{ld}\,(X)$.

   (b) $\mathrm{tr}\,(X^p)$, for any nonnegative even integer p.

   (c)
   $$\begin{cases} -\mathrm{tr}\,(X^{1/2}) & \text{if } X \in \mathbf{S}^n_+ \\ +\infty & \text{otherwise.} \end{cases}$$

(d)
$$\begin{cases} \operatorname{tr}(X^{-p}) & \text{if } X \in \mathbf{S}^n_{++} \\ +\infty & \text{otherwise} \end{cases}$$
for any nonnegative integer p.

(e)
$$\begin{cases} \operatorname{tr}(X^{1/2})^{-1} & \text{if } X \in \mathbf{S}^n_{++} \\ +\infty & \text{otherwise.} \end{cases}$$

(f)
$$\begin{cases} -(\det X)^{1/n} & \text{if } X \in \mathbf{S}^n_{+} \\ +\infty & \text{otherwise.} \end{cases}$$

Deduce from the sublinearity of the function in part (f) the property

$$0 \preceq X \preceq Y \Rightarrow 0 \leq \det X \leq \det Y$$

for matrices X and Y in $\mathbf{S}^n$.

5. Calculate the conjugate of each of the functions in Exercise 4.

6. Use formula (5.2.6) to calculate the gradients of the functions in Exercise 4.

7. For a matrix A in $\mathbf{S}^n_{++}$ and a real $b > 0$, use the Lagrangian sufficient conditions (3.2.3) to solve the problem

$$\inf\{f(X) \mid \operatorname{tr}(AX) \leq b,\ X \in \mathbf{S}^n\},$$

where f is one of the functions in Exercise 4.

8. * **(Orthogonal invariance)** A function $h : \mathbf{S}^n \to (\infty, +\infty]$ is *orthogonally invariant* if all matrices X in $\mathbf{S}^n$ and U in $\mathbf{O}^n$ satisfy the relation $h(U^T XU) = h(X)$; in other words, orthogonal similarity transformations do not change the value of h.

   (a) Prove h is orthogonally invariant if and only if there is a symmetric function $f : \mathbf{R}^n \to (\infty, +\infty]$ with $h = f \circ \lambda$.

   (b) Prove that an orthogonally invariant function h is closed and convex if and only if $h \circ \mathrm{Diag}$ is closed and convex.

9. * Suppose the function $f : \mathbf{R}^n \to (-\infty, +\infty]$ is symmetric.

   (a) Prove f^* is symmetric.

   (b) If vectors x and y in $\mathbf{R}^n$ satisfy $y \in \partial f(x)$, prove $[y] \in \partial f([x])$ using Proposition 1.2.4.

   (c) Finish the proof of the Spectral subgradients corollary (5.2.4).

   (d) Deduce $\partial(f \circ \lambda)(X) = \emptyset \Leftrightarrow \partial f(\lambda(X)) = \emptyset$.

   (e) Prove Corollary 5.2.7.

10. * **(Fillmore–Williams [78])** Suppose the set $C \subset \mathbf{R}^n$ is *symmetric*: that is, $PC = C$ holds for all permutation matrices P. Prove the set

$$\lambda^{-1}(C) = \{X \in \mathbf{S}^n \mid \lambda(X) \in C\}$$

is closed and convex if and only if C is closed and convex.

11. ** **(Semidefinite complementarity)** Suppose matrices X and Y lie in $\mathbf{S}^n_+$.

    (a) If $\mathrm{tr}\,(XY) = 0$, prove $-Y \in \partial\delta_{\mathbf{S}^n_+}(X)$.

    (b) Hence prove the following properties are equivalent:

      (i) $\mathrm{tr}\,(XY) = 0$.

      (ii) $XY = 0$.

      (iii) $XY + YX = 0$.

    (c) Using Exercise 5 in Section 1.2, prove for any matrices U and V in $\mathbf{S}^n$

$$(U^2 + V^2)^{1/2} = U + V \quad \Leftrightarrow \quad U, V \succeq 0 \text{ and } \mathrm{tr}\,(UV) = 0.$$

12. ** **(Eigenvalue sums)** Consider a vector μ in $\mathbf{R}^n_{\geq}$.

    (a) Prove the function $\mu^T\lambda(\cdot)$ is sublinear using Section 2.2, Exercise 9 (Schur-convexity).

    (b) Deduce the map λ is $(-\mathbf{R}^n_{\geq})^-$-sublinear. (See Section 3.3, Exercise 18 (Order convexity).)

    (c) Use Section 3.1, Exercise 10 to prove

$$\partial(\mu^T\lambda)(0) = \lambda^{-1}(\mathrm{conv}\,(\mathbf{P}^n\mu)).$$

13. ** **(Davis theorem)** Suppose the function $f : \mathbf{R}^n \to [-\infty, +\infty]$ is symmetric (but not necessarily closed). Use Exercise 12 (Eigenvalue sums) and Section 2.2, Exercise 9(d) (Schur-convexity) to prove that $f \circ \lambda$ is convex if and only if f is convex.

14. * **(DC functions)** We call a real function f on a convex set $C \subset \mathbf{E}$ a *DC function* if it can be written as the difference of two real convex functions on C.

    (a) Prove the set of DC functions is a vector space.

    (b) If f is a DC function, prove it is locally Lipschitz on $\mathrm{int}\, C$.

    (c) Prove λ_k is a DC function on $\mathbf{S}^n$ for all k, and deduce it is locally Lipschitz.

5.3 Duality for Linear and Semidefinite Programming

Linear programming (LP) is the study of optimization problems involving a linear objective function subject to linear constraints. This simple optimization model has proved enormously powerful in both theory and practice, so we devote this section to deriving linear programming duality theory from our convex-analytic perspective. We contrast this theory with the corresponding results for *semidefinite programming* (SDP), a class of matrix optimization problems analogous to linear programs but involving the positive semidefinite cone.

Linear programs are inherently polyhedral, so our main development follows directly from the polyhedrality section (Section 5.1). But to begin, we sketch an alternative development directly from the Farkas lemma (2.2.7). Given vectors $a^1, a^2, \ldots, a^m$, and c in $\mathbf{R}^n$ and a vector b in $\mathbf{R}^m$, consider the *primal linear program*

$$\left.\begin{array}{lrcl} \inf & \langle c, x\rangle & & \\ \text{subject to} & \langle a^i, x\rangle - b_i & \leq & 0 \ \text{ for } i = 1, 2, \ldots, m \\ & x & \in & \mathbf{R}^n. \end{array}\right\} \tag{5.3.1}$$

Denote the primal optimal value by $p \in [-\infty, +\infty]$. In the Lagrangian duality framework (Section 4.3), the dual problem is

$$\sup\Big\{ -b^T\mu \ \Big|\ \sum_{i=1}^m \mu_i a^i = -c,\ \mu \in \mathbf{R}^m_+\Big\} \tag{5.3.2}$$

with dual optimal value $d \in [-\infty, +\infty]$. From Section 4.3 we know the weak duality inequality $p \geq d$. If the primal problem (5.3.1) satisfies the Slater condition then the Dual attainment theorem (4.3.7) shows $p = d$ with dual attainment when the values are finite. However, as we shall see, the Slater condition is superfluous here.

Suppose the primal value p is finite. Then it is easy to see that the "homogenized" system of inequalities in $\mathbf{R}^{n+1}$,

$$\left.\begin{array}{rl} \langle a^i, x\rangle - b_i z \leq 0 & \text{for } i = 1, 2, \ldots, m \\ -z \leq 0 & \text{and} \\ \langle -c, x\rangle + pz > 0, & x \in \mathbf{R}^n,\ \ z \in \mathbf{R} \end{array}\right\} \tag{5.3.3}$$

has no solution. Applying the Farkas lemma (2.2.7) to this system, we deduce there is a vector $\bar{\mu}$ in $\mathbf{R}^n_+$ and a scalar β in $\mathbf{R}_+$ satisfying

$$\sum_{i=1}^m \bar{\mu}_i (a^i, -b_i) + \beta(0, -1) = (-c, p).$$

Thus $\bar{\mu}$ is a feasible solution for the dual problem (5.3.2) with objective value at least p. The weak duality inequality now implies $\bar{\mu}$ is optimal and $p = d$. We needed no Slater condition; the assumption of a finite primal optimal value alone implies zero duality gap and dual attainment.

We can be more systematic using our polyhedral theory. Suppose that $\mathbf{Y}$ is a Euclidean space, that the map $A : \mathbf{E} \to \mathbf{Y}$ is linear, and consider cones $H \subset \mathbf{Y}$ and $K \subset \mathbf{E}$. For given elements c of $\mathbf{E}$ and b of $\mathbf{Y}$, consider the primal *abstract linear program*

$$\inf\{\langle c, x\rangle \mid Ax - b \in H,\ x \in K\}. \tag{5.3.4}$$

As usual, denote the optimal value by p. We can write this problem in Fenchel form (3.3.6) if we define functions f on $\mathbf{E}$ and g on $\mathbf{Y}$ by $f(x) = \langle c, x\rangle + \delta_K(x)$ and $g(y) = \delta_H(y - b)$. Then the Fenchel dual problem (3.3.7) is

$$\sup\{\langle b, \phi\rangle \mid A^*\phi - c \in K^-,\ \phi \in -H^-\} \tag{5.3.5}$$

with dual optimal value d. If we now apply the Fenchel duality theorem (3.3.5) in turn to problem (5.3.4), and then to problem (5.3.5) (using the Bipolar cone theorem (3.3.14)), we obtain the following general result.

Corollary 5.3.6 (Cone programming duality) *Suppose the cones H and K in problem (5.3.4) are convex.*

(*a*) *If any of the conditions*

(*i*) $b \in \operatorname{int}(AK - H)$,

(*ii*) $b \in AK - \operatorname{int} H$, *or*

(*iii*) $b \in A(\operatorname{int} K) - H$, *and either H is polyhedral or A is surjective*

hold then there is no duality gap ($p = d$) and the dual optimal value d is attained if finite.

(*b*) *Suppose H and K are also closed. If any of the conditions*

(*i*) $-c \in \operatorname{int}(A^*H^- + K^-)$,

(*ii*) $-c \in A^*H^- + \operatorname{int} K^-$, *or*

(*iii*) $-c \in A^*(\operatorname{int} H^-) + K^-$, *and either K is polyhedral or A^* is surjective*

hold then there is no duality gap and the primal optimal value p is attained if finite.

In both parts (a) and (b), the sufficiency of condition (iii) follows by applying the Mixed Fenchel duality corollary (5.1.10), or the Open mapping theorem (Section 4.1, Exercise 9). In the fully polyhedral case we obtain the following result.

Corollary 5.3.7 (Linear programming duality) *Suppose the cones H and K in the dual pair of problems (5.3.4) and (5.3.5) are polyhedral. If either problem has finite optimal value then there is no duality gap and both problems have optimal solutions.*

Proof. We apply the Polyhedral Fenchel duality corollary (5.1.9) to each problem in turn. $\square$

Our earlier result for the linear program (5.3.1) is clearly just a special case of this corollary.

Linear programming has an interesting matrix analogue. Given matrices $A_1, A_2, \ldots, A_m$, and C in $\mathbf{S}^n_+$ and a vector b in $\mathbf{R}^m$, consider the primal *semidefinite program*

$$\left.\begin{array}{lrcl} \inf & \operatorname{tr}(CX) & & \\ \text{subject to} & \operatorname{tr}(A_iX) & = & b_i \ \text{ for } i = 1, 2, \ldots, m \\ & X & \in & \mathbf{S}^n_+. \end{array}\right\} \tag{5.3.8}$$

This is a special case of the abstract linear program (5.3.4), so the dual problem is

$$\sup \Big\{ b^T\phi \ \Big|\ C - \sum_{i=1}^{m} \phi_i A_i \in \mathbf{S}^n_+, \ \phi \in \mathbf{R}^m \Big\}, \tag{5.3.9}$$

since $(\mathbf{S}^n_+)^- = -\mathbf{S}^n_+$, by the Self-dual cones proposition (3.3.12), and we obtain the following duality theorem from the general result above.

Corollary 5.3.10 (Semidefinite programming duality) *If the primal problem (5.3.8) has a positive definite feasible solution, there is no duality gap and the dual optimal value is attained when finite. On the other hand, if there is a vector ϕ in $\mathbf{R}^m$ with $C - \sum_i \phi_i A_i$ positive definite then once again there is no duality gap and the primal optimal value is attained when finite.*

Unlike linear programming, we need a condition stronger than mere consistency to guarantee no duality gap. For example, if we consider the primal semidefinite program (5.3.8) with

$$n = 2, \ m = 1, \ C = \begin{bmatrix} 0 & 1 \\ 1 & 0 \end{bmatrix}, \ A_1 = \begin{bmatrix} 1 & 0 \\ 0 & 0 \end{bmatrix}, \ \text{and } b = 0,$$

the primal optimal value is 0 (and is attained), whereas the dual problem (5.3.9) is inconsistent.

Exercises and Commentary

The importance of linear programming duality was first emphasized by Dantzig [57] and that of semidefinite duality by Nesterov and Nemirovskii [148]. A good general reference for linear programming is [53]. A straightforward exposition of the central path (see Exercise 10) may be found in [187]. Semidefinite programming has wide application in control theory [46].

1. Check the form of the dual problem for the linear program (5.3.1).

2. If the optimal value of problem (5.3.1) is finite, prove system (5.3.3) has no solution.

3. **(Linear programming duality gap)** Give an example of a linear program of the form (5.3.1) which is inconsistent ($p = +\infty$) with the dual problem (5.3.2) *also* inconsistent ($d = -\infty$).

4. Check the form of the dual problem for the abstract linear program (5.3.4).

5. Fill in the details of the proof of the Cone programming duality corollary (5.3.6). In particular, when the cones H and K are closed, show how to interpret problem (5.3.4) as the dual of problem (5.3.5).

6. Fill in the details of the proof of the linear programming duality corollary (5.3.7).

7. **(Complementary slackness)** Suppose we know the optimal values of problems (5.3.4) and (5.3.5) are equal and the dual value is attained. Prove a feasible solution x for problem (5.3.4) is optimal if and only if there is a feasible solution ϕ for the dual problem (5.3.5) satisfying the conditions

$$\langle Ax - b, \phi \rangle = 0 = \langle x, A^*\phi - c \rangle.$$

8. **(Semidefinite programming duality)** Prove Corollary 5.3.10.

9. **(Semidefinite programming duality gap)** Check the details of the example after Corollary 5.3.10.

10. ** **(Central path)** Consider the dual pair of linear programs (5.3.1) and (5.3.2). Define a linear map $A : \mathbf{R}^n \to \mathbf{R}^m$ by $(Ax)_i = (a^i)^T x$ for each index i. Make the following assumptions:

    (i) There is a vector x in $\mathbf{R}^n$ satisfying $b - Ax \in \mathbf{R}^n_{++}$.

    (ii) There is a feasible solution μ in $\mathbf{R}^m_{++}$ for problem (5.3.2).

(iii) The set $\{a^1, a^2, \ldots, a^m\}$ is linearly independent.

Now consider the "penalized" problem (for real $\epsilon > 0$)

$$\inf_{x \in \mathbf{R}^n} \{c^T x + \epsilon \operatorname{lb}(b - Ax)\}. \tag{5.3.11}$$

(a) Write this problem as a Fenchel problem (3.3.6), and show the dual problem is

$$\sup \Big\{ -b^T \mu - \epsilon \operatorname{lb}(\mu) - k(\epsilon) \Big| \sum_{i=1}^m \mu_i a^i = -c,\ \mu \in \mathbf{R}^m_+ \Big\} \tag{5.3.12}$$

for some function $k : \mathbf{R}_+ \to \mathbf{R}$.

(b) Prove that both problems (5.3.11) and (5.3.12) have optimal solutions, with equal optimal values.

(c) By complementary slackness (Section 3.3, Exercise 9(f)), prove problems (5.3.11) and (5.3.12) have unique optimal solutions $x^\epsilon \in \mathbf{R}^n$ and $\mu^\epsilon \in \mathbf{R}^m$, characterized as the unique solution of the system

$$\begin{aligned} \sum_{i=1}^m \mu_i a^i &= -c \\ \mu_i (b_i - (a^i)^T x) &= \epsilon \text{ for each } i \\ b - Ax &\geq 0, \text{ and} \\ \mu \in \mathbf{R}^m_+, &\quad x \in \mathbf{R}^n. \end{aligned}$$

(d) Calculate $c^T x^\epsilon + b^T \mu^\epsilon$.

(e) Deduce that, as ϵ decreases to 0, the feasible solution x^ϵ approaches optimality in problem (5.3.1) and μ^ϵ approaches optimality in problem (5.3.2).

11. ** **(Semidefinite central path)** Imitate the development of Exercise 10 for the semidefinite programs (5.3.8) and (5.3.9).

12. ** **(Relativizing cone programming duality)** Prove other conditions guaranteeing part (a) of Corollary 5.3.6 are

(i) $b \in A(\operatorname{ri} K) - \operatorname{ri} H$ or

(ii) $b \in A(\operatorname{ri} K) - H$ and H polyhedral.

(Hint: Use Section 4.1, Exercise 20, and Section 5.1, Exercise 12.)

5.4 Convex Process Duality

In this section we introduce the idea of a "closed convex process". These are set-valued maps whose graphs are closed convex cones. As such, they provide a powerful unifying formulation for the study of linear maps, convex cones, and linear programming. The exercises show the elegance of this approach in a range of applications.

Throughout this section we fix a Euclidean space $\mathbf{Y}$. For clarity, we denote the closed unit balls in $\mathbf{E}$ and $\mathbf{Y}$ by $B_{\mathbf{E}}$ and $B_{\mathbf{Y}}$, respectively. A *multifunction* (or *set-valued map*) $\Phi : \mathbf{E} \to \mathbf{Y}$ is a map from $\mathbf{E}$ to the set of subsets of $\mathbf{Y}$. The *domain* of Φ is the set

$$D(\Phi) = \{x \in \mathbf{E} \mid \Phi(x) \neq \emptyset\}.$$

We say Φ *has nonempty images* if its domain is $\mathbf{E}$. For any subset C of $\mathbf{E}$ we write $\Phi(C)$ for the image $\cup_{x \in C} \Phi(x)$, and the *range* of Φ is the set $R(\Phi) = \Phi(\mathbf{E})$. We say Φ is *surjective* if its range is $\mathbf{Y}$. The *graph* of Φ is the set

$$G(\Phi) = \{(x, y) \in \mathbf{E} \times \mathbf{Y} \mid y \in \Phi(x)\},$$

and we define the *inverse* multifunction $\Phi^{-1} : \mathbf{Y} \to \mathbf{E}$ by the relationship

$$x \in \Phi^{-1}(y) \Leftrightarrow y \in \Phi(x) \quad \text{for } x \text{ in } \mathbf{E} \text{ and } y \text{ in } \mathbf{Y}.$$

A multifunction is *convex*, or *closed*, or *polyhedral* if its graph is likewise. A *process* is a multifunction whose graph is a cone. For example, we can interpret linear maps as closed convex processes in the obvious way.

Closure is one example of a variety of continuity properties of multifunctions we study in this section. We say the multifunction Φ is *LSC* at a point (x_0, y) in its graph if, for all neighbourhoods V of y, the image $\Phi(x)$ intersects V for all points x close to x_0. (In particular, x_0 must lie in $\operatorname{int}(D(\Phi))$.) Equivalently, for any sequence of points (x_n) approaching x_0 there is a sequence of points $y_n \in \Phi(x_n)$ approaching y. If, for x_0 in the domain, this property holds for all points y in $\Phi(x_0)$, we say Φ is *LSC at* x_0. (The notation comes from "lower semicontinuous", a name we avoid in this context because of incompatibility with the single-valued case; see Exercise 5.)

On the other hand, we say Φ is *open* at a point (x, y_0) in its graph if, for all neighbourhoods U of x, the point y_0 lies in $\operatorname{int}(\Phi(U))$. (In particular, y_0 must lie in $\operatorname{int}(R(\Phi))$.) Equivalently, for any sequence of points (y_n) approaching y_0 there is a sequence of points (x_n) approaching x such that $y_n \in \Phi(x_n)$ for all n. If, for y_0 in the range, this property holds for all points x in $\Phi^{-1}(y_0)$, we say Φ is *open* at y_0. These properties are inverse to each other in the following sense.

Proposition 5.4.1 (Openness and lower semicontinuity) *Any multifunction $\Phi : \mathbf{E} \to \mathbf{Y}$ is LSC at a point (x, y) in its graph if and only if Φ^{-1} is open at (y, x).*

We leave the proof as an exercise.

For convex multifunctions, openness at a point in the graph has strong global implications: the following result is another exercise.

Proposition 5.4.2 *If a convex multifunction is open at some point in its graph then it is open throughout the interior of its range.*

In particular, a convex process $\Phi : \mathbf{E} \to \mathbf{Y}$ is open at $(0, 0) \in \mathbf{E} \times \mathbf{Y}$ if and only if it is open at $0 \in \mathbf{Y}$; we just say Φ is *open at zero* (or, dually, Φ^{-1} is *LSC at zero*).

There is a natural duality for convex processes that generalizes the adjoint operation for linear maps. For a convex process $\Phi : \mathbf{E} \to \mathbf{Y}$, we define the *adjoint* process $\Phi^* : \mathbf{Y} \to \mathbf{E}$ by

$$G(\Phi^*) = \{(\mu, \nu) \mid (\nu, -\mu) \in G(\Phi)^-\}.$$

Then an easy consequence of the Bipolar cone theorem (3.3.14) is

$$G(\Phi^{**}) = -G(\Phi),$$

providing Φ is closed. (We could define a "lower" adjoint by the relationship $\Phi_*(\mu) = -\Phi^*(-\mu)$, in which case $(\Phi^*)_* = \Phi$.)

The language of adjoint processes is elegant and concise for many variational problems involving cones. A good example is the cone program (5.3.4). We can write this problem as

$$\inf_{x \in \mathbf{E}} \{\langle c, x \rangle \mid b \in \Psi(x)\},$$

where Ψ is the closed convex process defined by

$$\Psi(x) = \begin{cases} Ax - H & \text{if } x \in K \\ \emptyset & \text{otherwise} \end{cases} \tag{5.4.3}$$

for points c in $\mathbf{E}$, b in $\mathbf{Y}$, and closed convex cones $H \subset \mathbf{Y}$ and $K \subset \mathbf{E}$. An easy calculation shows the adjoint process is

$$\Psi^*(\mu) = \begin{cases} A^*\mu + K^- & \text{if } \mu \in H^- \\ \emptyset & \text{otherwise,} \end{cases} \tag{5.4.4}$$

so we can write the dual problem (5.3.5) as

$$\sup_{\mu \in \mathbf{Y}} \{\langle b, \mu \rangle \mid -c \in \Psi^*(-\mu)\}. \tag{5.4.5}$$

Furthermore, the constraint qualifications in the Cone programming duality corollary (5.3.6) become simply $b \in \operatorname{int} R(\Psi)$ and $-c \in \operatorname{int} R(\Psi^*)$.

In Section 1.1 we mentioned the fundamental linear-algebraic fact that the null space of any linear map A and the range of its adjoint satisfy the relationship

$$(A^{-1}(0))^- = R(A^*). \tag{5.4.6}$$

Our next step is to generalize this to processes. We begin with an easy lemma.

Lemma 5.4.7 *Any convex process $\Phi : \mathbf{E} \to \mathbf{Y}$ and subset C of $\mathbf{Y}$ satisfy $\Phi^*(C^\circ) \subset (\Phi^{-1}(C))^\circ$.*

Equality in this relationship requires more structure.

Theorem 5.4.8 (Adjoint process duality) *Let $\Phi : \mathbf{E} \to \mathbf{Y}$ be a convex process, and suppose the set $C \subset \mathbf{Y}$ is convex with $R(\Phi) \cap C$ nonempty.*

(*a*) *Either of the assumptions*

(*i*) *the multifunction $x \in \mathbf{E} \mapsto \Phi(x) - C$ is open at zero (or, in particular,* $\operatorname{int} C$ *contains zero), or*

(*ii*) *Φ and C are polyhedral*

implies

$$(\Phi^{-1}(C))^\circ = \Phi^*(C^\circ).$$

(*b*) *On the other hand, if C is compact and Φ is closed then*

$$(\Phi^{-1}(C))^\circ = \operatorname{cl}(\Phi^*(C^\circ)).$$

Proof. Suppose assumption (i) holds in part (a). For a fixed element ϕ of $(\Phi^{-1}(C))^\circ$, we can check that the "value function" $v : \mathbf{Y} \to [-\infty, +\infty]$ defined for elements y of $\mathbf{Y}$ by

$$v(y) = \inf_{x \in \mathbf{E}} \{-\langle \phi, x \rangle \mid y \in \Phi(x) - C\} \tag{5.4.9}$$

is convex. The assumption $\phi \in (\Phi^{-1}(C))^\circ$ is equivalent to $v(0) \geq -1$, while the openness assumption implies $0 \in \operatorname{core}(\operatorname{dom} v)$. Thus v is proper by Lemma 3.2.6, and so the Max formula (3.1.8) shows v has a subgradient $-\lambda \in \mathbf{Y}$ at 0. A simple calculation now shows $\lambda \in C^\circ$ and $\phi \in \Phi^*(\lambda)$, which, together with Lemma 5.4.7, proves the result.

If Φ and C are polyhedral, the Polyhedral algebra proposition (5.1.8) shows v is also polyhedral, so again has a subgradient, and our argument proceeds as before.

Turning to part (b), we can rewrite $\phi \in (\Phi^{-1}(C))^\circ$ as

$$(\phi, 0) \in (G(\Phi) \cap (\mathbf{E} \times C))^\circ$$

and apply the polarity formula in Section 4.1, Exercise 8 to deduce

$$(\phi, 0) \in \mathrm{cl}\,(G(\Phi)^- + (0 \times C^\circ)).$$

Hence there are sequences $(\phi_n, -\rho_n)$ in $G(\Phi)^-$ and μ_n in C° with ϕ_n approaching ϕ and $\mu_n - \rho_n$ approaching 0. We deduce

$$\phi_n \in \Phi^*(\rho_n) \subset \Phi^*(C^\circ + \epsilon_n B_{\mathbf{Y}}),$$

where the real sequence $\epsilon_n = \|\mu_n - \rho_n\|$ approaches 0. Since C is bounded we know $\mathrm{int}\,(C^\circ)$ contains 0 (by Section 4.1, Exercise 5), and the result follows using the positive homogeneity of Φ^*. $\square$

The null space/range formula (5.4.6) thus generalizes to a closed convex process Φ:

$$(\Phi^{-1}(0))^\circ = \mathrm{cl}\,(R(\Phi^*)),$$

and the closure is not required if Φ is open at zero.

We are mainly interested in using these polarity formulae to relate two "norms" for a convex process $\Phi : \mathbf{E} \to \mathbf{Y}$. The "lower norm"

$$\|\Phi\|_l = \inf\{r \in \mathbf{R}_{++} \mid \Phi(x) \cap rB_{\mathbf{Y}} \neq \emptyset, \ \forall x \in B_{\mathbf{E}}\}$$

quantifies Φ being LSC at zero; it is easy to check that Φ is LSC at zero if and only if its lower norm is finite. The "upper norm"

$$\|\Phi\|_u = \inf\{r \in \mathbf{R}_{++} \mid \Phi(B_{\mathbf{E}}) \subset rB_{\mathbf{Y}}\}$$

quantifies a form of "upper semicontinuity" (see Section 8.2). Clearly Φ is *bounded* (that is, bounded sets have bounded images) if and only if its upper norm is finite. Both norms generalize the *norm* of a linear map $A : \mathbf{E} \to \mathbf{Y}$, defined by

$$\|A\| = \sup\{\|Ax\| \mid \|x\| \leq 1\}.$$

Theorem 5.4.10 (Norm duality) *Any closed convex process* Φ *satisfies*

$$\|\Phi\|_l = \|\Phi^*\|_u.$$

Proof. For any real $r > \|\Phi\|_l$ we know $B_{\mathbf{E}} \subset \Phi^{-1}(rB_{\mathbf{Y}})$ by definition. Taking polars implies $B_{\mathbf{E}} \supset r^{-1}\Phi^*(B_{\mathbf{Y}})$, by the Adjoint process duality theorem (5.4.8), whence $\|\Phi^*\|_u < r$.

Conversely, $\|\Phi^*\|_u < r$ implies $\Phi^*(B_{\mathbf{Y}}) \subset rB_{\mathbf{E}}$. Taking polars and applying the Adjoint process duality theorem again followed by the Bipolar set theorem (4.1.5) shows $B_{\mathbf{E}} \subset r(\mathrm{cl}\,(\Phi^{-1}(B_{\mathbf{Y}})))$. But since $B_{\mathbf{Y}}$ is compact we can check $\Phi^{-1}(B_{\mathbf{Y}})$ is closed, and the result follows. □

The values of the upper and lower norms of course depend on the spaces **E** and **Y**. Our proof of the Norm duality theorem above shows that it remains valid when $B_{\mathbf{E}}$ and $B_{\mathbf{Y}}$ denote unit balls for arbitrary norms (see Section 4.1, Exercise 2), providing we replace them by their polars $B_{\mathbf{E}}^{\circ}$ and $B_{\mathbf{Y}}^{\circ}$ in the definition of $\|\Phi^*\|_u$.

The next result is an immediate consequence of the Norm duality theorem.

Corollary 5.4.11 *A closed convex process is LSC at zero if and only if its adjoint is bounded.*

We are now ready to prove the main result of this section.

Theorem 5.4.12 (Open mapping) *The following properties of a closed convex process Φ are equivalent:*

(i) *Φ is open at zero.*

(ii) *$(\Phi^*)^{-1}$ is bounded.*

(iii) *Φ is surjective.*

Proof. The equivalence of parts (i) and (ii) is just Corollary 5.4.11 (after taking inverses and observing the identity $G((\Phi^*)^{-1}) = -G((\Phi^{-1})^*)$. Part (i) clearly implies part (iii), so it remains to prove the converse. But if Φ is surjective then we know

$$Y = \bigcup_{n=1}^{\infty} \Phi(nB_{\mathbf{E}}) = \bigcup_{n=1}^{\infty} n\Phi(B_{\mathbf{E}}),$$

so zero lies in the core, and hence the interior, of the convex set $\Phi(B_{\mathbf{E}})$. Thus Φ is open at zero. □

Taking inverses gives the following equivalent result.

Theorem 5.4.13 (Closed graph) *The following properties of a closed convex process Φ are equivalent:*

(i) *Φ is LSC at zero.*

(ii) *Φ^* is bounded.*

(iii) *Φ has nonempty images.*

Exercises and Commentary

A classic reference for multifunctions is [13], and [113] is a good compendium which includes applications in mathematical economics. Convex processes were introduced by Rockafellar [166, 167]. The power of normed convex processes was highlighted by Robinson [161, 162]. Our development here follows [23, 24]. The importance of the "distance to inconsistency" (Exercise 21) was first made clear in [160]. For broad extensions, see [66].

1. **(Inverse multifunctions)** For any multifunction $\Phi : \mathbf{E} \to \mathbf{Y}$, prove

   (a) $R(\Phi^{-1}) = D(\Phi)$.

   (b) $G(\Phi^{-1}) = \{(y, x) \in \mathbf{Y} \times \mathbf{E} \,|\, (x, y) \in G(\Phi)\}$.

2. **(Convex images)** Prove the image of a convex set under a convex multifunction is convex.

3. For any proper closed convex function $f : \mathbf{E} \to (\infty, +\infty]$, prove $\partial(f^*) = (\partial f)^{-1}$.

4. Prove Proposition 5.4.1 (Openness and lower semicontinuity).

5. **(LSC and lower semicontinuity)** For a function $f : \mathbf{E} \to [-\infty, \infty]$, suppose f is finite at a point $z \in \mathbf{E}$.

   (a) Prove f is continuous at z if and only if the multifunction

   $$t \in \mathbf{R} \mapsto f^{-1}(t)$$

   is open at $(f(z), z)$.

   (b) Prove f is lower semicontinuous at z if and only if the multifunction whose graph is $\operatorname{epi}(-f)$ is LSC at $(z, f(z))$.

6. * Prove Proposition 5.4.2. (Hint: See Section 4.1, Exercise 1(b).)

7. **(Biconjugation)** Prove any closed convex process Φ satisfies

   $$G(\Phi^{**}) = -G(\Phi).$$

8. Check the adjoint formula (5.4.4).

9. Prove Lemma 5.4.7.

10. Prove the value function (5.4.9) is convex.

11. * Write a complete proof of the Adjoint process duality theorem (5.4.8).

12. If the multifunction $\Phi : \mathbf{E} \to \mathbf{Y}$ is closed and the set $C \subset \mathbf{Y}$ is compact, prove $\Phi^{-1}(C)$ is closed.

13. Prove $G((\Phi^*)^{-1}) = -G((\Phi^{-1})^*)$ for any closed convex process Φ.

14. **(Linear maps)** Consider a linear map $A : \mathbf{E} \to \mathbf{Y}$, and define a multifunction $\Phi : \mathbf{E} \to \mathbf{Y}$ by $\Phi(x) = \{Ax\}$ for all points x in $\mathbf{E}$.

    (a) Prove Φ is a closed convex process.

    (b) Prove Φ^* is the closed convex process $y \in \mathbf{Y} \mapsto \{A^*y\}$.

    (c) Prove $\|\Phi\|_l = \|\Phi\|_u = \|A\|$.

    (d) Prove A is an open map (that is, A maps open sets to open sets) if and only if Φ is open throughout $\mathbf{Y}$.

    (e) Hence deduce the Open mapping theorem for linear maps (Section 4.1, Exercise 9) as a special case of Theorem 5.4.12.

    (f) For any closed convex process $\Omega : \mathbf{E} \to \mathbf{Y}$, prove

    $$(\Omega + A)^* = \Omega^* + A^*.$$

15. * **(Normal cones)** A closed convex cone $K \subset \mathbf{E}$ is *generating* if it satisfies $K - K = \mathbf{E}$. For a point x in $\mathbf{E}$, the *order interval* $[0, x]_K$ is the set $K \cap (x - K)$. We say K is *normal* if there is a real $c > 0$ such that

    $$y \in [0, x]_K \Rightarrow \|y\| \leq c\|x\|.$$

    (a) Prove the multifunction $\Phi : \mathbf{E} \to \mathbf{E}$ defined by $\Phi(x) = [0, x]_K$ is a closed convex process.

    (b) Calculate $(\Phi^*)^{-1}$.

    (c) **(Krein–Grossberg)** Deduce K is normal if and only if K^- is generating.

    (d) Use Section 3.3, Exercise 20 (Pointed cones) to deduce K is normal if and only if it is pointed.

16. **(Inverse boundedness)** By considering the convex process (5.4.3), demonstrate the following equivalence for any linear map $A : \mathbf{E} \to \mathbf{Y}$ and closed cones $K \subset \mathbf{E}$ and $H \subset \mathbf{Y}$:

    $$AK - H = Y \quad \Leftrightarrow \quad \{y \in H^- \mid A^*y \in B_{\mathbf{E}} - K^-\} \text{ is bounded.}$$

17. ** **(Localization [24])** Given a closed convex process $\Phi : \mathbf{E} \to \mathbf{Y}$ and a point b in $\mathbf{Y}$, define the "homogenized" process $\Psi : \mathbf{E} \times \mathbf{R} \to \mathbf{Y} \times \mathbf{R}$ by

    $$\Psi(x, t) = \begin{cases} (\Phi(x) - tb) \times (t - \mathbf{R}_+) & \text{if } t \geq 0 \\ \emptyset & \text{if } t < 0. \end{cases}$$

(a) Prove Ψ is a closed convex process.
(b) Prove Ψ is surjective if and only if b lies in core $(R(\Phi))$.
(c) Prove Ψ is open at zero if and only if Φ is open at b.
(d) Calculate Ψ^*.
(e) Prove the following statements are equivalent:
   (i) Φ is open at b.
   (ii) b lies in core $(R(\Phi))$.
   (iii) The set

$$\{\mu \in \mathbf{Y} \mid \Phi^*(\mu) \cap B_{\mathbf{E}} \neq \emptyset \text{ and } \langle \mu, b \rangle \leq 1\}$$

   is bounded.
(f) If $R(\Phi)$ has nonempty core, use a separation argument to prove the statements in part (e) are equivalent to

$$\{\mu \in (\Phi^*)^{-1}(0) \mid \langle \mu, b \rangle \leq 0\} = \{0\}.$$

18. ** **(Cone duality)** By applying part (e) of Exercise 17 to example (5.4.3) with $A = 0$ and $K = \mathbf{E}$, deduce that a point b lies in the core of the closed convex cone $H \subset \mathbf{Y}$ if and only if the set

$$\{\mu \in H^- \mid -\langle \mu, b \rangle \leq 1\}$$

is bounded. Hence, give another proof that a closed convex cone has a bounded base if and only if its polar has nonempty interior (Section 3.3, Exercise 20).

19. ** **(Order epigraphs)**

(a) Suppose $C \subset \mathbf{E}$ is a convex cone, $S \subset \mathbf{Y}$ is a closed convex cone, and $F : C \to \mathbf{Y}$ is an S-sublinear function (Section 3.3, Exercise 18 (Order convexity)). Prove the multifunction $\Phi : \mathbf{E} \to \mathbf{Y}$ defined by

$$\Phi(x) = \begin{cases} F(x) + S & \text{if } x \in C \\ \emptyset & \text{otherwise,} \end{cases}$$

is a convex process, with adjoint

$$\Phi^*(\mu) = \begin{cases} \partial\langle \mu, F(\cdot)\rangle(0) & \text{if } \mu \in -S^- \\ \emptyset & \text{otherwise.} \end{cases}$$

(b) Use Section 5.2, Exercise 12 to prove the adjoint of the closed convex process

$$X \in \mathbf{S}^n \mapsto \lambda(X) - (\mathbf{R}^n_{\geq})^-$$

is the closed convex process with domain $\mathbf{R}^n_{\geq}$ defined by

$$\mu \mapsto \lambda^{-1}(\mathrm{conv}\,(\mathbf{P}^n \mu)).$$

20. ** **(Condition number [123])** Consider any closed convex process $\Phi : \mathbf{E} \to \mathbf{Y}$ and a linear map $G : \mathbf{E} \to \mathbf{Y}$.

    (a) If $\|G\|^{-1} > \|\Phi^{-1}\|_u$, prove the process $(\Phi + G)^{-1}$ is bounded.

    (b) If $\|G\|^{-1} > \|\Phi^{-1}\|_l$, use part (a) to prove the process $\Phi + G$ is surjective.

    (c) Suppose Φ is surjective and the point y lies on the boundary of the set $\Phi(B_{\mathbf{E}})$. By considering a supporting hyperplane, prove there is a *rank-one* linear map $G : \mathbf{E} \to \mathbf{Y}$ defined by

    $$Gx = \langle \mu, x \rangle y$$

    for some element μ of $\mathbf{E}$ such that $\Phi + G$ is not surjective.

    (d) Deduce

    $$\min\{\|G\| \mid \Phi + G \text{ not surjective}\} = \|\Phi^{-1}\|_l^{-1},$$

    where the minimum is attained by a rank-one map when finite.

21. ** **(Distance to inconsistency [123])** Consider a given linear map $A : \mathbf{E} \to \mathbf{Y}$ and an element b of $\mathbf{Y}$. Suppose the space $\mathbf{E} \times \mathbf{R}$ has the norm $\|(x,t)\| = \|x\| + |t|$.

    (a) Prove the linear map

    $$(x,t) \in \mathbf{E} \times \mathbf{R} \mapsto Ax - tb$$

    has norm $\|A\| \vee \|b\|$.

    Now consider closed convex cones $P \subset \mathbf{E}$ and $Q \subset \mathbf{Y}$, and systems

    $$\begin{array}{lll} (S) & b - Ax \in Q, & x \in P \quad \text{and} \\ (S_z) & z + tb - Ax \in Q, & x \in P, \quad t \in \mathbf{R}_+, \ \|x\| + |t| \le 1. \end{array}$$

    Let I denote the set of pairs (A, b) such that system (S) is inconsistent, and let I_0 denote the set of (A, b) such that the process

    $$(x,t) \in \mathbf{E} \times \mathbf{R} \mapsto \begin{cases} Ax - tb + Q & \text{if } x \in P,\ t \in \mathbf{R}_+ \\ \emptyset & \text{if otherwise} \end{cases}$$

    is not surjective.

    (b) Prove $I = \operatorname{cl} I_0$.

    (c) By applying Exercise 20 (Condition number), prove the distance of (A, b) from I is given by the formula

    $$d_I(A, b) = \inf\{\|z\| \mid (S_z) \text{ inconsistent}\}.$$

Chapter 6

Nonsmooth Optimization

6.1 Generalized Derivatives

From the perspective of optimization, the subdifferential $\partial f(\cdot)$ of a convex function f has many of the useful properties of the derivative. Some examples: it gives the necessary optimality condition $0 \in \partial f(x)$ when the point x is a (local) minimizer (Proposition 3.1.5); it reduces to $\{\nabla f(x)\}$ when f is differentiable at x (Corollary 3.1.10); and it often satisfies certain calculus rules such as $\partial(f+g)(x) = \partial f(x) + \partial g(x)$ (Theorem 3.3.5). For a variety of reasons, if the function f is not convex, the subdifferential $\partial f(\cdot)$ is not a particularly helpful idea. This makes it very tempting to look for definitions of the subdifferential for a nonconvex function. In this section we outline some examples; the most appropriate choice often depends on context.

For a convex function $f : \mathbf{E} \to (\infty, +\infty]$ with x in $\mathrm{dom}\, f$, we can characterize the subdifferential via the directional derivative: $\phi \in \partial f(x)$ if and only if $\langle \phi, \cdot \rangle \leq f'(x; \cdot)$ (Proposition 3.1.6). A natural approach is therefore to generalize the directional derivative. Henceforth in this section we make the simplifying assumption that the *real function* f (a real-valued function defined on some subset of $\mathbf{E}$) is locally Lipschitz around the point x in $\mathbf{E}$.

Partly motivated by the development of optimality conditions, a simple first try is the *Dini directional derivative*:

$$f^-(x; h) = \liminf_{t \downarrow 0} \frac{f(x+th) - f(x)}{t} .$$

A disadvantage of this idea is that $f^-(x; \cdot)$ is not usually sublinear (consider for example $f = -|\cdot|$ on $\mathbf{R}$), so we could not expect an analogue of the Max formula (3.1.9). With this in mind, we introduce the *Clarke directional derivative,*

$$f^\circ(x;h) = \limsup_{y\to x,\ t\downarrow 0} \frac{f(y+th)-f(y)}{t} = \inf_{\delta>0} \sup_{\|y-x\|\le\delta,\ 0<t<\delta} \frac{f(y+th)-f(y)}{t}$$

and the *Michel–Penot directional derivative,*

$$f^\diamond(x;h) = \sup_{u\in\mathbf{E}} \limsup_{t\downarrow 0} \frac{f(x+th+tu)-f(x+tu)}{t}.$$

Proposition 6.1.1 *If the real function f has Lipschitz constant K around the point x in $\mathbf{E}$ then the Clarke and Michel–Penot directional derivatives $f^\circ(x;\cdot)$ and $f^\diamond(x;\cdot)$ are sublinear and satisfy*

$$f^-(x;\cdot) \le f^\diamond(x;\cdot) \le f^\circ(x;\cdot) \le K\|\cdot\|.$$

Proof. The positive homogeneity and upper bound are straightforward, so let us prove subadditivity in the Clarke case. For any sequences $x^r \to x$ in $\mathbf{E}$ and $t_r \downarrow 0$ in $\mathbf{R}$, and any real $\epsilon > 0$, we have

$$\frac{f(x^r + t_r(u+v)) - f(x^r + t_r u)}{t_r} \le f^\circ(x;v) + \epsilon$$

and

$$\frac{f(x^r + t_r u) - f(x^r)}{t_r} \le f^\circ(x;u) + \epsilon$$

for all large r. Adding and letting r approach ∞ shows

$$f^\circ(x;u+v) \le f^\circ(x;u) + f^\circ(x;v) + 2\epsilon,$$

and the result follows. We leave the Michel–Penot case as an exercise. The inequalities are straightforward. □

Using our knowledge of support functions (Corollary 4.2.3), we can now define the *Clarke subdifferential*

$$\partial_\circ f(x) = \{\phi \in \mathbf{E} \mid \langle \phi, h\rangle \le f^\circ(x;h) \text{ for all } h \in \mathbf{E}\}$$

and the *Dini* and *Michel–Penot subdifferentials* $\partial_- f(x)$ and $\partial_\diamond f(x)$ analogously. Elements of the respective subdifferentials are called *subgradients.* We leave the proof of the following result as an exercise.

Corollary 6.1.2 (Nonsmooth max formulae) *If the real function f has Lipschitz constant K around the point x in $\mathbf{E}$ then the Clarke and*

Michel–Penot subdifferentials $\partial_\circ f(x)$ and $\partial_\diamond f(x)$ are nonempty, compact, and convex, and satisfy

$$\partial_- f(x) \subset \partial_\diamond f(x) \subset \partial_\circ f(x) \subset KB.$$

Furthermore, the Clarke and Michel–Penot directional derivatives are the support functions of the corresponding subdifferentials:

$$f^\circ(x;h) = \max\{\langle \phi, h\rangle \mid \phi \in \partial_\circ f(x)\} \tag{6.1.3}$$

and

$$f^\diamond(x;h) = \max\{\langle \phi, h\rangle \mid \phi \in \partial_\diamond f(x)\} \tag{6.1.4}$$

for any direction h in $\mathbf{E}$.

Notice the Dini subdifferential is also compact and convex, but may be empty.

Clearly if the point x is a local minimizer of f then any direction h in $\mathbf{E}$ satisfies $f^-(x;h) \geq 0$, and hence the necessary optimality conditions

$$0 \in \partial_- f(x) \subset \partial_\diamond f(x) \subset \partial_\circ f(x)$$

hold. If g is another real function which is locally Lipschitz around x then we would not typically expect $\partial_\circ(f+g)(x) = \partial_\circ f(x) + \partial_\circ g(x)$ (consider $f = -g = |\cdot|$ on $\mathbf{R}$ at $x = 0$, for example). On the other hand, if we are interested in an optimality condition like $0 \in \partial_\circ(f+g)(x)$, it is the *sum rule* $\partial_\circ(f+g)(x) \subset \partial_\circ f(x) + \partial_\circ g(x)$ that really matters. (A good example we see later is Corollary 6.3.9.) We address this in the next result, along with an analogue of the formula for the convex subdifferential of a max-function in Section 3.3, Exercise 17. We write $f \vee g$ for the function $x \mapsto \max\{f(x), g(x)\}$.

Theorem 6.1.5 (Nonsmooth calculus) *If the real functions f and g are locally Lipschitz around the point x in $\mathbf{E}$, then the Clarke subdifferential satisfies*

$$\partial_\circ(f+g)(x) \subset \partial_\circ f(x) + \partial_\circ g(x) \tag{6.1.6}$$

and

$$\partial_\circ(f \vee g)(x) \subset \operatorname{conv}(\partial_\circ f(x) \cup \partial_\circ g(x)). \tag{6.1.7}$$

Analogous results hold for the Michel–Penot subdifferential.

Proof. The Clarke directional derivative satisfies

$$(f+g)^\circ(x;\cdot) \leq f^\circ(x;\cdot) + g^\circ(x;\cdot),$$

since lim sup is a sublinear function. Using the Max formula (6.1.3) we deduce

$$\delta^*_{\partial_\circ(f+g)(x)} \leq \delta^*_{\partial_\circ f(x) + \partial_\circ g(x)}$$

and taking conjugates now gives the result using the Fenchel biconjugation theorem (4.2.1) and the fact that both sides of inclusion (6.1.6) are compact and convex.

To see inclusion (6.1.7), fix a direction h in $\mathbf{E}$ and choose sequences $x^r \to x$ in $\mathbf{E}$ and $t_r \downarrow 0$ in $\mathbf{R}$ satisfying

$$\frac{(f \vee g)(x^r + t_r h) - (f \vee g)(x^r)}{t_r} \to (f \vee g)^\circ(x; h).$$

Without loss of generality, suppose $(f \vee g)(x^r + t_r h) = f(x^r + t_r h)$ for all r in some subsequence R of $\mathbf{N}$, and now note

$$\begin{aligned} f^\circ(x; h) &\geq \limsup_{r \to \infty,\ r \in R} \frac{f(x^r + t_r h) - f(x^r)}{t_r} \\ &\geq \limsup_{r \to \infty,\ r \in R} \frac{(f \vee g)(x^r + t_r h) - (f \vee g)(x^r)}{t_r} \\ &= (f \vee g)^\circ(x; h). \end{aligned}$$

We deduce $(f \vee g)^\circ(x; \cdot) \leq f^\circ(x; \cdot) \vee g^\circ(x; \cdot)$, which, using the Max formula (6.1.3), we can rewrite as

$$\delta^*_{\partial_\circ (f \vee g)(x)} \leq \delta^*_{\partial_\circ f(x)} \vee \delta^*_{\partial_\circ g(x)} = \delta^*_{\mathrm{conv}(\partial_\circ f(x) \cup \partial_\circ g(x))}$$

using Exercise 9(b) (Support functions) in Section 4.2. Now the Fenchel biconjugation theorem again completes the proof. The Michel–Penot case is analogous. □

We now have the tools to derive a nonsmooth necessary optimality condition.

Theorem 6.1.8 (Nonsmooth necessary condition) *Suppose the point $\bar{x}$ is a local minimizer for the problem*

$$\inf\{f(x) \mid g_i(x) \leq 0 \ (i \in I)\}, \tag{6.1.9}$$

where the real functions f and g_i (for i in finite index set I) are locally Lipschitz around $\bar{x}$. Let $I(\bar{x}) = \{i \mid g_i(\bar{x}) = 0\}$ be the active set. Then there exist real $\lambda_0, \lambda_i \geq 0$ for i in $I(\bar{x})$, not all zero, satisfying

$$0 \in \lambda_0 \partial_\diamond f(\bar{x}) + \sum_{i \in I(\bar{x})} \lambda_i \partial_\diamond g_i(\bar{x}). \tag{6.1.10}$$

If, furthermore, some direction d in $\mathbf{E}$ satisfies

$$g_i^\diamond(\bar{x}; d) < 0 \ \ \textit{for all } i \textit{ in } I(\bar{x}) \tag{6.1.11}$$

then we can assume $\lambda_0 = 1$.

Proof. Imitating the approach of Section 2.3, we note that $\bar{x}$ is a local minimizer of the function

$$x \mapsto \max\{f(x) - f(\bar{x}),\ g_i(x) \text{ for } i \in I(\bar{x})\}.$$

We deduce

$$\begin{aligned} 0 &\in \partial_\diamond(\max\{f - f(\bar{x}),\ g_i \text{ for } i \in I(\bar{x})\})(\bar{x}) \\ &\subset \operatorname{conv}\Big(\partial_\diamond f(\bar{x}) \cup \bigcup_{i \in I(\bar{x})} \partial_\diamond g_i(\bar{x})\Big) \end{aligned}$$

by inclusion (6.1.7).

If condition (6.1.11) holds and λ_0 is zero in condition (6.1.10), we obtain the contradiction

$$0 \leq \max\Big\{\langle \phi, d\rangle \ \Big|\ \phi \in \sum_{i \in I(\bar{x})} \lambda_i \partial_\diamond g_i(\bar{x})\Big\} = \sum_{i \in I(\bar{x})} \lambda_i g_i^\diamond(\bar{x}; d) < 0.$$

Thus λ_0 is strictly positive, and hence without loss of generality equals one. □

Condition (6.1.10) is a Fritz John type condition analogous to Theorem 2.3.6. Assumption (6.1.11) is a Mangasarian–Fromovitz type constraint qualification like Assumption 2.3.7, and the conclusion is a Karush–Kuhn–Tucker condition analogous to Theorem 2.3.8. We used the Michel–Penot subdifferential in the above argument because it is in general smaller than the Clarke subdifferential, and hence provides stronger necessary conditions. By contrast to our approach here, the developments in Section 2.3 and Section 3.2 do not assume the local Lipschitz condition around the optimal point $\bar{x}$.

Exercises and Commentary

Dini derivatives were first used in [64]. The Clarke subdifferential appeared in [54]. A good reference is [55]. The Michel–Penot subdifferential was introduced in [138, 139]. A good general reference for this material is [5].

1. **(Examples of nonsmooth derivatives)** For the following functions $f : \mathbf{R} \to \mathbf{R}$ defined for each point x in $\mathbf{R}$ by

   (a) $f(x) = |x|$,

   (b) $f(x) = -|x|$,

   (c) $f(x) = \begin{cases} x^2 \sin(x^{-1}) & \text{if } x \neq 0 \\ 0 & \text{if } x = 0, \end{cases}$

(d) $f(x) = \begin{cases} 3^n & \text{if } 3^n \leq x \leq 2(3^n) \text{ for any integer } n \\ 2x - 3^{n+1} & \text{if } 2(3^n) \leq x \leq 3^{n+1} \text{ for any integer } n \\ 0 & \text{if } x \leq 0, \end{cases}$

compute the Dini, Michel–Penot and Clarke directional derivatives and subdifferentials at $x = 0$.

2. **(Continuity of Dini derivative)** For a point x in $\mathbf{E}$, prove the function $f^-(x;\cdot)$ is Lipschitz if f is locally Lipschitz around x.

3. Complete the proof of Proposition 6.1.1.

4. **(Surjective Dini subdifferential)** Suppose the continuous function $f : \mathbf{E} \to \mathbf{R}$ satisfies the growth condition

$$\lim_{\|x\| \to \infty} \frac{f(x)}{\|x\|} = +\infty.$$

For any element ϕ of $\mathbf{E}$, prove there is a point x in $\mathbf{E}$ with ϕ in $\partial_- f(x)$.

5. Prove Corollary 6.1.2 (Nonsmooth max formulae) using Corollary 4.2.3 (Support functions).

6. **(Failure of Dini calculus)** Show that the inclusion

$$\partial_-(f+g)(x) \subset \partial_- f(x) + \partial_- g(x)$$

can fail for locally Lipschitz functions f and g.

7. * Complete the details of the proof of the Nonsmooth calculus theorem (6.1.5).

8. * Prove the following results:

   (a) $f^\circ(x;-h) = (-f)^\circ(x;h)$.

   (b) $(\lambda f)^\circ(x;h) = \lambda f^\circ(x;h)$ for $0 \leq \lambda \in \mathbf{R}$.

   (c) $\partial_\circ(\lambda f)(x) = \lambda \partial_\circ f(x)$ for all λ in $\mathbf{R}$.

   Derive similar results for the Michel–Penot version.

9. * **(Mean value theorem [119])**

   (a) Suppose the function $f : \mathbf{E} \to \mathbf{R}$ is locally Lipschitz. For any points x and y in $\mathbf{E}$, prove there is a real t in (0,1) satisfying

$$f(x) - f(y) \in \langle x - y, \partial_\diamond f(tx + (1-t)y) \rangle.$$

   (Hint: Consider a local minimizer or maximizer of the function $g : [0,1] \to \mathbf{R}$ defined by $g(t) = f(tx + (1-t)y)$.)

(b) **(Monotonicity and convexity)** If the set C in $\mathbf{E}$ is open and convex and the function $f : C \to \mathbf{R}$ is locally Lipschitz, prove f is convex if and only if it satisfies

$$\langle x - y, \phi - \psi \rangle \geq 0 \text{ for all } x, y \in C,\ \phi \in \partial_\diamond f(x), \text{ and } \psi \in \partial_\diamond f(y).$$

(c) If $\partial_\diamond f(y) \subset kB$ for all points y near x, prove f has local Lipschitz constant k about x.

Prove similar results for the Clarke case.

10. * **(Max-functions)** Consider a compact set $T \subset \mathbf{R}^n$ and a continuous function $g : \mathbf{E} \times T \to \mathbf{R}$. For each element t of T define a function $g_t : \mathbf{E} \to \mathbf{R}$ by $g_t(x) = g(x, t)$ and suppose, for all t, that this function is locally Lipschitz around the point z. Define $G : \mathbf{E} \to \mathbf{R}$ by

$$G(x) = \max\{g(x, t) \mid t \in T\}$$

and let T_z be the set $\{t \in T \mid g(z, t) = G(z)\}$. Prove the inclusion

$$\partial_\circ G(z) \subset \mathrm{cl}\Big(\mathrm{conv} \bigcup_{t \in T_z} \{\lim \phi^r \mid z^r \to z,\ t_r \to t,\ \phi^r \in \partial_\circ g_{t_r}(z^r)\}\Big).$$

Specialize to the case where T is finite, and to the case where $\nabla g_t(x)$ is a continuous function of (x, t).

11. ** **(Order statistics [125])** Calculate the Dini, the Michel–Penot, and the Clarke directional derivatives and subdifferentials of the function

$$x \in \mathbf{R}^n \mapsto [x]_k.$$

6.2 Regularity and Strict Differentiability

We have outlined in Section 2.3 and Section 3.2 two very distinct versions of the necessary optimality conditions in constrained optimization. The first, culminating in the Karush–Kuhn–Tucker conditions (2.3.8), relied on Gâteaux differentiability, while the second, leading to the Lagrangian necessary conditions (3.2.8), used convexity. A primary aim of the nonsmooth theory of this chapter is to unify these types of results; in this section we show how this is possible.

A principal feature of the Michel–Penot subdifferential is that it coincides with the Gâteaux derivative when this exists.

Proposition 6.2.1 (Unique Michel–Penot subgradient) *A real function f which is locally Lipschitz around the point x in $\mathbf{E}$ has a unique Michel–Penot subgradient ϕ at x if and only if ϕ is the Gâteaux derivative $\nabla f(x)$.*

Proof. If f has a unique Michel–Penot subgradient ϕ at x, then all directions h in $\mathbf{E}$ satisfy

$$f^{\diamond}(x;h) = \sup_{u \in \mathbf{E}} \limsup_{t \downarrow 0} \frac{f(x+th+tu) - f(x+tu)}{t} = \langle \phi, h \rangle.$$

The cases $h = w$ with $u = 0$ and $h = -w$ with $u = w$ show

$$\limsup_{t \downarrow 0} \frac{f(x+tw) - f(x)}{t} \leq \langle \phi, w \rangle \leq \liminf_{t \downarrow 0} \frac{f(x+tw) - f(x)}{t},$$

so we deduce $f'(x,w) = \langle \phi, w \rangle$ as required.

Conversely, if f has Gâteaux derivative ϕ at x then any directions h and u in $\mathbf{E}$ satisfy

$$\begin{aligned}
&\limsup_{t \downarrow 0} \frac{f(x+th+tu) - f(x+tu)}{t} \\
&\leq \limsup_{t \downarrow 0} \frac{f(x+t(h+u)) - f(x)}{t} - \liminf_{t \downarrow 0} \frac{f(x+tu) - f(x)}{t} \\
&= f'(x;h+u) - f'(x;u) \\
&= \langle \phi, h+u \rangle - \langle \phi, u \rangle = \langle \phi, h \rangle = f'(x;h) \leq f^{\diamond}(x;h).
\end{aligned}$$

Now taking the supremum over u shows $f^{\diamond}(x;h) = \langle \phi, h \rangle$ for all h, as we claimed. □

Thus, for example, the Fritz John condition (6.1.10) reduces to Theorem 2.3.6 in the differentiable case (under the extra, locally Lipschitz assumption).

The above result shows that when f is Gâteaux differentiable at the point x, the Dini and Michel–Penot directional derivatives coincide. If they also equal the Clarke directional derivative then we say f is *regular* at x. Thus a real function f, locally Lipschitz around x, is regular at x exactly when the ordinary directional derivative $f'(x;\cdot)$ exists and equals the Clarke directional derivative $f^\circ(x;\cdot)$.

One of the reasons we are interested in regularity is that when the two functions f and g are regular at x, the nonsmooth calculus rules (6.1.6) and (6.1.7) hold with equality (assuming $f(x) = g(x)$ in the latter). The proof is a straightforward exercise.

We know that a convex function is locally Lipschitz around any point in the interior of its domain (Theorem 4.1.3). In fact such functions are also regular at such points: consequently our various subdifferentials are all generalizations of the convex subdifferential.

Theorem 6.2.2 (Regularity of convex functions) *Suppose the function $f : \mathbf{E} \to (\infty, +\infty]$ is convex. If the point x lies in* $\operatorname{int}(\operatorname{dom} f)$ *then f is regular at x, and hence the convex, Dini, Michel–Penot and Clarke subdifferentials all coincide:*

$$\partial_\circ f(x) = \partial_\diamond f(x) = \partial_- f(x) = \partial f(x).$$

Proof. Fix a direction h in $\mathbf{E}$ and choose a real $\delta > 0$. Denoting the local Lipschitz constant by K, we know

$$\begin{aligned} f^\circ(x;h) &= \lim_{\epsilon \downarrow 0} \sup_{\|y-x\| \le \epsilon\delta} \sup_{0<t<\epsilon} \frac{f(y+th) - f(y)}{t} \\ &= \lim_{\epsilon \downarrow 0} \sup_{\|y-x\| \le \epsilon\delta} \frac{f(y+\epsilon h) - f(y)}{\epsilon} \\ &\le \lim_{\epsilon \downarrow 0} \frac{f(x+\epsilon h) - f(x)}{\epsilon} + 2K\delta \\ &= f'(x;h) + 2K\delta, \end{aligned}$$

using the convexity of f. We deduce

$$f^\circ(x;h) \le f'(x;h) = f^-(x;h) \le f^\diamond(x;h) \le f^\circ(x;h),$$

and the result follows. □

Thus, for example, the Karush–Kuhn–Tucker type condition that we obtained at the end of Section 6.1 reduces exactly to the Lagrangian necessary conditions (3.2.8), written in the form $0 \in \partial f(\bar{x}) + \sum_{i \in I(\bar{x})} \lambda_i \partial g_i(\bar{x})$, assuming the convex functions f and g_i (for indices i in $I(\bar{x})$) are continuous at the optimal solution $\bar{x}$.

By analogy with Proposition 6.2.1 (Unique Michel–Penot subgradient), we might ask when the Clarke subdifferential of a function f at a point x is a singleton $\{\phi\}$? Clearly in this case f must be regular with Gâteaux derivative $\nabla f(x) = \phi$, although Gâteaux differentiability is not enough, as the example $x^2 \sin(1/x)$ shows (Exercise 1 in Section 6.1). To answer the question we need a stronger notion of differentiability.

For future reference we introduce three gradually stronger conditions for an arbitrary real function f. We say an element ϕ of $\mathbf{E}$ is the *Fréchet derivative* of f at x if it satisfies

$$\lim_{y \to x,\ y \neq x} \frac{f(y) - f(x) - \langle \phi, y - x \rangle}{\|y - x\|} = 0,$$

and we say ϕ is the *strict derivative* of f at x if it satisfies

$$\lim_{y,z \to x,\ y \neq z} \frac{f(y) - f(z) - \langle \phi, y - z \rangle}{\|y - z\|} = 0.$$

In either case, it is easy to see $\nabla f(x)$ is ϕ. For locally Lipschitz functions on $\mathbf{E}$, a straightforward exercise shows Gâteaux and Fréchet differentiability coincide, but notice that the function $x^2 \sin(1/x)$ is *not* strictly differentiable at zero. Finally, if f is Gâteaux differentiable close to x with gradient map $\nabla f(\cdot)$ continuous then we say f is *continuously differentiable* around x. In the case $\mathbf{E} = \mathbf{R}^n$ we see in elementary calculus that this is equivalent to the partial derivatives of f being continuous around x. We make analogous definitions of Gâteaux, Fréchet, strict and continuous differentiability for a function $F : \mathbf{E} \to \mathbf{Y}$ (where $\mathbf{Y}$ is another Euclidean space). The derivative $\nabla f(x)$ is in this case a linear map from $\mathbf{E}$ to $\mathbf{Y}$.

The following result clarifies the idea of a strict derivative and suggests its connection with the Clarke directional derivative; we leave the proof as another exercise.

Theorem 6.2.3 (Strict differentiability) *A real function f has strict derivative ϕ at a point x in $\mathbf{E}$ if and only if it is locally Lipschitz around x with*

$$\lim_{y \to x,\ t \downarrow 0} \frac{f(y + th) - f(y)}{t} = \langle \phi, h \rangle$$

for all directions h in $\mathbf{E}$. In particular this holds if f is continuously differentiable around x with $\nabla f(x) = \phi$.

We can now answer our question about the Clarke subdifferential.

Theorem 6.2.4 (Unique Clarke subgradient) *A real function f which is locally Lipschitz around the point x in $\mathbf{E}$ has a unique Clarke subgradient ϕ at x if and only if ϕ is the strict derivative of f at x. In this case f is regular at x.*

Proof. One direction is clear, so let us assume $\partial_\circ f(x) = \{\phi\}$. Then we deduce

$$\begin{aligned}\liminf_{y\to x,\ t\downarrow 0} \frac{f(y+th)-f(y)}{t} &= -\limsup_{y\to x,\ t\downarrow 0} \frac{f((y+th)-th)-f(y+th)}{t} \\ &= -f^\circ(x;-h) = \langle \phi, h\rangle = f^\circ(x;h) \\ &= \limsup_{y\to x,\ t\downarrow 0} \frac{f(y+th)-f(y)}{t},\end{aligned}$$

and the result now follows using Theorem 6.2.3 (Strict differentiability). □

The Clarke subdifferential has a remarkable alternative description, often more convenient for computation. It is a reasonably straightforward measure-theoretic consequence of Rademacher's theorem (9.1.2), which states that locally Lipschitz functions are almost everywhere differentiable.

Theorem 6.2.5 (Intrinsic Clarke subdifferential) *Suppose that the real function f is locally Lipschitz around the point x in $\mathbf{E}$ and that the set $S \subset \mathbf{E}$ has measure zero. Then the Clarke subdifferential of f at x is*

$$\partial_\circ f(x) = \operatorname{conv}\{\lim_r \nabla f(x^r) \mid x^r \to x,\ x^r \notin S\}.$$

Exercises and Commentary

Again, references for this material are [55, 138, 139, 5]. A nice proof of Theorem 6.2.5 (Intrinsic Clarke subdifferential) appears in [14]. For some related ideas applied to distance functions, see [33]. Rademacher's theorem can be found in [71], for example. See also Section 9.1. For more details on the functions of eigenvalues appearing in Exercise 15, see [121, 124].

1. Which of the functions in Section 6.1, Exercise 1 are regular at zero?

2. **(Regularity and nonsmooth calculus)** If the functions f and g are regular at the point x, prove that the nonsmooth calculus rules (6.1.6) and (6.1.7) hold with equality (assuming $f(x) = g(x)$ in the latter) and that the resulting functions are also regular at x.

3. Show by a direct calculation that the function $x \in \mathbf{R} \mapsto x^2 \sin(1/x)$ is not strictly differentiable at the point $x = 0$.

4. Prove the special case of the Lagrangian necessary conditions we claim after Theorem 6.2.2.

5. * Prove that the notions of Gâteaux and Fréchet differentiability coincide for locally Lipschitz real functions.

6. Without using Theorem 6.2.4, prove that a unique Clarke subgradient implies regularity.

7. * Prove the Strict differentiability theorem (6.2.3).

8. Write out a complete proof of the unique Clarke subgradient theorem (6.2.4).

9. **(Mixed sum rules)** Suppose that the real function f is locally Lipschitz around the point x in $\mathbf{E}$ and that the function $g : \mathbf{E} \to (-\infty, +\infty]$ is convex with x in $\operatorname{int}(\operatorname{dom} g)$. Prove:

   (a) $\partial_\diamond(f+g)(x) = \nabla f(x) + \partial g(x)$ if f is Gâteaux differentiable at x.

   (b) $\partial_\circ(f+g)(x) = \nabla f(x) + \partial g(x)$ if f is strictly differentiable at x.

10. **(Types of differentiability)** Consider the function $f : \mathbf{R}^2 \to \mathbf{R}$, defined for $(x,y) \neq 0$ by

$$f(x,y) = \frac{x^a y^b}{x^p + y^q}$$

   with $f(0) = 0$, in the five cases:

   (i) $a = 2$, $b = 3$, $p = 2$, and $q = 4$.

   (ii) $a = 1$, $b = 3$, $p = 2$, and $q = 4$.

   (iii) $a = 2$, $b = 4$, $p = 4$, and $q = 8$.

   (iv) $a = 1$, $b = 2$, $p = 2$, and $q = 2$.

   (v) $a = 1$, $b = 2$, $p = 2$, and $q = 4$.

   In each case determine if f is continuous, Gâteaux, Fréchet, or continuously differentiable at zero.

11. Construct a function $f : \mathbf{R} \to \mathbf{R}$ which is strictly differentiable at zero but not continuously differentiable around zero.

12. * **(Closed subdifferentials)**

   (a) Suppose the function $f : \mathbf{E} \to (\infty, +\infty]$ is convex, and the point x lies in $\operatorname{int}(\operatorname{dom} f)$. Prove the convex subdifferential $\partial f(\cdot)$ is *closed* at x; in other words, $x^r \to x$ and $\phi^r \to \phi$ in $\mathbf{E}$ with ϕ^r in $\partial f(x^r)$ implies $\phi \in \partial f(x)$. (See Exercise 8 in Section 4.2.)

   (b) Suppose the real function f is locally Lipschitz around the point x in $\mathbf{E}$.

(i) For any direction h in $\mathbf{E}$, prove the Clarke directional derivative has the property that $-f^\circ(\cdot; h)$ is lower semicontinuous at x.

(ii) Deduce the Clarke subdifferential is closed at x.

(iii) Deduce further the inclusion $\subset$ in the Intrinsic Clarke subdifferential theorem (6.2.5).

(c) Show that the Dini and Michel–Penot subdifferentials are not necessarily closed.

13. * **(Dense Dini subgradients)** Suppose the real function f is locally Lipschitz around the point x in $\mathbf{E}$. By considering the closest point in epi f to the point $(x, f(x) - \delta)$ (for a small real $\delta > 0$), prove there are Dini subgradients at points arbitrarily close to x.

14. ** **(Regularity of order statistics [125])** At which points is the function

$$x \in \mathbf{R}^n \mapsto [x]_k$$

regular? (See Section 6.1, Exercise 11.)

15. ** **(Subdifferentials of eigenvalues)** Define a function $\gamma_k : \mathbf{R}^n \to \mathbf{R}$ by $\gamma_k(x) = \sum_{i=1}^k [x]_i$ for $k = 1, 2, \ldots, n$. (See Section 2.2, Exercise 9 (Schur-convexity).)

(a) For any point x in $\mathbf{R}^n_\geq$ satisfying $x_k > x_{k+1}$, prove $\nabla \gamma_k(x) = \sum_1^k e^i$ (where e^i is the ith standard unit vector in $\mathbf{R}^n$).

Now define a function $\sigma_k : \mathbf{S}^n \to \mathbf{R}$ by $\sigma_k = \sum_1^k \lambda_i$.

(b) Prove $\sigma_k = \gamma_k \circ \lambda$.

(c) Deduce σ_k is convex and hence locally Lipschitz.

(d) Deduce λ_k is locally Lipschitz.

(e) If the matrix X in $\mathbf{S}^n$ satisfies $\lambda_k(X) > \lambda_{k+1}(X)$, prove σ_k is Gâteaux differentiable at X and calculate the derivative. (Hint: Use formula (5.2.6).)

(f) If the matrix X in $\mathbf{S}^n$ satisfies $\lambda_{k-1}(X) > \lambda_k(X) > \lambda_{k+1}(X)$, prove

$$\nabla \lambda_k(X) = uu^T$$

for any unit vector u in $\mathbf{R}^n$ satisfying $\lambda_k(X)u = Xu$.

(g) Using the Intrinsic Clarke subdifferential theorem (6.2.5), deduce the formula

$$\partial_\circ \lambda_k(X) = \operatorname{conv}\{uu^T \mid Xu = \lambda_k(X)u,\ \|u\| = 1\}.$$

(h) **(Isotonicity of λ)** Using the Mean value theorem (Section 6.1, Exercise 9), deduce for any matrices X and Y in $\mathbf{S}^n$

$$X \succeq Y \;\Rightarrow\; \lambda(X) \geq \lambda(Y).$$

6.3 Tangent Cones

We simplified our brief outline of some of the fundamental ideas of nonsmooth analysis by restricting attention to locally Lipschitz functions. By contrast, the convex analysis we have developed lets us study the optimization problem $\inf\{f(x) \mid x \in S\}$ via the function $f + \delta_S$, even though the indicator function δ_S is not locally Lipschitz on the boundary of the set S. The following simple but very important idea circumvents this difficulty. We define the *distance function* to the nonempty set $S \subset \mathbf{E}$ by

$$d_S(x) = \inf\{\|y - x\| \mid y \in S\} \tag{6.3.1}$$

(see Section 3.3, Exercise 12 (Infimal convolution).) We can easily check that d_S has Lipschitz constant equal to one on $\mathbf{E}$, and is convex if and only if S has convex closure.

Proposition 6.3.2 (Exact penalization) *For a point x in a set $S \subset \mathbf{E}$, suppose the real function f is locally Lipschitz around x. If x is a local minimizer of f on S then for real L sufficiently large, x is a local minimizer of $f + Ld_S$.*

Proof. Suppose the Lipschitz constant is no larger than L. Fix a point z close to x. Clearly $d_S(z)$ is the infimum of $\|z - y\|$ over points y close to x in S, and such points satisfy

$$f(z) + Ld_S(z) \geq f(y) + L(d_S(z) - \|z - y\|) \geq f(x) + L(d_S(z) - \|z - y\|).$$

The result follows by taking the supremum over y. □

With the assumptions of the previous proposition, we know that any direction h in $\mathbf{E}$ satisfies

$$0 \leq (f + Ld_S)^\circ(x; h) \leq f^\circ(x; h) + Ld_S^\circ(x; h),$$

and hence the Clarke directional derivative satisfies $f^\circ(x; h) \geq 0$ whenever h lies in the set

$$T_S(x) = \{h \mid d_S^\circ(x; h) = 0\}. \tag{6.3.3}$$

Since $d_S^\circ(x; \cdot)$ is finite and sublinear (and an easy exercise shows it is nonnegative) it follows that $T_S(x)$ is a closed convex cone. We call it the *Clarke tangent cone*.

Tangent cones are "conical" approximations to sets in an analogous way to directional derivatives being sublinear approximations to functions. Different directional derivatives give rise to different tangent cones. For example, the Dini directional derivative leads to the cone

$$K_S(x) = \{h \mid d_S^-(x; h) = 0\}, \tag{6.3.4}$$

a (nonconvex) closed cone containing $T_S(x)$ called the *contingent cone.* If the set S is convex then we can use the ordinary directional derivative to define the cone

$$T_S(x) = \{h \mid d'_S(x;h) = 0\}, \tag{6.3.5}$$

which again will be a closed convex cone called the *(convex) tangent cone.* We can use the same notation as the Clarke cone because finite convex functions are regular at every point (Theorem 6.2.2). We also show below that our notation agrees in the convex case with that of Section 3.3.

Our definitions of the Clarke and contingent cones do not reveal that these cones are topological objects, independent of the choice of norm. The following are more intrinsic descriptions. We leave the proofs as exercises.

Theorem 6.3.6 (Tangent cones) *Suppose the point x lies in a set S in $\mathbf{E}$.*

(a) *The contingent cone $K_S(x)$ consists of those vectors h in $\mathbf{E}$ for which there are sequences $t_r \downarrow 0$ in $\mathbf{R}$ and $h^r \to h$ in $\mathbf{E}$ such that $x + t_r h^r$ lies in S for all r.*

(b) *The Clarke tangent cone $T_S(x)$ consists of those vectors h in $\mathbf{E}$ such that for any sequences $t_r \downarrow 0$ in $\mathbf{R}$ and $x^r \to x$ in S, there is a sequence $h^r \to h$ in $\mathbf{E}$ such that $x^r + t_r h^r$ lies in S for all r.*

Intuitively, the contingent cone $K_S(x)$ consists of limits of directions to points near x in S, while the Clarke tangent cone $T_S(x)$ "stabilizes" this tangency idea by allowing perturbations of the base point x.

We call the set S *tangentially regular* at the point $x \in S$ if the contingent and Clarke tangent cones coincide (which clearly holds if the distance function d_S is regular at x). The convex case is an example.

Corollary 6.3.7 (Convex tangent cone) *If the point x lies in the convex set $C \subset \mathbf{E}$, then C is tangentially regular at x with*

$$T_C(x) = K_C(x) = \operatorname{cl} \mathbf{R}_+(C - x).$$

Proof. The regularity follows from Theorem 6.2.2 (Regularity of convex functions). The identity $K_C(x) = \operatorname{cl} \mathbf{R}_+(C - x)$ follows easily from the contingent cone characterization in Theorem 6.3.6. □

Our very first optimality result (Proposition 2.1.1) required the condition $-\nabla f(x) \in N_C(x)$ if the point x is a local minimizer of a differentiable function f on a convex set $C \subset \mathbf{E}$. If the function $f : \mathbf{E} \to (\infty, +\infty]$ is convex and continuous at $x \in C$, then in fact a necessary and sufficient condition for global minimality is

$$0 \in \partial(f + \delta_C)(x) = \partial f(x) + N_C(x),$$

using the sum formula in Theorem 3.3.5. This suggests transforming the result of our earlier discussion in this section into an analogous form. We use the following idea.

Theorem 6.3.8 *For a point x in a set $S \subset \mathbf{E}$, the* **Clarke normal cone,** *defined by $N_S(x) = T_S(x)^-$, is* $\mathrm{cl}\,(\mathbf{R}_+\partial_\circ d_S(x))$.

Proof. By the Bipolar cone theorem (3.3.14), we need only show that $(\partial_\circ d_S(x))^- = T_S(x)$, and this follows from the Max formula (6.1.3). □

Notice that our notation for the normal cone is again consistent with the convex case we discussed in Section 3.3.

Corollary 6.3.9 (Nonsmooth necessary conditions) *For a point x in a set $S \subset \mathbf{E}$, suppose the real function f is locally Lipschitz around x. Any local minimizer x of f on S must satisfy the condition*

$$0 \in \partial_\diamond f(x) + N_S(x).$$

Proof. For large real L, the point x is a local minimizer of $f + Ld_S$ by the Exact penalization proposition (6.3.2), so it satisfies

$$0 \in \partial_\diamond(f + Ld_S)(x) \subset \partial_\diamond f(x) + L\partial_\diamond d_S(x) \subset \partial_\diamond f(x) + N_S(x),$$

using the nonsmooth sum rule (6.1.6). □

In particular in the above result, if f is Gâteaux differentiable at x then $-\nabla f(x) \in N_S(x)$, and when S is convex we recover the first order necessary condition (2.1.1). However, we can obtain a more useful, and indeed, fundamental, geometric necessary condition by using the contingent cone.

Proposition 6.3.10 (Contingent necessary condition) *If a point x is a local minimizer of the real function f on the set $S \subset \mathbf{E}$, and if f is Fréchet differentiable at x, then the condition*

$$-\nabla f(x) \in K_S(x)^-$$

must hold.

Proof. If the condition fails then there is a direction h in $K_S(x)$ which satisfies $\langle \nabla f(x), h\rangle < 0$. By Theorem 6.3.6 (Tangent cones) there are sequences $t_r \downarrow 0$ in $\mathbf{R}$ and $h^r \to h$ in $\mathbf{E}$ satisfying $x + t_r h^r$ in S for all r. But then, since we know

$$\lim_{r\to\infty} \frac{f(x + t_r h^r) - f(x) - \langle \nabla f(x), t_r h^r\rangle}{t_r\|h^r\|} = 0,$$

we deduce $f(x + t_r h^r) < f(x)$ for all large r, contradicting the local minimality of x. □

Precisely because of this result, our aim in the next chapter will be to identify concrete circumstances where we can calculate the contingent cone $K_S(x)$.

Exercises and Commentary

Our philosophy in this section is guided by [55]. The contingent cone was introduced by Bouligand [45]. Scalarization (see Exercise 12) is a central tool in vector (or multicriteria) optimization [104]. For the background to Exercise 13 (Boundary properties), see [39, 40, 41].

1. **(Exact penalization)** For a set $U \subset \mathbf{E}$, suppose that the function $f : U \to \mathbf{R}$ has Lipschitz constant L', and that the set $S \subset U$ is closed. For any real $L > L'$, if the point x minimizes $f + Ld_S$ on U, prove $x \in S$.

2. **(Distance function)** For any nonempty set $S \subset \mathbf{E}$, prove the distance function d_S has Lipschitz constant equal to one on $\mathbf{E}$, and it is convex if and only if $\mathrm{cl}\, S$ is convex.

3. **(Examples of tangent cones)** For the following sets $S \subset \mathbf{R}^2$, calculate $T_S(0)$ and $K_S(0)$:

   (a) $\{(x, y) \mid y \geq x^3\}$.

   (b) $\{(x, y) \mid x \geq 0 \text{ or } y \geq 0\}$.

   (c) $\{(x, y) \mid x = 0 \text{ or } y = 0\}$.

   (d) $\left\{r(\cos\theta, \sin\theta) \,\middle|\, 0 \leq r \leq 1,\ \frac{\pi}{4} \leq \theta \leq \frac{7\pi}{4}\right\}$.

4. * **(Topology of contingent cone)** Prove that the contingent cone is closed, and derive the topological description given in Theorem 6.3.6.

5. * **(Topology of Clarke cone)** Suppose the point x lies in the set $S \subset \mathbf{E}$.

   (a) Prove $d_S^\circ(x; \cdot) \geq 0$.

   (b) Prove

   $$d_S^\circ(x; h) = \limsup_{y \to x \text{ in } S,\ t \downarrow 0} \frac{d_S(y + th)}{t}.$$

   (c) Deduce the topological description of $T_S(x)$ given in Theorem 6.3.6.

6. * **(Intrinsic tangent cones)** Prove directly from the intrinsic description of the Clarke and contingent cones (Theorem 6.3.6) that the Clarke cone is convex and the contingent cone is closed.

7. Write a complete proof of the Convex tangent cone corollary (6.3.7).

8. **(Isotonicity)** Suppose $x \in U \subset V \subset \mathbf{E}$. Prove $K_U(x) \subset K_V(x)$, but give an example where $T_U(x) \not\subset T_V(x)$.

9. **(Products)** Let $\mathbf{Y}$ be a Euclidean space. Suppose $x \in U \subset \mathbf{E}$ and $y \in V \subset \mathbf{Y}$. Prove $T_{U \times V}(x,y) = T_U(x) \times T_V(y)$, but give an example where $K_{U\times V}(x,y) \neq K_U(x) \times K_V(y)$.

10. **(Tangents to graphs)** Suppose the function $F : \mathbf{E} \to \mathbf{Y}$ is Fréchet differentiable at the point x in $\mathbf{E}$. Prove
$$K_{G(F)}(x, F(x)) = G(\nabla F).$$

11. * **(Graphs of Lipschitz functions)** Given a Euclidean space $\mathbf{Y}$, suppose the function $F : \mathbf{E} \to \mathbf{Y}$ is locally Lipschitz around the point x in $\mathbf{E}$.

    (a) For elements μ of $\mathbf{E}$ and ν of $\mathbf{Y}$, prove
    $$(\mu, -\nu) \in (K_{G(F)}(x, F(x)))^\circ \ \Leftrightarrow\ \mu \in \partial_-\langle \nu, F(\cdot)\rangle(x).$$

    (b) In the case $\mathbf{Y} = \mathbf{R}$, deduce
    $$\mu \in \partial_- F(x) \ \Leftrightarrow\ (\mu, -1) \in (K_{G(F)}(x, F(x)))^\circ$$

12. ** **(Proper Pareto minimization)** We return to the notation of Section 4.1, Exercise 12 (Pareto minimization), but dropping the assumption that the cone S has nonempty interior. Recall that S is pointed, and hence has a compact base by Section 3.3, Exercise 20. We say the point y in D is a *proper Pareto minimum (with respect to S)* if it satisfies
$$-K_D(y) \cap S = \{0\},$$
and the point $\bar{x}$ in C is a *proper Pareto minimum* of the vector optimization problem
$$\inf\{F(x) \mid x \in C\} \tag{6.3.11}$$
if $F(\bar{x})$ is a proper Pareto minimum of $F(C)$.

    (a) If D is a polyhedron, use Section 5.1, Exercise 6 to prove any Pareto minimum is proper. Show this can fail for a general convex set D.

(b) For any point y in D, prove

$$K_{D+S}(y) = \mathrm{cl}\,(K_D(y) + S).$$

(c) **(Scalarization)** Suppose $\bar{x}$ is as above. By separating the cone $-K_{F(C)+S}(F(\bar{x}))$ from a compact base for S, prove there is an element ϕ of $-\mathrm{int}\,S^-$ such that $\bar{x}$ solves the convex problem

$$\inf\{\langle \phi, F(x)\rangle \mid x \in C\}.$$

Conversely, show any solution of this problem is a proper Pareto minimum of the original problem (6.3.11).

13. ** **(Boundary properties)** For points x and y in $\mathbf{E}$, define the *line segments*

$$[x, y] = x + [0, 1](y - x), \quad (x, y) = x + (0, 1)(y - x).$$

Suppose the set $S \subset \mathbf{E}$ is nonempty and closed. Define a subset

$$\mathrm{star}\, S = \{x \in S \mid [x, y] \subset S \text{ for all } y \text{ in } S\}.$$

(a) Prove S is convex if and only if $\mathrm{star}\, S = S$.

(b) For all points x in S, prove $\mathrm{star}\, S \subset (T_S(x) + x)$.

The *pseudotangent cone* to S at a point x in S is

$$P_S(x) = \mathrm{cl}\,(\mathrm{conv}\, K_S(x)).$$

We say x is a *proper point of S* if $P_S(x) \neq \mathbf{E}$.

(c) If S is convex, prove the boundary points of S coincide with the proper points.

(d) Prove the proper points of S are dense in the boundary of S.

We say S is *pseudoconvex* at x if $P_S(x) \supset S - x$.

(e) Prove any convex set is pseudoconvex at every element.

(f) **(Nonconvex separation)** Given points x in S and y in $\mathbf{E}$ satisfying $[x, y] \not\subset S$ and any real $\epsilon > 0$, prove there exists a point z in S such that

$$y \notin P_S(z) + z \quad \text{and} \quad \|z - x\| \le \|y - x\| + \epsilon.$$

(Complete the following steps: Fix a real δ in $(0, \epsilon)$ and a point w in (x, y) such that the ball $w + \delta B$ is disjoint from S. For each real t, define a point $x_t = w + t(x - w)$ and a real

$$\tau = \sup\{t \in [0, 1] \mid S \cap (x_t + \delta B) = \emptyset\}.$$

Now pick any point z in $S \cap (x_\tau + \delta B)$ and deduce the result from the properties

$$P_S(x) \subset \{u \in \mathbf{E} \mid \langle u, z - x_\tau \rangle \geq 0\} \text{ and}$$
$$0 \geq \langle y - x_\tau, z - x_\tau \rangle.)$$

(g) Explain why the nonconvex separation principle in part (f) generalizes the Basic separation theorem (2.1.6).

(h) Deduce $\cap_{x \in S}(P_S(x) + x) \subset \text{star}\, S$.

(i) Deduce

$$\bigcap_{x \in S}(P_S(x) + x) = \text{star}\, S = \bigcap_{x \in S}(T_S(x) + x)$$

(and hence star S is closed). Verify this formula for the set in Exercise 3(d).

(j) Prove a set is convex if and only if it is pseudoconvex at every element.

(k) If star S is nonempty, prove its recession cone (see Section 1.1, Exercise 6) is given by

$$\bigcap_{x \in S} P_S(x) = 0^+(\text{star}\, S) = \bigcap_{x \in S} T_S(x).$$

14. **(Pseudoconvexity and sufficiency)** Given a set $S \subset \mathbf{E}$ and a real function f which is Gâteaux differentiable at a point x in S, we say f is *pseudoconvex at x on S* if

$$\langle \nabla f(x), y - x \rangle \geq 0,\ y \in S \ \Rightarrow\ f(y) \geq f(x).$$

(a) Suppose S is convex, the function $g : S \to \mathbf{R}_+$ is convex, the function $h : S \to \mathbf{R}_{++}$ is concave, and both g and h are Fréchet differentiable at the point x in S. Prove the function g/h is pseudoconvex at x.

(b) If the contingent necessary condition $-\nabla f(x) \in K_S(x)^-$ holds and f and S are pseudoconvex at x, prove x is a global minimizer of f on S (see Exercise 13).

(c) If the point x is a local minimizer of the convex function f on the set S, prove x minimizes f on $x + P_S(x)$ (see Exercise 13).

15. **(No ideal tangent cone exists)** Consider a convex set $Q_S(x)$ defined for sets $S \subset \mathbf{R}^2$ and points x in S and satisfying the properties

(i) **(isotonicity)** $x \in R \subset S \Rightarrow Q_R(x) \subset Q_S(x)$.

(ii) **(convex tangents)** $x \in$ closed convex $S \Rightarrow Q_S(x) = T_S(x)$.

Deduce $Q_{\{(u,v) \mid u \text{ or } v=0\}}(0) = \mathbf{R}^2$.

16. ** **(Distance function [32])** We can define the distance function (6.3.1) with respect to *any* norm $\|\cdot\|$. Providing the norm is continuously differentiable away from zero, prove that for any nonempty closed set S and any point x outside S, we have

$$(-d_S)^\circ(x;\cdot) = (-d_S)^\diamond(x;\cdot).$$

6.4 The Limiting Subdifferential

In this chapter we have seen a variety of subdifferentials. As we have observed, the smaller the subdifferential, the stronger the necessary optimality conditions we obtain by using it. On the other hand, the smallest of our subdifferentials, the Dini subdifferential, is in some sense *too* small. It may be empty, it is not a closed multifunction, and it may not always satisfy a sum rule:

$$\partial_-(f+g)(x) \not\subset \partial_- f(x) + \partial_- g(x)$$

in general. In this section we show how to enlarge it somewhat to construct what is, in many senses, the smallest adequate closed subdifferential.

Consider for the moment a real function f that is locally Lipschitz around the point x in $\mathbf{E}$. Using a construction analogous to the Intrinsic Clarke subdifferential theorem (6.2.5), we can construct a nonempty subdifferential incorporating the local information from the Dini subdifferential. Specifically, we define the *limiting subdifferential* by closing the graph of the Dini subdifferential:

$$\partial_a f(x) = \{\lim_r \phi^r \mid x^r \to x,\ \phi^r \in \partial_- f(x^r)\}.$$

(Recall $\partial_- f(z)$ is nonempty at points z arbitrarily close to x by Section 6.2, Exercise 13.) We sketch some of the properties of the limiting subdifferential in the exercises. In particular, it is nonempty and compact, it coincides with $\partial f(x)$ when f is convex and continuous at the point x, and any local minimizer x of f must satisfy $0 \in \partial_a f(x)$. Often the limiting subdifferential is not convex; in fact its convex hull is exactly the Clarke subdifferential. A harder fact is that if the real function g is also locally Lipschitz around x then a sum rule holds:

$$\partial_a(f+g)(x) \subset \partial_a f(x) + \partial_a g(x).$$

We prove a more general version of this rule below.

We first extend our definitions beyond locally Lipschitz functions. As in the convex case, the additional possibilities of studying extended-real-valued functions are very powerful. For a function $f : \mathbf{E} \to [-\infty, +\infty]$ that is finite at the point $x \in \mathbf{E}$, we define the *Dini directional derivative* of f at x in the direction $v \in \mathbf{E}$ by

$$f^-(x;v) = \liminf_{t \downarrow 0,\ u \to v} \frac{f(x+tu) - f(x)}{t},$$

and the *Dini subdifferential* of f at x is the set

$$\partial_- f(x) = \{\phi \in \mathbf{E} \mid \langle \phi, v \rangle \le f^-(x;v) \text{ for all } v \text{ in } \mathbf{E}\}.$$

If $f(x)$ is infinite we define $\partial_- f(x) = \emptyset$. These definitions agree with our previous notions by Section 6.1, Exercise 2 (Continuity of Dini derivative).

For real $\delta > 0$, we define a subset of $\mathbf{E}$ by

$$U(f, x, \delta) = \{z \in \mathbf{E} \mid \|z - x\| < \delta,\ |f(z) - f(x)| < \delta\}.$$

The *limiting subdifferential* of f at x is the set

$$\partial_a f(x) = \bigcap_{\delta > 0} \mathrm{cl}\,(\partial_- f(U(f, x, \delta))).$$

Thus an element ϕ of $\mathbf{E}$ belongs to $\partial_a f(x)$ if and only if there is a sequence of points (x^r) in $\mathbf{E}$ approaching x with $f(x^r)$ approaching $f(x)$, and a sequence of Dini subgradients $\phi^r \in \partial_- f(x^r)$ approaching ϕ.

The case of an indicator function is particularly important. Recall that if the set $C \subset \mathbf{E}$ is convex and the point x lies in C then $\partial \delta_C(x) = N_C(x)$. By analogy, we define the *limiting normal cone* to a set $S \subset \mathbf{E}$ at a point x in $\mathbf{E}$ by

$$N_S^a(x) = \partial_a \delta_S(x).$$

We first prove an "inexact" or "fuzzy" sum rule: point and subgradients are all allowed to move a little. Since such rules are central to modern nonsmooth analysis, we give the proof in detail.

Theorem 6.4.1 (Fuzzy sum rule) *If the functions*

$$f_1, f_2, \ldots, f_n : \mathbf{E} \to [-\infty, +\infty]$$

are lower semicontinuous near the point $z \in \mathbf{E}$ then the inclusion

$$\partial_- \Big(\sum_i f_i\Big)(z) \subset \delta B + \sum_i \partial_- f_i(U(f_i, z, \delta)).$$

holds for any real $\delta > 0$.

Proof. Assume without loss of generality that $z = 0$ and $f_i(0) = 0$ for each i. We assume zero belongs to the left hand side of our desired inclusion and deduce it belongs to the right hand side, or, in other words,

$$\delta B \cap \sum_i \partial_- f_i(U(f_i, 0, \delta)) \neq \emptyset. \tag{6.4.2}$$

(The general case follows by adding a linear function to f_1.)

Since $0 \in \partial_- (\sum_i f_i)(0)$, Exercise 3 shows zero is a strict local minimizer of the function $g = \delta \|\cdot\| + \sum_i f_i$. Choose a real ϵ from the interval $(0, \delta)$ such that

$$0 \neq x \in \epsilon B \ \Rightarrow\ g(x) > 0 \text{ and } f_i(x) \geq -\frac{1}{n} \text{ for each } i$$

(using the lower semicontinuity of each f_i). Define a sequence of functions $p_r : \mathbf{E}^{n+1} \to [-\infty, +\infty]$ by

$$p_r(x_0, x_1, \ldots, x_n) = \delta\|x_0\| + \sum_i \Big(f_i(x_i) + \frac{r}{2}\|x_i - x_0\|^2 \Big)$$

for $r = 1, 2, \ldots$, and for each r choose a minimizer $(x_0^r, x_1^r, \ldots, x_n^r)$ of p_r on $(\epsilon B)^{n+1}$. Since $p_r(0, 0, \ldots, 0) = 0$, we deduce

$$p_r(x_0^r, x_1^r, \ldots, x_n^r) \le 0 \tag{6.4.3}$$

for each r.

Our choice of ϵ implies $\sum_i f_i(x_i^r) \ge -1$, so

$$\delta\|x_0^r\| + \frac{r}{2}\sum_i \|x_i^r - x_0^r\|^2 \le p_r(x_0^r, x_1^r, \ldots, x_n^r) + 1 \le 1$$

for each r. Hence, for each index i the sequence (x_i^r) is bounded, so there is a subsequence S of $\mathbf{N}$ such that $\lim_{r \in S} x_i^r$ exists for each i. The above inequality also shows this limit must be independent of i; call the limit $\bar{x}$, and note it lies in ϵB.

From inequality (6.4.3) we see $\delta\|x_0^r\| + \sum_i f_i(x_i^r) \le 0$ for all r, and using lower semicontinuity shows

$$g(\bar{x}) = \delta\|\bar{x}\| + \sum_i f_i(\bar{x}) \le 0,$$

so our choice of ϵ implies $\bar{x} = 0$. We have thus shown

$$\lim_{r \in S} x_i^r = 0 \ \text{ for each } i.$$

Inequality (6.4.3) implies $\sum_i f_i(x_i^r) \le 0$ for all r, and since

$$\liminf_{r \in S} f_i(x_i^r) \ge f_i(0) = 0 \ \text{ for each } i$$

by lower semicontinuity, we deduce

$$\lim_{r \in S} f_i(x_i^r) = 0$$

for each i.

Fix an index r in S large enough to ensure $\|x_0^r\| < \epsilon$, $\|x_i^r\| < \epsilon$ and $|f_i(x_i^r)| < \delta$ for each $i = 1, 2, \ldots, n$. For this r, the function p_r has a local minimum at $(x_0^r, x_1^r, \ldots, x_n^r)$, so its Dini directional derivative in every direction $(v_0, v_1, \ldots, v_n) \in \mathbf{E}^{n+1}$ is nonnegative. Define vectors

$$\phi_i = r(x_0^r - x_i^r) \ \text{ for } i = 1, 2, \ldots, n.$$

Then for any nonzero i, setting $v_j = 0$ for all $j \neq i$ shows

$$f_i^-(x_i^r; v_i) - \langle \phi_i, v_i \rangle \geq 0 \text{ for all } v_i \text{ in } \mathbf{E},$$

whence

$$\phi_i \in \partial_- f_i(x_i^r) \text{ for } i = 1, 2, \ldots, n.$$

On the other hand, setting $v_i = 0$ for all nonzero i shows

$$\delta \|v_0\| + \langle \textstyle\sum_i \phi_i, v_0 \rangle \geq 0 \text{ for all } v_0 \text{ in } \mathbf{E},$$

whence $\sum_i \phi_i \in \delta B$, and the desired relationship (6.4.2) now follows. $\square$

It is not difficult to construct examples where the above result fails if $\delta = 0$ (Exercise 4). In fact there are also examples where

$$\partial_a(f_1 + f_2)(z) \not\subset \partial_a f_1(z) + \partial_a f_2(z).$$

In general the following result is the best we can expect.

Theorem 6.4.4 (Limiting subdifferential sum rule) *If one of the functions $f, g : \mathbf{E} \to [-\infty, +\infty]$ is locally Lipschitz and the other is lower semicontinuous near the point z in $\mathbf{E}$ then*

$$\partial_a(f + g)(z) \subset \partial_a f(z) + \partial_a g(z).$$

Proof. For any element ϕ of $\partial_a(f+g)(z)$ there is a sequence of points (z^r) approaching z in $\mathbf{E}$ with $(f+g)(z^r)$ approaching $(f+g)(z)$, and a sequence of Dini subgradients $\phi^r \in \partial_-(f+g)(z^r)$ approaching ϕ. By the Fuzzy sum rule (6.4.1), there exist points w^r and y^r in $\mathbf{E}$ satisfying

$$\|w^r - z^r\|, \ \|y^r - z^r\|, \ |f(w^r) - f(z^r)|, \ |g(y^r) - g(z^r)| < \frac{1}{r},$$

and elements μ^r of $\partial_- f(w^r)$ and ρ^r of $\partial_- g(y^r)$ satisfying

$$\|\mu^r + \rho^r - \phi^r\| \leq \frac{1}{r}$$

for each $r = 1, 2, \ldots$.

Now since f is locally Lipschitz, the sequence (μ^r) is bounded, so it has a subsequence converging to some element μ of $\partial f_a(z)$. The corresponding subsequence of (ρ^r) converges to an element ρ of $\partial_a g(z)$, and since these elements satisfy $\mu + \rho = \phi$, the result follows. $\square$

Exercises and Commentary

The limiting subdifferential was first studied by Mordukhovich in [143], followed by joint work with Kruger in [116], and by work of Ioffe [102, 103]. For a very complete development see [168]. A comprehensive survey of the infinite-dimensional literature (including some background to Exercise 11 (Viscosity subderivatives)) may be found in [42]. Somewhat surprisingly, on the real line the limiting and Clarke subdifferentials may only differ at countably many points, and at these points the limiting subdifferential is the union of two (possibly degenerate) intervals [31].

1. For the functions in Section 6.1, Exercise 1, compute the limiting subdifferential $\partial_a f(0)$ in each case.

2. Prove that the convex, Dini, and limiting subdifferential all coincide for convex functions.

3. **(Local minimizers)** Consider a function $f : \mathbf{E} \to [-\infty, +\infty]$ which is finite at the point $x \in \mathbf{E}$.

   (a) If x is a local minimizer, prove $0 \in \partial_- f(x)$.

   (b) If $0 \in \partial_- f(x)$, prove for any real $\delta > 0$ that x is a strict local minimizer of the function $f(\cdot) + \delta\|\cdot - x\|$.

4. **(Failure of sum rule)** Construct two lower semicontinuous functions $f, g : \mathbf{R} \to [-\infty, +\infty]$ satisfying $\partial_a f(0) = \partial_a g(0) = \emptyset$ and $\partial_a (f + g)(0) \neq \emptyset$.

5. If the real function f is continuous at x, prove the multifunction $\partial_a f$ is closed at x (see Section 6.2, Exercise 12 (Closed subdifferentials)).

6. Prove a limiting subdifferential sum rule for a finite number of lower semicontinuous functions, with all but one being locally Lipschitz.

7. * **(Limiting and Clarke subdifferentials)** Suppose the real function f is locally Lipschitz around the point x in $\mathbf{E}$.

   (a) Use the fact that the Clarke subdifferential is a closed multifunction to show $\partial_a f(x) \subset \partial_\circ f(x)$.

   (b) Deduce from the Intrinsic Clarke subdifferential theorem (6.2.5) the property $\partial_\circ f(x) = \operatorname{conv} \partial_a f(x)$.

   (c) Prove $\partial_a f(x) = \{\phi\}$ if and only if ϕ is the strict derivative of f at x.

8. * **(Topology of limiting subdifferential)** Suppose the real function f is locally Lipschitz around the point $x \in \mathbf{E}$.

(a) Prove $\partial_a f(x)$ is compact.

(b) Use the Fuzzy sum rule (6.4.1) to prove $\partial_- f(z)$ is nonempty at points z in $\mathbf{E}$ arbitrarily close to x (c.f. Section 6.2, Exercise 13).

(c) Deduce $\partial_a f(x)$ is nonempty.

9. * **(Tangents to graphs)** Consider a point z in a set $S \subset \mathbf{E}$, and a direction v in $\mathbf{E}$.

(a) Prove $\delta_S^-(z; v) = \delta_{K_S(z)}(v)$.

(b) Deduce $\partial_- \delta_S(z) = (K_S(z))^\circ$.

Now consider a Euclidean space $\mathbf{Y}$, a function $F : \mathbf{E} \to \mathbf{Y}$ which is locally Lipschitz around the point x in $\mathbf{E}$, and elements μ of $\mathbf{E}$ and ν of $\mathbf{Y}$.

(c) Use Section 6.3, Exercise 11 (Graphs of Lipschitz functions) to prove

$$(\mu, -\nu) \in \partial_- \delta_{G(F)}(x, F(x)) \iff \mu \in \partial_- \langle \nu, F(\cdot) \rangle (x).$$

(d) Deduce

$$(\mu, -\nu) \in N^a_{G(F)}(x, F(x)) \iff \mu \in \partial_a \langle \nu, F(\cdot) \rangle (x).$$

(e) If $\mathbf{Y} = \mathbf{R}$, deduce

$$(\mu, -1) \in N^a_{G(F)}(x, F(x)) \iff \mu \in \partial_a F(x).$$

(e) If F is strictly differentiable at x, deduce

$$N^a_{G(F)}(x, F(x)) = G(-(\nabla F(x))^*).$$

10. ** **(Composition)** Given a Euclidean space $\mathbf{Y}$ and two functions, $F : \mathbf{E} \to \mathbf{Y}$ and $f : \mathbf{Y} \to [-\infty, +\infty]$, define a function $p : \mathbf{E} \times \mathbf{Y} \to [-\infty, +\infty]$ by $p(x, y) = f(y)$ for points x in $\mathbf{E}$ and y in $\mathbf{Y}$.

(a) Prove $\partial_a p(x, y) = \{0\} \times \partial_a f(y)$.

(b) Prove $\partial_- (f \circ F)(x) \times \{0\} \subset \partial_- (p + \delta_{G(F)})(x, F(x))$.

(c) Deduce $\partial_a (f \circ F)(x) \times \{0\} \subset \partial_a (p + \delta_{G(F)})(x, F(x))$.

Now suppose F is continuous near a point z in $\mathbf{E}$ and f is locally Lipschitz around $F(z)$.

(d) Use the Limiting subdifferential sum rule (6.4.4) to deduce

$$\partial_a (f \circ F)(z) \times \{0\} \subset (\{0\} \times \partial_a f(F(z))) + N^a_{G(F)}(z, F(z)).$$

(e) **(Composition rule)** If F is strictly differentiable at z, use Exercise 9 (Tangents to graphs) to deduce

$$\partial_a(f \circ F)(z) \subset (\nabla F(z))^* \partial_a f(z).$$

Derive the corresponding formula for the Clarke subdifferential using Exercise 7(b).

(f) **(Mean value theorem)** If f is locally Lipschitz on $\mathbf{Y}$ then for any points u and v in $\mathbf{Y}$, prove there is a point z in the line segment (u, v) such that

$$f(u) - f(v) \in \langle \partial_a f(z) \cup -\partial_a(-f)(z), u - v \rangle.$$

(Hint: Consider the functions $t \mapsto \pm f(v + t(u - v))$.)

(g) **(Max rule)** Consider two real functions g and h which are locally Lipschitz around z and satisfy $g(z) = h(z)$. Using the functions

$$x \in \mathbf{E} \mapsto F(x) = (g(x), h(x)) \in \mathbf{R}^2$$

and

$$(u, v) \in \mathbf{R}^2 \mapsto f(u, v) = \max\{u, v\} \in \mathbf{R}$$

in part (d), apply Exercise 9 to prove

$$\partial_a(g \vee h)(z) \subset \bigcup_{\gamma \in [0,1]} \partial_a(\gamma g + (1 - \gamma)h)(z).$$

Derive the corresponding formula for the Clarke subdifferential, using Exercise 7(b)

(h) Use the Max rule in part (g) to strengthen the Nonsmooth necessary condition (6.1.8) for inequality-constrained optimization.

11. * **(Viscosity subderivatives)** Consider a real function f which is locally Lipschitz around zero and satisfies $f(0) = 0$ and $0 \in \partial_- f(0)$. Define a function $\rho : \mathbf{R}_+ \to \mathbf{R}$ by

$$\rho(r) = \min\{f(x) \mid \|x\| = r\}.$$

(a) Prove ρ is locally Lipschitz around zero.

(b) Prove $\rho^-(0; 1) \geq 0$.

(c) Prove the function $\gamma = \min\{0, \rho\}$ is locally Lipschitz and satisfies

$$f(x) \geq \gamma(\|x\|) \text{ for all } x \text{ in } \mathbf{E}$$

and

$$\lim_{t \downarrow 0} \frac{\gamma(t)}{t} = 0.$$

(d) Consider a real function g which is locally Lipschitz around a point $x \in \mathbf{E}$. If ϕ is any element of $\partial_- g(x)$ then prove ϕ is a *viscosity subderivative* of g: there is a real function h which is locally Lipschitz around x, minorizes g near x, and satisfies $h(x) = g(x)$ and has Fréchet derivative $\nabla h(x) = \phi$. Prove the converse is also true.

(e)** Prove the function h in part (d) can be assumed continuously differentiable near x.

12. ** **(Order statistic [125])** Consider the function $x \in \mathbf{R}^n \mapsto [x]_k$ (for some index $k = 1, 2, \ldots, n$).

(a) Calculate $\partial_-[\cdot]_k(0)$.

(b) Hence calculate $\partial_-[\cdot]_k(x)$ at an arbitrary point x in $\mathbf{R}^n$.

(c) Hence calculate $\partial_a[\cdot]_k(x)$.

Chapter 7

Karush–Kuhn–Tucker Theory

7.1 An Introduction to Metric Regularity

Our main optimization models so far are inequality-constrained. A little thought shows our techniques are not useful for equality-constrained problems like

$$\inf\{f(x) \mid h(x) = 0\}.$$

In this section we study such problems by linearizing the feasible region $h^{-1}(0)$ using the contingent cone.

Throughout this section we consider an open set $U \subset \mathbf{E}$, a closed set $S \subset U$, a Euclidean space $\mathbf{Y}$, and a continuous map $h : U \to \mathbf{Y}$. The restriction of h to S we denote $h|_S$. The following easy result (Exercise 1) suggests our direction.

Proposition 7.1.1 *If h is Fréchet differentiable at the point $x \in U$ then*

$$K_{h^{-1}(h(x))}(x) \subset N(\nabla h(x)).$$

Our aim in this section is to find conditions guaranteeing equality in this result.

Our key tool is the next result. It states that if a closed function attains a value close to its infimum at some point then a nearby point minimizes a slightly perturbed function.

Theorem 7.1.2 (Ekeland variational principle) *Suppose the function $f : \mathbf{E} \to (\infty, +\infty]$ is closed and the point $x \in \mathbf{E}$ satisfies $f(x) \leq \inf f + \epsilon$ for some real $\epsilon > 0$. Then for any real $\lambda > 0$ there is a point $v \in \mathbf{E}$ satisfying the conditions* *(a)* $\|x - v\| \leq \lambda$,

(b) $f(v) \leq f(x)$, *and*

(c) v *is the unique minimizer of the function* $f(\cdot) + (\epsilon/\lambda)\| \cdot - v\|$.

Proof. We can assume f is proper, and by assumption it is bounded below. Since the function

$$f(\cdot) + \frac{\epsilon}{\lambda}\| \cdot - x\|$$

therefore has compact level sets, its set of minimizers $M \subset \mathbf{E}$ is nonempty and compact. Choose a minimizer v for f on M. Then for points $z \neq v$ in M we know

$$f(v) \leq f(z) < f(z) + \frac{\epsilon}{\lambda}\|z - v\|,$$

while for z not in M we have

$$f(v) + \frac{\epsilon}{\lambda}\|v - x\| < f(z) + \frac{\epsilon}{\lambda}\|z - x\|.$$

Part (c) follows by the triangle inequality. Since v lies in M we have

$$f(z) + \frac{\epsilon}{\lambda}\|z - x\| \geq f(v) + \frac{\epsilon}{\lambda}\|v - x\| \text{ for all } z \text{ in } \mathbf{E}.$$

Setting $z = x$ shows the inequalities

$$f(v) + \epsilon \geq \inf f + \epsilon \geq f(x) \geq f(v) + \frac{\epsilon}{\lambda}\|v - x\|.$$

Properties (a) and (b) follow. □

As we shall see, precise calculation of the contingent cone $K_{h^{-1}(h(x))}(x)$ requires us first to bound the distance of a point z to the set $h^{-1}(h(x))$ in terms of the function value $h(z)$. This leads us to the notion of "metric regularity". In this section we present a somewhat simplified version of this idea, which suffices for most of our purposes; we defer a more comprehensive treatment to a later section. We say h is *weakly metrically regular* on S at the point x in S if there is a real constant k such that

$$d_{S \cap h^{-1}(h(x))}(z) \leq k\|h(z) - h(x)\| \text{ for all } z \text{ in } S \text{ close to } x.$$

Lemma 7.1.3 *Suppose* $0 \in S$ *and* $h(0) = 0$. *If* h *is not weakly metrically regular on* S *at zero then there is a sequence* $v_r \to 0$ *in* S *such that* $h(v_r) \neq 0$ *for all* r, *and a strictly positive sequence* $\delta_r \downarrow 0$ *such that the function*

$$\|h(\cdot)\| + \delta_r\| \cdot - v_r\|$$

is minimized on S *at* v_r.

Proof. By definition there is a sequence $x_r \to 0$ in S such that

$$d_{S \cap h^{-1}(0)}(x_r) > r\|h(x_r)\| \text{ for all } r. \tag{7.1.4}$$

For each index r we apply the Ekeland principle with

$$f = \|h\| + \delta_S, \quad \epsilon = \|h(x_r)\|, \quad \lambda = \min\{r\epsilon, \sqrt{\epsilon}\}, \quad \text{and } x = x_r$$

to deduce the existence of a point v_r in S such that

(a) $\|x_r - v_r\| \le \min\left\{r\|h(x_r)\|, \sqrt{\|h(x_r)\|}\right\}$ and

(c) v_r minimizes the function

$$\|h(\cdot)\| + \max\left\{r^{-1}, \sqrt{\|h(x_r)\|}\right\} \|\cdot - v_r\|$$

on S.

Property (a) shows $v_r \to 0$, while (c) reveals the minimizing property of v_r. Finally, inequality (7.1.4) and property (a) prove $h(v_r) \neq 0$. □

We can now present a convenient condition for weak metric regularity.

Theorem 7.1.5 (Surjectivity and metric regularity) *If h is strictly differentiable at the point x in S and*

$$\nabla h(x)(T_S(x)) = \mathbf{Y}$$

then h is weakly metrically regular on S at x.

Proof. Notice first h is locally Lipschitz around x (see Theorem 6.2.3). Without loss of generality, suppose $x = 0$ and $h(0) = 0$. If h is not weakly metrically regular on S at zero then by Lemma 7.1.3 there is a sequence $v_r \to 0$ in S such that $h(v_r) \neq 0$ for all r, and a real sequence $\delta_r \downarrow 0$ such that the function

$$\|h(\cdot)\| + \delta_r \|\cdot - v_r\|$$

is minimized on S at v_r. Denoting the local Lipschitz constant by L, we deduce from the sum rule (6.1.6) and the Exact penalization proposition (6.3.2) the condition

$$0 \in \partial_\circ(\|h\|)(v_r) + \delta_r B + L\partial_\circ d_S(v_r).$$

Hence there are elements u_r of $\partial_\circ(\|h\|)(v_r)$ and w_r of $L\partial_\circ d_S(v_r)$ such that $u_r + w_r$ approaches zero.

By choosing a subsequence we can assume

$$\|h(v_r)\|^{-1} h(v_r) \to y \neq 0$$

and an exercise then shows $u_r \to (\nabla h(0))^* y$. Since the Clarke subdifferential is closed at zero (Section 6.2, Exercise 12), we deduce

$$-(\nabla h(0))^* y \in L\partial_\circ d_S(0) \subset N_S(0).$$

However, by assumption there is a nonzero element p of $T_S(0)$ such that $\nabla h(0)p = -y$, so we arrive at the contradiction

$$0 \geq \langle p, -(\nabla h(0))^* y\rangle = \langle \nabla h(0)p, -y\rangle = \|y\|^2 > 0,$$

which completes the proof. □

We can now prove the main result of this section.

Theorem 7.1.6 (Liusternik) *If h is strictly differentiable at the point x and $\nabla h(x)$ is surjective then the set $h^{-1}(h(x))$ is tangentially regular at x and*

$$K_{h^{-1}(h(x))}(x) = N(\nabla h(x)).$$

Proof. Assume without loss of generality that $x = 0$ and $h(0) = 0$. In light of Proposition 7.1.1, it suffices to prove

$$N(\nabla h(0)) \subset T_{h^{-1}(0)}(0).$$

Fix any element p of $N(\nabla h(0))$ and consider a sequence $x^r \to 0$ in $h^{-1}(0)$ and $t_r \downarrow 0$ in $\mathbf{R}_{++}$. The previous result shows h is weakly metrically regular at zero, so there is a constant k such that

$$d_{h^{-1}(0)}(x^r + t_r p) \leq k\|h(x^r + t_r p)\|$$

holds for all large r, and hence there are points z^r in $h^{-1}(0)$ satisfying

$$\|x^r + t_r p - z^r\| \leq k\|h(x^r + t_r p)\|.$$

If we define directions $p^r = t_r^{-1}(z^r - x^r)$ then clearly the points $x^r + t_r p^r$ lie in $h^{-1}(0)$ for large r, and since

$$\begin{aligned} \|p - p^r\| &= \frac{\|x^r + t_r p - z^r\|}{t_r} \\ &\leq \frac{k\|h(x^r + t_r p) - h(x^r)\|}{t_r} \\ &\to k\|(\nabla h(0))p\| \\ &= 0, \end{aligned}$$

we deduce $p \in T_{h^{-1}(0)}(0)$. □

Exercises and Commentary

Liusternik's original study of tangent spaces appeared in [130]. Closely related ideas were pursued by Graves [85] (see [65] for a good survey). The Ekeland principle first appeared in [69], motivated by the study of infinite-dimensional problems where techniques based on compactness might be unavailable. As we see in this section, it is a powerful idea even in finite dimensions; the simplified version we present here was observed in [94]. See also Exercise 14 in Section 9.2. The inversion technique we use (Lemma 7.1.3) is based on the approach in [101]. The recognition of "metric" regularity (a term perhaps best suited to nonsmooth analysis) as a central idea began largely with Robinson; see [162, 163] for example. Many equivalences are discussed in [5, 168].

1. Suppose h is Fréchet differentiable at the point $x \in S$.

   (a) Prove for any set $D \supset h(S)$ the inclusion

   $$\nabla h(x) K_S(x) \subset K_D(h(x)).$$

   (b) If h is constant on S, deduce

   $$K_S(x) \subset N(\nabla h(x)).$$

   (c) If h is a real function and x is a local minimizer of h on S, prove

   $$-\nabla h(x) \in (K_S(x))^-.$$

2. **(Lipschitz extension)** Suppose the real function f has Lipschitz constant k on the set $C \subset \mathbf{E}$. By considering the infimal convolution of the functions $f + \delta_C$ and $k\|\cdot\|$, prove there is a function $\tilde{f} : \mathbf{E} \to \mathbf{R}$ with Lipschitz constant k that agrees with f on C. Prove furthermore that if f and C are convex then $\tilde{f}$ can be assumed convex.

3. * **(Closure and the Ekeland principle)** Given a subset S of $\mathbf{E}$, suppose the conclusion of Ekeland's principle holds for all functions of the form $g + \delta_S$ where the function g is continuous on S. Deduce S is closed. (Hint: For any point x in $\operatorname{cl} S$, let $g = \|\cdot - x\|$.)

4. ** Suppose h is strictly differentiable at zero and satisfies

   $$h(0) = 0, \; v_r \to 0, \; \|h(v_r)\|^{-1} h(v_r) \to y, \text{ and } u_r \in \partial_\circ(\|h\|)(v_r).$$

   Prove $u_r \to (\nabla h(0))^* y$. Write out a shorter proof when h is continuously differentiable at zero.

5. ** Interpret Exercise 27 (Conical open mapping) in Section 4.2 in terms of metric regularity.

6. ** **(Transversality)** Suppose the set $V \subset \mathbf{Y}$ is open and the set $R \subset V$ is closed. Suppose furthermore h is strictly differentiable at the point x in S with $h(x)$ in R and

$$\nabla h(x)(T_S(x)) - T_R(h(x)) = \mathbf{Y}. \tag{7.1.7}$$

   (a) Define the function $g : U \times V \to \mathbf{Y}$ by $g(z, y) = h(z) - y$. Prove g is weakly metrically regular on $S \times R$ at the point $(x, h(x))$.

   (b) Deduce the existence of a constant k' such that the inequality

$$d_{(S \times R) \cap g^{-1}(g(x,h(x)))}(z, y) \le k' \|h(z) - y\|$$

   holds for all points (z, y) in $S \times R$ close to $(x, h(x))$.

   (c) Apply Proposition 6.3.2 (Exact penalization) to deduce the existence of a constant k such that the inequality

$$d_{(S \times R) \cap g^{-1}(g(x,h(x)))}(z, y) \le k(\|h(z) - y\| + d_S(z) + d_R(y))$$

   holds for all points (z, y) in $U \times V$ close to $(x, h(x))$.

   (d) Deduce the inequality

$$d_{S \cap h^{-1}(R)}(z) \le k(d_S(z) + d_R(h(z)))$$

   holds for all points z in U close to x.

   (e) Imitate the proof of Liusternik's theorem (7.1.6) to deduce the inclusions

$$T_{S \cap h^{-1}(R)}(x) \supset T_S(x) \cap (\nabla h(x))^{-1} T_R(h(x))$$

   and

$$K_{S \cap h^{-1}(R)}(x) \supset K_S(x) \cap (\nabla h(x))^{-1} T_R(h(x)).$$

   (f) Suppose h is the identity map, so

$$T_S(x) - T_R(x) = \mathbf{E}.$$

   If either R or S is tangentially regular at x, prove

$$K_{R \cap S}(x) = K_R(x) \cap K_S(x).$$

   (g) **(Guignard)** By taking polars and applying the Krein–Rutman polar cone calculus (3.3.13) and condition (7.1.7) again, deduce

$$N_{S \cap h^{-1}(R)}(x) \subset N_S(x) + (\nabla h(x))^* N_R(h(x)).$$

(h) If C and D are convex subsets of $\mathbf{E}$ satisfying $0 \in \operatorname{core}(C - D)$ (or $\operatorname{ri} C \cap \operatorname{ri} D \neq \emptyset$), and the point x lies in $C \cap D$, use part (e) to prove

$$T_{C \cap D}(x) = T_C(x) \cap T_D(x).$$

7. ** **(Liusternik via inverse functions)** We first fix $\mathbf{E} = \mathbf{R}^n$. The classical inverse function theorem states that if the map $g : U \to \mathbf{R}^n$ is continuously differentiable then at any point x in U at which $\nabla g(x)$ is invertible, x has an open neighbourhood V whose image $g(V)$ is open, and the restricted map $g|_V$ has a continuously differentiable inverse satisfying the condition

$$\nabla (g|_V)^{-1} (g(x)) = (\nabla g(x))^{-1}.$$

Consider now a continuously differentiable map $h : U \to \mathbf{R}^m$, and a point x in U with $\nabla h(x)$ surjective, and fix a direction d in the null space $N(\nabla h(x))$. Choose any $(n \times (n - m))$ matrix D making the matrix $A = (\nabla h(x), D)$ invertible, define a function $g : U \to \mathbf{R}^n$ by $g(z) = (h(z), Dz)$, and for a small real $\delta > 0$ define a function $p : (-\delta, \delta) \to \mathbf{R}^n$ by

$$p(t) = g^{-1}(g(x) + tAd).$$

(a) Prove p is well-defined providing δ is small.

(b) Prove the following properties:

(i) p is continuously differentiable.

(ii) $p(0) = x$.

(iii) $p'(0) = d$.

(iv) $h(p(t)) = h(x)$ for all small t.

(c) Deduce that a direction d lies in $N(\nabla h(x))$ if and only if there is a function $p : (-\delta, \delta) \to \mathbf{R}^n$ for some $\delta > 0$ in $\mathbf{R}$ satisfying the four conditions in part (b).

(d) Deduce $K_{h^{-1}(h(x))}(x) = N(\nabla h(x))$.

7.2 The Karush–Kuhn–Tucker Theorem

The central result of optimization theory describes first order necessary optimality conditions for the general nonlinear problem

$$\inf\{f(x) \mid x \in S\}, \tag{7.2.1}$$

where, given an open set $U \subset \mathbf{E}$, the objective function is $f : U \to \mathbf{R}$ and the feasible region S is described by equality and inequality constraints:

$$S = \{x \in U \mid g_i(x) \leq 0 \text{ for } i = 1, 2, \ldots, m,\ h(x) = 0\}. \tag{7.2.2}$$

The equality constraint map $h : U \to \mathbf{Y}$ (where $\mathbf{Y}$ is a Euclidean space) and the inequality constraint functions $g_i : U \to \mathbf{R}$ (for $i = 1, 2, \ldots, m$) are all continuous. In this section we derive necessary conditions for the point $\bar{x}$ in S to be a local minimizer for the problem (7.2.1).

In outline, the approach takes three steps. We first extend Liusternik's theorem (7.1.6) to describe the contingent cone $K_S(\bar{x})$. Next we calculate this cone's polar cone using the Farkas lemma (2.2.7). Finally, we apply the Contingent necessary condition (6.3.10) to derive the result.

As in our development for the inequality-constrained problem in Section 2.3, we need a regularity condition. Once again, we denote the set of indices of the active inequality constraints by $I(\bar{x}) = \{i \mid g_i(\bar{x}) = 0\}$.

Assumption 7.2.3 (The Mangasarian–Fromovitz constraint qualification) *The active constraint functions g_i (for i in $I(\bar{x})$) are Fréchet differentiable at the point $\bar{x}$, the equality constraint map h is strictly differentiable, with a surjective gradient, at $\bar{x}$, and the set*

$$\{p \in N(\nabla h(\bar{x})) \mid \langle \nabla g_i(\bar{x}), p \rangle < 0 \text{ for } i \text{ in } I(\bar{x})\} \tag{7.2.4}$$

is nonempty.

Notice in particular that the set (7.2.4) is nonempty in the case where the map $h : U \to \mathbf{R}^q$ has components $h_1, h_2, \ldots, h_q$ and the set of gradients

$$\{\nabla h_j(\bar{x}) \mid j = 1, 2, \ldots, q\} \cup \{\nabla g_i(\bar{x}) \mid i \in I(\bar{x})\} \tag{7.2.5}$$

is linearly independent (Exercise 1).

Theorem 7.2.6 *Suppose the Mangasarian–Fromovitz constraint qualification (7.2.3) holds. Then the contingent cone to the feasible region S defined by equation (7.2.2) is given by*

$$K_S(\bar{x}) = \{p \in N(\nabla h(\bar{x})) \mid \langle \nabla g_i(\bar{x}), p \rangle \leq 0 \text{ for } i \text{ in } I(\bar{x})\}. \tag{7.2.7}$$

Proof. Denote the set (7.2.4) by $\widetilde{K}$ and the right hand side of formula (7.2.7) by K. The inclusion

$$K_S(\bar{x}) \subset K$$

is a straightforward exercise. Furthermore, since $\widetilde{K}$ is nonempty, it is easy to see $K = \operatorname{cl} \widetilde{K}$. If we can show $\widetilde{K} \subset K_S(\bar{x})$ then the result will follow since the contingent cone is always closed.

To see $\widetilde{K} \subset K_S(\bar{x})$, fix an element p of $\widetilde{K}$. Since p lies in $N(\nabla h(\bar{x}))$, Liusternik's theorem (7.1.6) shows $p \in K_{h^{-1}(0)}(\bar{x})$. Hence there are sequences $t_r \downarrow 0$ in $\mathbf{R}_{++}$ and $p^r \to p$ in $\mathbf{E}$ satisfying $h(\bar{x} + t_r p^r) = 0$ for all r. Clearly $\bar{x} + t_r p^r \in U$ for all large r, and we claim $g_i(\bar{x} + t_r p^r) < 0$. For indices i not in $I(\bar{x})$ this follows by continuity, so we suppose $i \in I(\bar{x})$ and $g_i(\bar{x} + t_r p^r) \geq 0$ for all r in some subsequence R of $\mathbf{N}$. We then obtain the contradiction

$$\begin{aligned} 0 &= \lim_{r \to \infty \text{ in } R} \frac{g_i(\bar{x} + t_r p^r) - g_i(\bar{x}) - \langle \nabla g_i(\bar{x}), t_r p^r \rangle}{t_r \|p^r\|} \\ &\geq -\frac{\langle \nabla g_i(\bar{x}), p \rangle}{\|p\|} \\ &> 0. \end{aligned}$$

The result now follows. □

Lemma 7.2.8 *Any linear maps $A : \mathbf{E} \to \mathbf{R}^q$ and $G : \mathbf{E} \to \mathbf{Y}$ satisfy*

$$\{x \in N(G) \mid Ax \leq 0\}^- = A^* \mathbf{R}^q_+ + G^* \mathbf{Y}.$$

Proof. This is an immediate application of Section 5.1, Exercise 9 (Polyhedral cones). □

Theorem 7.2.9 (Karush–Kuhn–Tucker conditions) *Suppose $\bar{x}$ is a local minimizer for problem (7.2.1) and the objective function f is Fréchet differentiable at $\bar{x}$. If the Mangasarian–Fromovitz constraint qualification (7.2.3) holds then there exist multipliers λ_i in $\mathbf{R}_+$ (for i in $I(\bar{x})$) and μ in $\mathbf{Y}$ satisfying*

$$\nabla f(\bar{x}) + \sum_{i \in I(\bar{x})} \lambda_i \nabla g_i(\bar{x}) + \nabla h(\bar{x})^* \mu = 0. \tag{7.2.10}$$

Proof. The Contingent necessary condition (6.3.10) shows

$$\begin{aligned} -\nabla f(\bar{x}) &\in K_S(\bar{x})^- \\ &= \{p \in N(\nabla h(\bar{x})) \mid \langle \nabla g_i(\bar{x}), p \rangle \leq 0 \text{ for } i \text{ in } I(\bar{x})\}^- \\ &= \sum_{i \in I(\bar{x})} \mathbf{R}_+ \nabla g_i(\bar{x}) + \nabla h(\bar{x})^* \mathbf{Y} \end{aligned}$$

using Theorem 7.2.6 and Lemma 7.2.8. □

Exercises and Commentary

A survey of the history of these results may be found in [158]. The Mangasarian–Fromovitz condition originated with [133], while the Karush–Kuhn–Tucker conditions first appeared in [111] and [117]. The idea of penalty functions (see Exercise 11 (Quadratic penalties)) is a common technique in optimization. The related notion of a barrier penalty is crucial for interior point methods; examples include the penalized linear and semidefinite programs we considered in Section 4.3, Exercise 4 (Examples of duals).

1. **(Linear independence implies Mangasarian–Fromovitz)** If the set of gradients (7.2.5) is linearly independent, then by considering the equations

$$\begin{aligned} \langle \nabla g_i(\bar{x}), p \rangle &= -1 \;\; \text{for } i \text{ in } I(\bar{x}) \\ \langle \nabla h_j(\bar{x}), p \rangle &= 0 \;\; \text{for } j = 1, 2, \ldots, q, \end{aligned}$$

   prove the set (7.2.4) is nonempty.

2. Consider the proof of Theorem 7.2.6.

   (a) Prove $K_S(\bar{x}) \subset K$.

   (b) If $\widetilde{K}$ is nonempty, prove $K = \operatorname{cl} \widetilde{K}$.

3. **(Linear constraints)** If the functions g_i (for i in $I(\bar{x})$) and h are affine, prove the contingent cone formula (7.2.7) holds.

4. **(Bounded multipliers)** In Theorem 7.2.9 (Karush–Kuhn–Tucker conditions), prove the set of multiplier vectors (λ, μ) satisfying equation (7.2.10) is compact.

5. **(Slater condition)** Suppose the set U is convex, the functions

$$g_1, g_2, \ldots, g_m : U \to \mathbf{R}$$

   are convex and Fréchet differentiable, and the function $h : \mathbf{E} \to \mathbf{Y}$ is affine and surjective. Suppose further there is a point $\hat{x}$ in $h^{-1}(0)$ satisfying $g_i(\hat{x}) < 0$ for $i = 1, 2, \ldots, m$. For any feasible point $\bar{x}$ for problem (7.2.1), prove the Mangasarian–Fromovitz constraint qualification holds.

6. **(Largest eigenvalue)** For a matrix A in $\mathbf{S}^n$, use the Karush–Kuhn–Tucker theorem to calculate

$$\sup\{x^T A x \mid \|x\| = 1,\; x \in \mathbf{R}^n\}.$$

7. * **(Largest singular value [100, p. 135])** Given any $m \times n$ matrix A, consider the optimization problem

$$\alpha = \sup\{x^T A y \mid \|x\|^2 = 1,\ \|y\|^2 = 1\} \tag{7.2.11}$$

and the matrix

$$\widetilde{A} = \begin{bmatrix} 0 & A \\ A^T & 0 \end{bmatrix}.$$

   (a) If μ is an eigenvalue of $\widetilde{A}$, prove $-\mu$ is also.
   (b) If μ is a nonzero eigenvalue of $\widetilde{A}$, use a corresponding eigenvector to construct a feasible solution to problem (7.2.11) with objective value μ.
   (c) Deduce $\alpha \geq \lambda_1(\widetilde{A})$.
   (d) Prove problem (7.2.11) has an optimal solution.
   (e) Use the Karush–Kuhn–Tucker theorem to prove any optimal solution of problem (7.2.11) corresponds to an eigenvector of $\widetilde{A}$.
   (f) **(Jordan [108])** Deduce $\alpha = \lambda_1(\widetilde{A})$. (This number is called the *largest singular value of* A.)

8. ** **(Hadamard's inequality [88])** The matrix with columns $x^1, x^2, \ldots, x^n$ in $\mathbf{R}^n$ we denote by $(x^1, x^2, \ldots, x^n)$. Prove $(\bar{x}^1, \bar{x}^2, \ldots, \bar{x}^n)$ solves the problem

$$\begin{array}{lrcl} \inf & -\det(x^1, x^2, \ldots, x^n) & & \\ \text{subject to} & \|x^i\|^2 & = & 1 \ \text{ for } i = 1, 2, \ldots, n \\ & x^1, x^2, \ldots, x^n & \in & \mathbf{R}^n \end{array}$$

if and only if the matrix $(\bar{x}^1, \bar{x}^2, \ldots, \bar{x}^n)$ has determinant equal to one and has columns forming an orthonormal basis, and deduce the inequality

$$\det(x^1, x^2, \ldots, x^n) \leq \prod_{i=1}^{n} \|x^i\|.$$

9. **(Nonexistence of multipliers [77])** Define a function $\mathrm{sgn} : \mathbf{R} \to \mathbf{R}$ by

$$\mathrm{sgn}(v) = \begin{cases} 1 & \text{if } v > 0 \\ 0 & \text{if } v = 0 \\ -1 & \text{if } v < 0 \end{cases}$$

and a function $h : \mathbf{R}^2 \to \mathbf{R}$ by

$$h(u, v) = v - \mathrm{sgn}(v)(u^+)^2.$$

   (a) Prove h is Fréchet differentiable at $(0, 0)$ with derivative $(0, 1)$.

(b) Prove h is not continuous on any neighbourhood of $(0,0)$, and deduce it is not strictly differentiable at $(0,0)$.

(c) Prove $(0,0)$ is optimal for the problem

$$\inf\{f(u,v) \mid h(u,v) = 0\},$$

where $f(u,v) = u$, and yet there is no real λ satisfying

$$\nabla f(0,0) + \lambda \nabla h(0,0) = (0,0).$$

(Exercise 14 in Section 8.1 gives an approach to weakening the conditions required in this section.)

10. * **(Guignard optimality conditions [87])** Suppose the point $\bar{x}$ is a local minimizer for the optimization problem

$$\inf\{f(x) \mid h(x) \in R,\ x \in S\}$$

where $R \subset \mathbf{Y}$. If the functions f and h are strictly differentiable at $\bar{x}$ and the transversality condition

$$\nabla h(\bar{x}) T_S(\bar{x}) - T_R(h(\bar{x})) = \mathbf{Y}$$

holds, use Section 7.1, Exercise 6 (Transversality) to prove the optimality condition

$$0 \in \nabla f(\bar{x}) + \nabla h(\bar{x})^* N_R(h(\bar{x})) + N_S(\bar{x}).$$

11. ** **(Quadratic penalties [136])** Take the nonlinear program (7.2.1) in the case $\mathbf{Y} = \mathbf{R}^q$ and now let us assume all the functions

$$f, g_1, g_2, \ldots, g_m, h_1, h_2, \ldots, h_q : U \to \mathbf{R}$$

are continuously differentiable on the set U. For positive integers k we define a function $p_k : U \to \mathbf{R}$ by

$$p_k(x) = f(x) + k\Big(\sum_{i=1}^{m} (g_i^+(x))^2 + \sum_{j=1}^{q} (h_j(x))^2\Big).$$

Suppose the point $\bar{x}$ is a local minimizer for the problem (7.2.1). Then for some compact neighbourhood W of $\bar{x}$ in U we know $f(x) \geq f(\bar{x})$ for all feasible points x in W. Now define a function $r_k : W \to \mathbf{R}$ by

$$r_k(x) = p_k(x) + \|x - \bar{x}\|^2,$$

and for each $k = 1, 2, \ldots$ choose a point x^k minimizing r_k on W.

(a) Prove $r_k(x^k) \leq f(\bar{x})$ for each $k = 1, 2, \ldots$.

(b) Deduce

$$\lim_{k\to\infty} g_i^+(x^k) = 0 \ \text{ for } i = 1, 2, \ldots, m$$

and

$$\lim_{k\to\infty} h_j(x^k) = 0 \ \text{ for } j = 1, 2, \ldots, q.$$

(c) Hence show $x^k \to \bar{x}$ as $k \to \infty$.

(d) Calculate $\nabla r_k(x)$.

(e) Deduce

$$-2(x^k - \bar{x}) = \nabla f(x^k) + \sum_{i=1}^{m} \lambda_i^k \nabla g_i(x^k) + \sum_{j=1}^{q} \mu_j^k \nabla h_j(x^k)$$

for some suitable choice of vectors λ^k in $\mathbf{R}_+^m$ and μ^k in $\mathbf{R}^q$.

(f) By taking a convergent subsequence of the vectors

$$\|(1, \lambda^k, \mu^k)\|^{-1}(1, \lambda^k, \mu^k) \in \mathbf{R} \times \mathbf{R}_+^m \times \mathbf{R}^q,$$

show from parts (c) and (e) the existence of a nonzero vector $(\lambda_0, \lambda, \mu)$ in $\mathbf{R} \times \mathbf{R}_+^m \times \mathbf{R}^q$ satisfying the *Fritz John conditions*:

(i) $\lambda_i g_i(\bar{x}) = 0$ for $i = 1, 2, \ldots, m$.

(ii) $\lambda_0 \nabla f(\bar{x}) + \sum_{i=1}^m \lambda_i \nabla g_i(\bar{x}) + \sum_{j=1}^q \mu_j \nabla h_j(\bar{x}) = 0$.

(g) Under the assumption of the Mangasarian–Fromovitz constraint qualification (7.2.3), show that the Fritz John conditions in part (f) imply the Karush–Kuhn–Tucker conditions.

7.3 Metric Regularity and the Limiting Subdifferential

In Section 7.1 we presented a convenient test for the weak metric regularity of a function at a point in terms of the surjectivity of its strict derivative there (Theorem 7.1.5). This test, while adequate for most of our purposes, can be richly refined using the limiting subdifferential.

As before, we consider an open set $U \subset \mathbf{E}$, a Euclidean space $\mathbf{Y}$, a closed set $S \subset U$, and a function $h : U \to \mathbf{Y}$ which we assume throughout this section is locally Lipschitz. We begin with the full definition of metric regularity, strengthening the notion of Section 7.1. We say h is *metrically regular on S at* the point x in S if there is a real constant k such that the estimate

$$d_{S \cap h^{-1}(y)}(z) \leq k\|h(z) - y\|$$

holds for all points z in S close to x and all vectors y in $\mathbf{Y}$ close to $h(x)$. (Before we only required this to be true when $y = h(x)$.)

Lemma 7.3.1 *If h is not metrically regular on S at x then there are sequences (v_r) in S converging to x, (y_r) in $\mathbf{Y}$ converging to $h(x)$, and (ϵ_r) in $\mathbf{R}_{++}$ decreasing to zero such that, for each index r, we have $h(v_r) \neq y_r$ and the function*

$$\|h(\cdot) - y_r\| + \epsilon_r\| \cdot - v_r\|$$

is minimized on S at v_r.

Proof. The proof is completely analogous to that of Lemma 7.1.3: we leave it as an exercise. □

We also need the following chain-rule-type result; we leave the proof as an exercise.

Lemma 7.3.2 *At any point x in $\mathbf{E}$ where $h(x) \neq 0$ we have*

$$\partial_a \|h(\cdot)\|(x) = \partial_a \langle \|h(x)\|^{-1} h(x), h(\cdot)\rangle(x).$$

Using this result and a very similar proof to Theorem 7.1.5, we can now extend the surjectivity and metric regularity result.

Theorem 7.3.3 (Limiting subdifferential and regularity) *If a point x lies in S and no nonzero element w of $\mathbf{Y}$ satisfies the condition*

$$0 \in \partial_a \langle w, h(\cdot)\rangle(x) + N_S^a(x)$$

then h is metrically regular on S at x.

Proof. If h is not metrically regular, we can apply Lemma 7.3.1, so with that notation the function

$$\|h(\cdot) - y_r\| + \epsilon_r\| \cdot - v_r\|$$

is minimized on S at v_r. By Proposition 6.3.2 (Exact penalization) we deduce for large enough real L

$$\begin{aligned} 0 &\in \partial_a(\|h(\cdot) - y_r\| + \epsilon_r\| \cdot - v_r\| + Ld_S(\cdot))(v_r) \\ &\subset \partial_a\|h(\cdot) - y_r\|(v_r) + \epsilon_r B + L\partial_a d_S(v_r) \end{aligned}$$

for all r, using the Limiting subdifferential sum rule (6.4.4). If we write $w_r = \|h(v_r) - y_r\|^{-1}(h(v_r) - y_r)$, we obtain by Lemma 7.3.2

$$0 \in \partial_a\langle w_r, h(\cdot)\rangle(v_r) + \epsilon_r B + L\partial_a d_S(v_r),$$

so there are elements u_r in $\partial_a\langle w_r, h(\cdot)\rangle(v_r)$ and z_r in $L\partial_a d_S(v_r)$ such that $\|u_r + z_r\| \leq \epsilon_r$. The sequences (w_r), (u_r), and (z_r) are all bounded, so by taking subsequences we can assume w_r approaches some nonzero vector w, z_r approaches some vector z, and u_r approaches $-z$.

Now, using the sum rule again we observe

$$u_r \in \partial_a\langle w, h(\cdot)\rangle(v_r) + \partial_a\langle w_r - w, h(\cdot)\rangle(v_r)$$

for each r. The local Lipschitz constant of the function $\langle w_r - w, h(\cdot)\rangle$ tends to zero, so since $\partial_a\langle w, h(\cdot)\rangle$ is a closed multifunction at x (by Section 6.4, Exercise 5) we deduce

$$-z \in \partial_a\langle w, h(\cdot)\rangle(x).$$

Similarly, since $\partial_a d_S(\cdot)$ is closed at x, we see

$$z \in L\partial_a d_S(x) \subset N_S^a(x)$$

by Exercise 4, and this contradicts the assumption of the theorem. □

This result strengthens and generalizes the elegant test of Theorem 7.1.5, as the next result shows.

Corollary 7.3.4 (Surjectivity and metric regularity) *If h is strictly differentiable at the point x in S and*

$$(\nabla h(x)^*)^{-1}(N_S^a(x)) = \{0\}$$

or, in particular,

$$\nabla h(x)(T_S(x)) = \mathbf{Y}$$

then h is metrically regular on S at x.

Proof. Since it is easy to check for any element w of $\mathbf{Y}$ the function $\langle w, h(\cdot)\rangle$ is strictly differentiable at x with derivative $\nabla h(x)^* w$, the first condition implies the result by Theorem 7.3.3. On the other hand, the second condition implies the first, since for any element w of $(\nabla h(x)^*)^{-1}(N_S^a(x))$ there is an element z of $T_S(x)$ satisfying $\nabla h(x)z = w$, and now we deduce

$$\|w\|^2 = \langle w, w\rangle = \langle w, \nabla h(x)z\rangle = \langle \nabla h(x)^* w, z\rangle \le 0$$

using Exercise 4, so $w = 0$. □

As a final extension to the idea of metric regularity, consider now a closed set $D \subset \mathbf{Y}$ containing $h(x)$. We say h is *metrically regular on S at x with respect to D* if there is a real constant k such that

$$d_{S\cap h^{-1}(y+D)}(z) \le k d_D(h(z) - y)$$

for all points z in S close to x and vectors y close to 0. Our previous definition was the case $D = \{h(x)\}$. This condition estimates how far a point $z \in S$ is from feasibility for the system

$$h(z) \in y + D, \quad z \in S,$$

in terms of the constraint error $d_D(h(z) - y)$.

Corollary 7.3.5 *If the point x lies in the closed set $S \subset \mathbf{E}$ with $h(x)$ in the closed set $D \subset \mathbf{Y}$, and no nonzero element w of $N_D^a(h(x))$ satisfies the condition*

$$0 \in \partial_a \langle w, h(\cdot)\rangle(x) + N_S^a(x),$$

then h is metrically regular on S at x with respect to D.

Proof. Define a function $\widetilde{h} : U \times \mathbf{Y} \to \mathbf{Y}$ by $\widetilde{h}(z, y) = h(z) - y$, a set $\widetilde{S} = S \times D$, and a point $\widetilde{x} = (x, h(x))$. Since by Exercise 5 we have

$$N_{\widetilde{S}}^a(\widetilde{x}) = N_S^a(x) \times N_D^a(h(x))$$

and

$$\partial_a \langle w, \widetilde{h}(\cdot)\rangle(\widetilde{x}) = \partial_a \langle w, h(\cdot)\rangle(x) \times \{-w\}$$

for any element w of $\mathbf{Y}$, there is no nonzero w satisfying the condition

$$0 \in \partial_a \langle w, \widetilde{h}(\cdot)\rangle(\widetilde{x}) + N_{\widetilde{S}}^a(\widetilde{x}),$$

so $\widetilde{h}$ is metrically regular on $\widetilde{S}$ at $\widetilde{x}$ by Theorem 7.3.3 (Limiting subdifferential and regularity). Some straightforward manipulation now shows h is metrically regular on S at x with respect to D. □

The case $D = \{h(x)\}$ recaptures Theorem 7.3.3.

A nice application of this last result estimates the distance to a level set under a Slater-type assumption, a typical illustration of the power of metric regularity.

Corollary 7.3.6 (Distance to level sets) *If the function $g : U \to \mathbf{R}$ is locally Lipschitz around a point x in U satisfying*

$$g(x) = 0 \quad \text{and} \quad 0 \notin \partial_a g(x)$$

then there is a real constant $k > 0$ such that the estimate

$$d_{g^{-1}(-\mathbf{R}_+)}(z) \le k g(z)^+$$

holds for all points z in $\mathbf{E}$ close to x.

Proof. Let $S \subset U$ be any closed neighbourhood of x and apply Corollary 7.3.5 with $h = g$ and $D = -\mathbf{R}_+$. □

Exercises and Commentary

In many circumstances, metric regularity is in fact equivalent to weak metric regularity (see [25]). The power of the limiting subdifferential as a tool in recognizing metric regularity was first observed by Mordukhovich [144]; there is a comprehensive discussion in [145, 168].

1. * Prove Lemma 7.3.1.

2. * Assume $h(x) \neq 0$.

   (a) Prove
   $$\partial_- \|h(\cdot)\|(x) = \partial_- \langle \|h(x)\|^{-1} h(x), h(\cdot)\rangle(x).$$

   (b) Prove the analogous result for the limiting subdifferential. (You may use the Limiting subdifferential sum rule (6.4.4).)

3. **(Metric regularity and openness)** If h is metrically regular on S at x, prove h is *open* on S at x; that is, for any neighbourhood U of x we have $h(x) \in \operatorname{int} h(U \cap S)$.

4. ** **(Limiting normals and distance functions)** Given a point z in $\mathbf{E}$, suppose y is a nearest point to z in S.

   (a) If $0 \le \alpha < 1$, prove the unique nearest point to $\alpha z + (1-\alpha)y$ in S is y.

   (b) For z not in S, deduce every element of $\partial_- d_S(z)$ has norm one.

   (c) For any element w of $\mathbf{E}$, prove
   $$d_S(z + w) \le d_S(z) + d_S(y + w).$$

   (d) Deduce $\partial_- d_S(z) \subset \partial_- d_S(y)$.

Now consider a point x in S.

(e) Prove ϕ is an element of $\partial_a d_S(x)$ if and only if there are sequences (x^r) in S approaching x, and (ϕ^r) in $\mathbf{E}$ approaching ϕ satisfying $\phi^r \in \partial_- d_S(x^r)$ for all r.

(f) Deduce $\mathbf{R}_+\partial_a d_S(x) \subset N_S^a(x)$.

(g) Suppose ϕ is an element of $\partial_- \delta_S(x)$. For any real $\epsilon > 0$, apply Section 6.4, Exercise 3 (Local minimizers) and the Limiting subdifferential sum rule to prove

$$\phi \in (\|\phi\| + \epsilon)\partial_a d_S(x) + \epsilon B.$$

(h) By taking limits, deduce

$$N_S^a(x) = \mathbf{R}_+\partial_a d_S(x).$$

(i) Deduce

$$N_S(x) = \mathrm{cl}\,(\mathrm{conv}\, N_S^a(x)),$$

and hence

$$T_S(x) = N_S^a(x)^-.$$

(Hint: Use Section 6.4, Exercise 7 (Limiting and Clarke subdifferentials).)

(j) Hence prove the following properties are equivalent:

(i) $T_S(x) = \mathbf{E}$.

(ii) $N_S^a(x) = \{0\}$.

(iii) $x \in \mathrm{int}\, S$.

5. **(Normals to products)** For closed sets $S \subset \mathbf{E}$ and $D \subset \mathbf{Y}$ and points x in S and y in D, prove

$$N_{S\times D}^a(x, y) = N_S^a(x) \times N_D^a(y).$$

6. * Complete the remaining details of the proof of Corollary 7.3.5.

7. Prove Corollary 7.3.6 (Distance to level sets).

8. **(Limiting versus Clarke conditions)** Define a set

$$S = \{(u, v) \in \mathbf{R}^2 \mid u \leq 0 \text{ or } v \leq 0\}$$

and a function $h : \mathbf{R}^2 \to \mathbf{R}$ by $h(u, v) = u + v$. In Corollary 7.3.4 (Surjectivity and metric regularity), prove the limiting normal cone condition holds at the point $x = 0$, and yet the Clarke tangent cone condition fails.

9. ** **(Normals to level sets)** Under the hypotheses of Corollary 7.3.6 (Distance to level sets), prove

$$N^a_{g^{-1}(-\mathbf{R}_+)}(x) = \mathbf{R}_+ \partial_a g(x).$$

(Hint: Use Exercise 4 and the Max rule (Section 6.4, Exercise 10(g).)

7.4 Second Order Conditions

Optimality conditions can be refined using second order information; we saw an early example in Theorem 2.1.5 (Second order conditions). Because of the importance of curvature information for Newton-type methods in numerical optimization, second order conditions are widely useful.

In this section we present prototypical second order conditions for constrained optimization. Our approach is a simple and elegant blend of convex analysis and metric regularity.

Consider an open set $U \subset \mathbf{E}$, a Euclidean space $\mathbf{Y}$. Given any function $h : U \to \mathbf{Y}$ that is Fréchet differentiable on U, the gradient map ∇h is a function from U to the vector space $L(\mathbf{E}, \mathbf{Y})$ of all linear maps from $\mathbf{E}$ to $\mathbf{Y}$ with the operator norm

$$\|A\| = \max_{x \in B_{\mathbf{E}}} \|Ax\| \quad (A \in L(\mathbf{E}, \mathbf{Y})).$$

If this map ∇h is itself Fréchet differentiable at the point $\bar{x}$ in U then we say h is *twice Fréchet differentiable* at $\bar{x}$: the gradient $\nabla^2 h(\bar{x})$ is a linear map from $\mathbf{E}$ to $L(\mathbf{E}, \mathbf{Y})$, and for any element v of $\mathbf{E}$ we write

$$(\nabla^2 h(\bar{x})v)(v) = \nabla^2 h(\bar{x})(v, v).$$

In this case h has the following *quadratic approximation* at $\bar{x}$:

$$h(\bar{x} + v) = h(\bar{x}) + \nabla h(\bar{x})v + \frac{1}{2}\nabla^2 h(\bar{x})(v, v) + o(\|v\|^2) \quad \text{for small } v.$$

We suppose throughout this section that the functions $f : U \to \mathbf{R}$ and h are twice Fréchet differentiable at $\bar{x}$, and that the closed convex set S contains $\bar{x}$. We consider the nonlinear optimization problem

$$\inf\{f(x) \mid h(x) = 0,\ x \in S\}, \tag{7.4.1}$$

and we define the *narrow critical cone at* $\bar{x}$ by

$$C(\bar{x}) = \{d \in \mathbf{R}_+(S - \bar{x}) \mid \nabla f(\bar{x})d \le 0,\ \nabla h(\bar{x})d = 0\}.$$

Theorem 7.4.2 (Second order necessary conditions) *Suppose that the point $\bar{x}$ is a local minimum for the problem (7.4.1), that the direction d lies in the narrow critical cone $C(\bar{x})$, and that the condition*

$$0 \in \operatorname{core}(\nabla h(\bar{x})(S - \bar{x})) \tag{7.4.3}$$

holds. Then there exists a multiplier λ in $\mathbf{Y}$ such that the Lagrangian

$$L(\cdot) = f(\cdot) + \langle \lambda, h(\cdot) \rangle \tag{7.4.4}$$

satisfies the conditions

$$\nabla L(\bar{x}) \in -N_S(\bar{x}) \tag{7.4.5}$$

and

$$\nabla^2 L(\bar{x})(d,d) \geq 0. \tag{7.4.6}$$

Proof. Consider first the convex program

$$\inf\{\nabla f(\bar{x})z \mid \nabla h(\bar{x})z = -\nabla^2 h(\bar{x})(d,d), \ z \in \mathbf{R}_+(S-\bar{x})\}. \tag{7.4.7}$$

Suppose the point z is feasible for problem (7.4.7). It is easy to check for small real $t \geq 0$ the path

$$x(t) = \bar{x} + td + \frac{t^2}{2}z$$

lies in S. Furthermore, the quadratic approximation shows this path almost satisfies the original constraint for small t:

$$\begin{aligned} h(x(t)) &= h(\bar{x}) + t\nabla h(\bar{x})d + \frac{t^2}{2}(\nabla h(\bar{x})z + \nabla^2 h(\bar{x})(d,d)) + o(t^2) \\ &= o(t^2). \end{aligned}$$

But condition (7.4.3) implies in particular that $\nabla h(\bar{x})T_S(\bar{x}) = \mathbf{Y}$; in fact these conditions are equivalent, since the only convex set whose closure is $\mathbf{Y}$ is $\mathbf{Y}$ itself (see Section 4.1, Exercise 20(a) (Properties of the relative interior)). So, by Theorem 7.1.5 (Surjectivity and metric regularity), h is (weakly) metrically regular on S at $\bar{x}$. Hence the path above is close to feasible for the original problem: there is a real constant k such that, for small $t \geq 0$, we have

$$d_{S \cap h^{-1}(0)}(x(t)) \leq k\|h(x(t))\| = o(t^2).$$

Thus we can perturb the path slightly to obtain a set of points

$$\{\tilde{x}(t) \mid t \geq 0\} \subset S \cap h^{-1}(0)$$

satisfying $\|\tilde{x}(t) - x(t)\| = o(t^2)$.

Since $\bar{x}$ is a local minimizer for the original problem (7.4.1), we know

$$f(\bar{x}) \leq f(\tilde{x}(t)) = f(\bar{x}) + t\nabla f(\bar{x})d + \frac{t^2}{2}(\nabla f(\bar{x})z + \nabla^2 f(\bar{x})(d,d)) + o(t^2)$$

using the quadratic approximation again. Hence $\nabla f(\bar{x})d \geq 0$, so in fact $\nabla f(\bar{x})d = 0$, since d lies in $C(\bar{x})$. We deduce

$$\nabla f(\bar{x})z + \nabla^2 f(\bar{x})(d,d) \geq 0.$$

We have therefore shown the optimal value of the convex program (7.4.7) is at least $-\nabla^2 f(\bar{x})(d,d)$.

For the final step in the proof, we rewrite problem (7.4.7) in Fenchel form:

$$\inf_{z \in \mathbf{E}} \left\{ \left(\langle \nabla f(\bar{x}), z \rangle + \delta_{\mathbf{R}_+(S-\bar{x})}(z) \right) + \delta_{\{-\nabla^2 h(\bar{x})(d,d)\}}(\nabla h(\bar{x}) z) \right\}.$$

Since condition (7.4.3) holds, we can apply Fenchel duality (3.3.5) to deduce there exists $\lambda \in \mathbf{Y}$ satisfying

$$\begin{aligned} -\nabla^2 f(\bar{x})(d,d) &\leq -\delta^*_{\mathbf{R}_+(S-\bar{x})}(-\nabla h(\bar{x})^* \lambda - \nabla f(\bar{x})) - \delta^*_{\{-\nabla^2 h(\bar{x})(d,d)\}}(\lambda) \\ &= -\delta_{N_S(\bar{x})}(-\nabla h(\bar{x})^* \lambda - \nabla f(\bar{x})) + \langle \lambda, \nabla^2 h(\bar{x})(d,d) \rangle, \end{aligned}$$

whence the result. □

Under some further conditions we can guarantee that for *any* multiplier λ satisfying the first order condition (7.4.5), the second order condition (7.4.6) holds for *all* directions d in the narrow critical cone (see Exercises 2 and 3).

We contrast the necessary condition above with a rather elementary second order *sufficient* condition. For this we use the *broad critical cone at* $\bar{x}$:

$$\overline{C}(\bar{x}) = \{ d \in K_S(\bar{x}) \mid \nabla f(\bar{x}) d \leq 0,\ \nabla h(\bar{x}) d = 0 \}.$$

Theorem 7.4.8 (Second order sufficient condition) *Suppose for each nonzero direction d in the broad critical cone $\overline{C}(\bar{x})$ there exist multipliers μ in $\mathbf{R}_+$ and λ in $\mathbf{Y}$ such that the Lagrangian*

$$\overline{L}(\cdot) = \mu f(\cdot) + \langle \lambda, h(\cdot) \rangle$$

satisfies the conditions

$$\nabla \overline{L}(\bar{x}) \in -N_S(\bar{x}) \quad \text{and} \quad \nabla^2 \overline{L}(\bar{x})(d,d) > 0.$$

Then for all small real $\delta > 0$ the point $\bar{x}$ is a strict local minimizer for the perturbed problem

$$\inf\{ f(x) - \delta \|x - \bar{x}\|^2 \mid h(x) = 0,\ x \in S \}. \tag{7.4.9}$$

Proof. Suppose there is no such δ, so there is a sequence of feasible solutions (x_r) for problem (7.4.9) converging to $\bar{x}$ and satisfying

$$\limsup_{r \to \infty} \frac{f(x_r) - f(\bar{x})}{\|x_r - \bar{x}\|^2} \leq 0. \tag{7.4.10}$$

By taking a subsequence, we can assume

$$\lim_{r \to \infty} \frac{x_r - \bar{x}}{\|x_r - \bar{x}\|} = d,$$

and it is easy to check the nonzero direction d lies in $\overline{C}(\bar{x})$. Hence by assumption there exist the required multipliers μ and λ.

From the first order condition we know

$$\nabla \overline{L}(\bar{x})(x_r - \bar{x}) \geq 0,$$

so by the quadratic approximation we deduce as $r \to \infty$

$$\begin{aligned}\mu(f(x_r) - f(\bar{x})) &= \overline{L}(x_r) - \overline{L}(\bar{x}) \\ &\geq \frac{1}{2}\nabla^2 \overline{L}(\bar{x})(x_r - \bar{x}, x_r - \bar{x}) + o(\|x_r - \bar{x}\|^2).\end{aligned}$$

Dividing by $\|x_r - \bar{x}\|^2$ and taking limits shows

$$\mu \liminf_{r\to\infty} \frac{f(x_r) - f(\bar{x})}{\|x_r - \bar{x}\|^2} \geq \frac{1}{2}\nabla^2 \overline{L}(\bar{x})(d, d) > 0,$$

which contradicts inequality (7.4.10). □

Notice this result is of Fritz John type (like Theorem 2.3.6): we do not assume the multiplier μ is nonzero. Furthermore, we can easily weaken the assumption that the set S is convex to the condition

$$(S - \bar{x}) \cap \epsilon B \subset K_S(\bar{x}) \quad \text{for some } \epsilon > 0.$$

Clearly the narrow critical cone may be smaller than the broad critical cone, even when S is convex. They are equal if S is *quasipolyhedral* at $\bar{x}$:

$$K_S(\bar{x}) = \mathbf{R}_+(S - \bar{x})$$

(as happens in particular when S is polyhedral). However, even for unconstrained problems there is an intrinsic gap between the second order necessary conditions and the sufficient conditions.

Exercises and Commentary

Our approach here is from [25] (see also [12]). There are higher order analogues [11]. Problems of the form (7.4.11) where all the functions involved are quadratic are called *quadratic programs.* Such problems are particularly well-behaved: the optimal value is attained when finite, and in this case the second order necessary conditions developed in Exercise 3 are also *sufficient* (see [21]). For a straightforward exposition of the standard second order conditions, see [132], for example.

1. **(Higher order conditions)** By considering the function

$$\operatorname{sgn}(x)\exp\Big(-\frac{1}{x^2}\Big)$$

on $\mathbf{R}$, explain why there is no necessary and sufficient nth order optimality condition.

2. * **(Uniform multipliers)** With the assumptions of Theorem 7.4.2 (Second order necessary conditions), suppose in addition that for all directions d in the narrow critical cone $C(\bar{x})$ there exists a solution z in $\mathbf{E}$ to the system

$$\nabla h(\bar{x})z = -\nabla^2 h(\bar{x})(d,d) \text{ and } z \in \operatorname{span}(S - \bar{x}).$$

By considering problem (7.4.7), prove that if the multiplier λ satisfies the first order condition (7.4.5) then the second order condition (7.4.6) holds for all d in $C(\bar{x})$. Observe this holds in particular if $S = \mathbf{E}$ and $\nabla h(\bar{x})$ is surjective.

3. ** **(Standard second order necessary conditions)** Consider the problem

$$\left.\begin{array}{lrcl} \inf & f(x) & & \\ \text{subject to} & g_i(x) & \leq & 0 \text{ for } i = 1, 2, \ldots, m \\ & h_j(x) & = & 0 \text{ for } j = 1, 2, \ldots, q \\ & x & \in & \mathbf{R}^n, \end{array}\right\} \quad (7.4.11)$$

where all the functions are twice Fréchet differentiable at the local minimizer $\bar{x}$ and the set of gradients

$$A = \{\nabla g_i(\bar{x}) \mid i \in I(\bar{x})\} \cup \{\nabla h_j(\bar{x}) \mid j = 1, 2, \ldots, q\}$$

is linearly independent (where we denote the set of indices of the active inequality constraints by $I(\bar{x}) = \{i \mid g_i(\bar{x}) = 0\}$, as usual). By writing this problem in the form (7.4.1) and applying Exercise 2, prove there exist multipliers μ_i in $\mathbf{R}_+$ (for i in $I(\bar{x})$) and $\lambda_1, \lambda_2, \ldots, \lambda_q$ in $\mathbf{R}$ such that the Lagrangian

$$L(\cdot) = f(\cdot) + \sum_{i \in I(\bar{x})} \mu_i g_i + \sum_{j=1}^{q} \lambda_j h_j$$

satisfies the conditions

$$\nabla L(\bar{x}) = 0 \text{ and } \nabla^2 L(\bar{x})(d,d) \geq 0 \text{ for all } d \text{ in } A^\perp.$$

4. **(Narrow and broad critical cones are needed)** By considering the set

$$S = \{x \in \mathbf{R}^2 \mid x_2 \geq x_1^2\}$$

and the problem

$$\inf\{x_2 - \alpha x_1^2 \mid x \in S\}$$

for various values of the real parameter α, explain why the narrow and broad critical cones cannot be interchanged in either the Second order necessary conditions (7.4.2) or the sufficient conditions (7.4.8).

5. **(Standard second order sufficient conditions)** Write down the second order sufficient optimality conditions for the general nonlinear program in Exercise 3.

6. * **(Guignard-type conditions)** Consider the problem of Section 7.2, Exercise 10,

$$\inf\{f(x) \mid h(x) \in R,\ x \in S\},$$

where the set $R \subset \mathbf{Y}$ is closed and convex. By rewriting this problem in the form (7.4.1), derive second order optimality conditions.

Chapter 8

Fixed Points

8.1 The Brouwer Fixed Point Theorem

Many questions in optimization and analysis reduce to solving a nonlinear equation $h(x) = 0$, for some function $h : \mathbf{E} \to \mathbf{E}$. Equivalently, if we define another map $f = I - h$ (where I is the identity map), we seek a point x in $\mathbf{E}$ satisfying $f(x) = x$; we call x a *fixed point* of f.

The most potent fixed point existence theorems fall into three categories: "geometric" results, devolving from the Banach contraction principle (which we state below), "order-theoretic" results (to which we briefly return in Section 8.3), and "topological" results, for which the prototype is the theorem of Brouwer forming the main body of this section. We begin with Banach's result.

Given a set $C \subset \mathbf{E}$ and a continuous *self map* $f : C \to C$, we ask whether f has a fixed point. We call f a *contraction* if there is a real constant $\gamma_f < 1$ such that

$$\|f(x) - f(y)\| \le \gamma_f \|x - y\| \text{ for all } x, y \in C. \tag{8.1.1}$$

Theorem 8.1.2 (Banach contraction) *Any contraction on a closed subset of* $\mathbf{E}$ *has a unique fixed point.*

Proof. Suppose the set $C \subset \mathbf{E}$ is closed and the function $f : C \to C$ satisfies the contraction condition (8.1.1). We apply the Ekeland variational principle (7.1.2) to the function

$$z \in \mathbf{E} \mapsto \begin{cases} \|z - f(z)\| & \text{if } z \in C \\ +\infty & \text{otherwise} \end{cases}$$

at an arbitrary point x in C, with the choice of constants

$$\epsilon = \|x - f(x)\| \text{ and } \lambda = \frac{\epsilon}{1 - \gamma_f}.$$

This shows there is a point v in C satisfying

$$\|v - f(v)\| < \|z - f(z)\| + (1 - \gamma_f)\|z - v\|$$

for all points $z \neq v$ in C. Hence v is a fixed point, since otherwise choosing $z = f(v)$ gives a contradiction. The uniqueness is easy. □

What if the map f is not a contraction? A very useful weakening of the notion is the idea of a *nonexpansive* map, which is to say a self map f satisfying

$$\|f(x) - f(y)\| \leq \|x - y\| \quad \text{for all } x, y$$

(see Exercise 2). A nonexpansive map on a nonempty compact set or a nonempty closed convex set may not have a fixed point, as simple examples like translations on $\mathbf{R}$ or rotations of the unit circle show. On the other hand, a straightforward argument using the Banach contraction theorem shows this cannot happen if the set is nonempty, compact, *and* convex. However, in this case we have the following more fundamental result.

Theorem 8.1.3 (Brouwer) *Any continuous self map of a nonempty compact convex subset of* $\mathbf{E}$ *has a fixed point.*

In this section we present an "analyst's approach" to Brouwer's theorem. We use the two following important analytic tools concerning $C^{(1)}$ (continuously differentiable) functions on the closed unit ball $B \subset \mathbf{R}^n$.

Theorem 8.1.4 (Stone–Weierstrass) *For any continuous map* $f : B \to \mathbf{R}^n$*, there is a sequence of* $C^{(1)}$ *maps* $f_r : B \to \mathbf{R}^n$ *converging uniformly to* f*.*

An easy exercise shows that, in this result, if f is a self map then we can assume each f_r is also a self map.

Theorem 8.1.5 (Change of variable) *Suppose that the set* $W \subset \mathbf{R}^n$ *is open and that the* $C^{(1)}$ *map* $g : W \to \mathbf{R}^n$ *is one-to-one with* ∇g *invertible throughout* W*. Then the set* $g(W)$ *is open with measure*

$$\int_W |\det \nabla g|.$$

We also use the elementary topological fact that the open unit ball $\text{int}\, B$ is *connected*; that is, it cannot be written as the disjoint union of two nonempty open sets.

The key step in our argument is the following topological result.

Theorem 8.1.6 (Retraction) *The unit sphere* S *is not a* $C^{(1)}$ **retract** *of the unit ball* B*; that is, there is no* $C^{(1)}$ *map from* B *to* S *whose restriction to* S *is the identity.*

Proof. Suppose there is such a retraction map $p : B \to S$. For real t in $[0,1]$, define a self map of B by $p_t = tp + (1-t)I$. As a function of the variables $x \in B$ and t, the function $\det \nabla p_t(x)$ is continuous and hence strictly positive for small t. Furthermore, p_t is one-to-one for small t (Exercise 7).

If we denote the open unit ball $B \setminus S$ by U, then the change of variables theorem above shows for small t that $p_t(U)$ is open with measure

$$\nu(t) = \int_U \det \nabla p_t. \tag{8.1.7}$$

On the other hand, by compactness, $p_t(B)$ is a closed subset of B, and we also know $p_t(S) = S$. A little manipulation now shows we can write U as a disjoint union of two open sets:

$$U = (p_t(U) \cap U) \cup (p_t(B)^c \cap U). \tag{8.1.8}$$

The first set is nonempty, since $p_t(0) = tp(0) \in U$. But as we observed, U is connected, so the second set must be empty, which shows $p_t(B) = B$. Thus the function $\nu(t)$ defined by equation (8.1.7) equals the volume of the unit ball B for all small t.

However, as a function of $t \in [0,1]$, $\nu(t)$ is a polynomial, so it must be constant. Since p is a retraction we know that all points x in U satisfy $\|p(x)\|^2 = 1$. Differentiating implies $(\nabla p(x))p(x) = 0$, from which we deduce $\det \nabla p(x) = 0$, since $p(x)$ is nonzero. Thus $\nu(1)$ is zero, which is a contradiction. $\square$

Proof of Brouwer's theorem. Consider first a $C^{(1)}$ self map f on the unit ball B. Suppose f has no fixed point. A straightforward exercise shows there are unique functions $\alpha : B \to \mathbf{R}_+$ and $p : B \to S$ satisfying the relationship

$$p(x) = x + \alpha(x)(x - f(x)) \quad \text{for all } x \text{ in } B. \tag{8.1.9}$$

Geometrically, $p(x)$ is the point where the line extending from the point $f(x)$ through the point x meets the unit sphere S. In fact p must then be a $C^{(1)}$ retraction, contradicting the retraction theorem above. Thus we have proved that any $C^{(1)}$ self map of B has a fixed point.

Now suppose the function f is just continuous. The Stone–Weierstrass theorem (8.1.4) implies there is a sequence of $C^{(1)}$ maps $f_r : B \to \mathbf{R}^n$ converging uniformly to f, and by Exercise 4 we can assume each f_r is a self map. Our argument above shows each f_r has a fixed point x^r. Since B is compact, the sequence (x^r) has a subsequence converging to some point x in B, which it is easy to see must be a fixed point of f. So any continuous self map of B has a fixed point.

Finally, consider a nonempty compact convex set $C \subset \mathbf{E}$ and a continuous self map g on C. Just as in our proof of Minkowski's theorem (4.1.8), we may as well assume C has nonempty interior. Thus there is a *homeomorphism* (a continuous onto map with continuous inverse) $h : C \to B$ (see Exercise 11). Since the function $h \circ g \circ h^{-1}$ is a continuous self map of B, our argument above shows this function has a fixed point x in B, and therefore $h^{-1}(x)$ is a fixed point of g. □

Exercises and Commentary

Good general references on fixed point theory are [68, 174, 83]. The Banach contraction principle appeared in [7]. Brouwer proved the three-dimensional case of his theorem in 1909 [49] and the general case in 1912 [50], with another proof by Hadamard in 1910 [89]. A nice exposition of the Stone–Weierstrass theorem may be found in [16], for example. The Change of variable theorem (8.1.5) we use can be found in [177]; a beautiful proof of a simplified version, also sufficient to prove Brouwer's theorem, appeared in [118]. Ulam conjectured and Borsuk proved their result in 1933 [17].

1. **(Banach iterates)** Consider a closed subset $C \subset \mathbf{E}$ and a contraction $f : C \to C$ with fixed point x^f. Given any point x_0 in C, define a sequence of points inductively by

$$x_{r+1} = f(x_r) \text{ for } r = 0, 1, \ldots.$$

   (a) Prove $\lim_{r,s\to\infty} \|x_r - x_s\| = 0$. Since $\mathbf{E}$ is *complete*, the sequence (x_r) converges. (Another approach first shows (x_r) is bounded.) Hence prove in fact x_r approaches x^f. Deduce the Banach contraction theorem.

   (b) Consider another contraction $g : C \to C$ with fixed point x^g. Use part (a) to prove the inequality

$$\|x^f - x^g\| \leq \frac{\sup_{z\in C} \|f(z) - g(z)\|}{1 - \gamma_f}.$$

2. **(Nonexpansive maps)**

   (a) If the $n \times n$ matrix U is orthogonal, prove the map $x \in \mathbf{R}^n \to Ux$ is nonexpansive.

   (b) If the set $S \subset \mathbf{E}$ is closed and convex then for any real λ in the interval $[0, 2]$ prove the *relaxed projection*

$$x \in \mathbf{E} \mapsto (1 - \lambda)x + \lambda P_S(x)$$

   is nonexpansive. (Hint: Use the nearest point characterization in Section 2.1, Exercise 8(c).)

(c) **(Browder–Kirk [51, 112])** Suppose the set $C \subset \mathbf{E}$ is compact and convex and the map $f : C \to C$ is nonexpansive. Prove f has a fixed point. (Hint: Choose an arbitrary point x in C and consider the contractions

$$z \in C \mapsto (1-\epsilon)f(z) + \epsilon x$$

for small real $\epsilon > 0$.)

(d)* In part (c), prove the fixed points form a nonempty compact convex set.

3. **(Non-uniform contractions)**

(a) Consider a nonempty compact set $C \subset \mathbf{E}$ and a self map f on C satisfying the condition

$$\|f(x) - f(y)\| < \|x - y\| \quad \text{for all distinct } x, y \in C.$$

By considering $\inf \|x - f(x)\|$, prove f has a unique fixed point.

(b) Show the result in part (a) can fail if C is unbounded.

(c) Prove the map $x \in [0,1] \mapsto xe^{-x}$ satisfies the condition in part (a).

4. In the Stone–Weierstrass theorem, prove that if f is a self map then we can assume each f_r is also a self map.

5. Prove the interval $(-1, 1)$ is connected. Deduce the open unit ball in $\mathbf{R}^n$ is connected.

6. In the Change of variable theorem (8.1.5), use metric regularity to prove the set $g(W)$ is open.

7. In the proof of the Retraction theorem (8.1.6), prove the map p is Lipschitz, and deduce that the map p_t is one-to-one for small t. Also prove that if t is small then $\det \nabla p_t$ is strictly positive throughout B.

8. In the proof of the Retraction theorem (8.1.6), prove the partition (8.1.8), and deduce $p_t(B) = B$.

9. In the proof of the Retraction theorem (8.1.6), prove $\nu(t)$ is a polynomial in t.

10. In the proof of Brouwer's theorem, prove the relationship (8.1.9) defines a $C^{(1)}$ retraction $p : B \to S$.

11. **(Convex sets homeomorphic to the ball)** Suppose the compact convex set $C \subset \mathbf{E}$ satisfies $0 \in \operatorname{int} C$. Prove that the map $h : C \to B$ defined by

$$h(x) = \begin{cases} \gamma_C(x)\|x\|^{-1}x & \text{if } x \neq 0 \\ 0 & \text{if } x = 0 \end{cases}$$

(where γ_C is the gauge function we defined in Section 4.1) is a homeomorphism.

12. * **(A nonclosed nonconvex set with the fixed point property)** Let Z be the subset of the unit disk in $\mathbf{R}^2$ consisting of all lines through the origin with rational slope. Prove every continuous self map of Z has a fixed point.

13. * **(Change of variable and Brouwer)** A very simple proof may be found in [118] of the formula

$$\int (f \circ g)|\nabla g| = \int f$$

when the function f is continuous with bounded support and the function g is differentiable, equaling the identity outside a large ball. Prove any such g is surjective by considering an f supported outside the range of g (which is closed). Deduce Brouwer's theorem.

14. ** **(Brouwer and inversion)** The central tool of the last chapter, the Surjectivity and metric regularity theorem (7.1.5), considers a function h whose *strict* derivative at a point satisfies a certain surjectivity condition. In this exercise, which comes out of a long tradition, we use Brouwer's theorem to consider functions h which are merely *Fréchet* differentiable. This exercise proves the following result.

Theorem 8.1.10 *Consider an open set $U \subset \mathbf{E}$, a closed convex set $S \subset U$, and a Euclidean space $\mathbf{Y}$, and suppose the continuous function $h : U \to \mathbf{Y}$ has Fréchet derivative at the point $x \in S$ satisfying the surjectivity condition*

$$\nabla h(x) T_S(x) = \mathbf{Y}.$$

Then there is a neighbourhood V of $h(x)$, a continuous, piecewise linear function $F : \mathbf{Y} \to \mathbf{E}$, and a function $g : V \to \mathbf{Y}$ that is Fréchet differentiable at $h(x)$ and satisfies $(F \circ g)(V) \subset S$ and

$$h((F \circ g)(y)) = y \ \textit{ for all } y \in V.$$

Proof. We can assume $x = 0$ and $h(0) = 0$.

(a) Use Section 4.1, Exercise 20 (Properties of the relative interior) to prove $\nabla h(0)(\mathbf{R}_+ S) = \mathbf{Y}$.

(b) Deduce that there exists a basis $y_1, y_2, \ldots, y_n$ of $\mathbf{Y}$ and points $u_1, u_2, \ldots, u_n$ and $w_1, w_2, \ldots, w_n$ in S satisfying

$$\nabla h(0)u_i = y_i = -\nabla h(0)w_i \quad \text{for } i = 1, 2, \ldots, n.$$

(c) Prove the set

$$B_1 = \Big\{ \sum_1^n t_i y_i \,\Big|\, t \in \mathbf{R}^n,\ \sum_1^n |t_i| \le 1 \Big\}$$

and the function F defined by

$$F\Big(\sum_1^n t_i y_i\Big) = \sum_1^n \big(t_i^+ u_i + (-t_i)^+ w_i\big)$$

satisfy $F(B_1) \subset S$ and $\nabla(h \circ F)(0) = I$.

(d) Deduce there exists a real $\epsilon > 0$ such that $\epsilon B_{\mathbf{Y}} \subset B_1$ and

$$\|h(F(y)) - y\| \le \frac{\|y\|}{2} \quad \text{whenever } \|y\| \le 2\epsilon.$$

(e) For any point v in the neighbourhood $V = (\epsilon/2)B_{\mathbf{Y}}$, prove the map

$$y \in V \mapsto v + y - h(F(y))$$

is a continuous self map of V.

(f) Apply Brouwer's theorem to deduce the existence of a fixed point $g(v)$ for the map in part (e). Prove $\nabla g(0) = I$, and hence complete the proof of the result.

(g) If x lies in the interior of S, prove F can be assumed linear.

(Exercise 9 (Nonexistence of multipliers) in Section 7.2 suggests the importance here of assuming h continuous.)

15. * **(Knaster–Kuratowski–Mazurkiewicz principle [114])** In this exercise we show the equivalence of Brouwer's theorem with the following result.

Theorem 8.1.11 (KKM) *Suppose for every point x in a nonempty set $X \subset \mathbf{E}$ there is an associated closed subset $M(x) \subset X$. Assume the property*

$$\operatorname{conv} F \subset \bigcup_{x \in F} M(x)$$

holds for all finite subsets $F \subset X$. Then for any finite subset $F \subset X$ we have

$$\bigcap_{x \in F} M(x) \neq \emptyset.$$

Hence if some subset $M(x)$ is compact we have

$$\bigcap_{x \in X} M(x) \neq \emptyset.$$

(a) Prove that the final assertion follows from the main part of the theorem using Theorem 8.2.3 (General definition of compactness).

(b) **(KKM implies Brouwer)** Given a continuous self map f on a nonempty compact convex set $C \subset \mathbf{E}$, apply the KKM theorem to the family of sets

$$M(x) = \{y \in C \mid \langle y - f(y), y - x\rangle \leq 0\} \quad \text{for } x \in C$$

to deduce f has a fixed point.

(c) **(Brouwer implies KKM)** With the hypotheses of the KKM theorem, assume $\cap_{x \in F} M(x)$ is empty for some finite set F. Consider a fixed point z of the self map

$$y \in \operatorname{conv} F \mapsto \frac{\sum_{x \in F} d_{M(x)}(y) x}{\sum_{x \in F} d_{M(x)}(y)}$$

and define $F' = \{x \in F \mid z \notin M(x)\}$. Show $z \in \operatorname{conv} F'$ and derive a contradiction.

16. ** **(Hairy ball theorem [140])** Let S_n denote the Euclidean sphere

$$\{x \in \mathbf{R}^{n+1} \mid \|x\| = 1\}.$$

A *tangent vector field* on S_n is a function $w : S_n \to \mathbf{R}^{n+1}$ satisfying $\langle x, w(x)\rangle = 0$ for all points x in S_n. This exercise proves the following result.

Theorem 8.1.12 *For every even n, any continuous tangent vector field on S_n must vanish somewhere.*

Proof. Consider a nonvanishing continuous tangent vector field u on S_n.

(a) Prove there is a nonvanishing $C^{(1)}$ tangent vector field on S_n, by using the Stone–Weierstrass theorem (8.1.4) to approximate u by a $C^{(1)}$ function p and then considering the vector field

$$x \in S_n \mapsto p(x) - \langle x, p(x)\rangle x.$$

(b) Deduce the existence of a positively homogeneous $C^{(1)}$ function $w : \mathbf{R}^{n+1} \to \mathbf{R}^{n+1}$ whose restriction to S_n is a *unit norm* $C^{(1)}$ tangent vector field: $\|w(x)\| = 1$ for all x in S_n.

Define a set

$$A = \{x \in \mathbf{R}^{n+1} \mid 1 < 2\|x\| < 3\}$$

and use the field w in part (b) to define functions $w_t : \mathbf{R}^{n+1} \to \mathbf{R}^{n+1}$ for real t by

$$w_t(x) = x + tw(x).$$

(c) Imitate the proof of Brouwer's theorem to prove the measure of the image set $w_t(A)$ is a polynomial in t when t is small.

(d) Prove directly the inclusion $w_t(A) \subset \sqrt{1+t^2}A$.

(e) For any point y in $\sqrt{1+t^2}A$, apply the Banach contraction theorem to the function $x \in kB \mapsto y - tw(x)$ (for large real k) to deduce in fact

$$w_t(A) = \sqrt{1+t^2}A \quad \text{for small } t.$$

(f) Complete the proof by combining parts (c) and (e). □

(g) If f is a continuous self map of S_n where n is even, prove either f or $-f$ has a fixed point.

(h) **(Hedgehog theorem)** Prove for even n that any nonvanishing continuous vector field must be somewhere *normal*: $|\langle x, f(x)\rangle| = \|f(x)\|$ for some x in S_n.

(i) Find examples to show the Hairy ball theorem fails for all odd n.

17. * **(Borsuk–Ulam theorem)** Let S_n denote the Euclidean sphere

$$\{x \in \mathbf{R}^{n+1} \mid \|x\| = 1\}.$$

We state the following result without proof.

Theorem 8.1.13 (Borsuk–Ulam) *For any positive integers $m \leq n$, if the function $f : S_n \to \mathbf{R}^m$ is continuous then there is a point x in S_n satisfying $f(x) = f(-x)$.*

(a) If $m \leq n$ and the map $f : S_n \to \mathbf{R}^m$ is continuous and odd, prove f vanishes somewhere.

(b) Prove any odd continuous self map f on S_n is surjective. (Hint: For any point u in S_n, consider the function

$$x \in S_n \mapsto f(x) - \langle f(x), u \rangle u$$

and apply part (a).)

(c) Prove the result in part (a) is equivalent to the following result:

Theorem 8.1.14 *For positive integers $m < n$ there is no continuous odd map from S_n to S_m.*

(d) **(Borsuk–Ulam implies Brouwer [178])** Let B denote the unit ball in $\mathbf{R}^n$, and let S denote the boundary of $B \times [-1, 1]$:

$$S = \{(x, t) \in B \times [-1, 1] \mid \|x\| = 1 \text{ or } |t| = 1\}.$$

(i) If the map $g : S \to \mathbf{R}^n$ is continuous and odd, use part (a) to prove g vanishes somewhere on S.

(ii) Consider a continuous self map f on B. By applying part (i) to the function

$$(x, t) \in S \mapsto (2 - |t|)x - tf(tx),$$

prove f has a fixed point.

18. ** **(Generalized Riesz lemma)** Consider a smooth norm $|||\cdot|||$ on $\mathbf{E}$ (that is, a norm which is continuously differentiable except at the origin) and linear subspaces $U, V \subset \mathbf{E}$ satisfying $\dim U > \dim V = n$. Denote the unit sphere in U (in this norm) by $S(U)$.

(a) By choosing a basis $v_1, v_2, \ldots, v_n$ of V and applying the Borsuk–Ulam theorem (see Exercise 17) to the map

$$x \in S(U) \mapsto (\langle \nabla|||\cdot|||(x), v_i \rangle)_{i=1}^n \in \mathbf{R}^n,$$

prove there is a point x in $S(U)$ satisfying $\nabla|||\cdot|||(x) \perp V$.

(b) Deduce the origin is the nearest point to x in V (in this norm).

(c) With this norm, deduce there is a unit vector in U whose distance from V is equal to one.

(d) Use the fact that any norm can be uniformly approximated arbitrarily well by a smooth norm to extend the result of part (c) to arbitrary norms.

(e) Find a simpler proof when $V \subset U$.

19. ** **(Riesz implies Borsuk)** In this question we use the generalized Riesz lemma, Exercise 18, to prove the Borsuk–Ulam result, Exercise 17(a). To this end, suppose the map $f : S_n \to \mathbf{R}^n$ is continuous and odd. Define functions

$$\begin{aligned} u_i : S_n &\to \mathbf{R} \text{ for } i = 1, 2, \ldots, n+1 \\ v_i : \mathbf{R}^n &\to \mathbf{R} \text{ for } i = 1, 2, \ldots, n \end{aligned}$$

by $u_i(x) = x_i$ and $v_i(x) = x_i$ for each index i. Define spaces of continuous odd functions on S_n by

$$\begin{aligned} U &= \operatorname{span}\{u_1, u_2, \ldots . u_{n+1}\} \\ V &= \operatorname{span}\{v_1 \circ f, v_2 \circ f, \ldots, v_n \circ f\} \\ \mathbf{E} &= U + V, \end{aligned}$$

with norm $\|u\| = \max u(S_n)$ (for u in $\mathbf{E}$).

(a) Prove there is a function u in U satisfying $\|u\| = 1$ and whose distance from V is equal to one.

(b) Prove u attains its maximum on S_n at a unique point y.

(c) Use the fact that for any function w in $\mathbf{E}$, we have

$$(\nabla \| \cdot \| (u))w = w(y)$$

to deduce $f(y) = 0$.

8.2 Selection and the Kakutani–Fan Fixed Point Theorem

The Brouwer fixed point theorem in the previous section concerns functions from a nonempty compact convex set to itself. In optimization, as we have already seen in Section 5.4, it may be convenient to broaden our language to consider *multifunctions* Ω from the set to itself and seek a *fixed point*—a point x satisfying $x \in \Omega(x)$. To begin this section we summarize some definitions for future reference.

We consider a subset $K \subset \mathbf{E}$, a Euclidean space $\mathbf{Y}$, and a multifunction $\Omega : K \to \mathbf{Y}$. We say Ω is *USC* at a point x in K if every open set U containing $\Omega(x)$ also contains $\Omega(z)$ for all points z in K close to x.

Thus a multifunction Ω is USC if for any sequence of points (x_n) approaching x, any sequence of elements $y_n \in \Omega(x_n)$ is eventually close to $\Omega(x)$. If Ω is USC at every point in K we simply call it *USC*. On the other hand, as in Section 5.4, we say Ω is *LSC* if, for every x in K, every neighbourhood V of any point in $\Omega(x)$ intersects $\Omega(z)$ for all points z in K close to x.

We refer to the sets $\Omega(x)$ $(x \in K)$ as the *images* of Ω. The multifunction Ω is a *cusco* if it is USC with nonempty compact convex images. Clearly such multifunctions are *locally bounded*: any point in K has a neighbourhood whose image is bounded. Cuscos appear in several important optimization contexts. For example, the Clarke subdifferential of a locally Lipschitz function is a cusco (Exercise 5).

To see another important class of examples we need a further definition. We say a multifunction $\Phi : \mathbf{E} \to \mathbf{E}$ is *monotone* if it satisfies the condition

$$\langle u - v, x - y \rangle \geq 0 \text{ whenever } u \in \Phi(x) \text{ and } v \in \Phi(y).$$

In particular, any (not necessarily self-adjoint) positive semidefinite linear operator is monotone, as is the subdifferential of any convex function. One multifunction *contains* another if the graph of the first contains the graph of the second. We say a monotone multifunction is *maximal* if the only monotone multifunction containing it is itself. The subdifferentials of closed proper convex functions are examples (see Exercise 16). Zorn's lemma (which lies outside our immediate scope) shows any monotone multifunction is contained in a maximal monotone multifunction.

Theorem 8.2.1 (Maximal monotonicity) *Maximal monotone multifunctions are cuscos on the interiors of their domains.*

Proof. See Exercise 16. □

Maximal monotone multifunctions in fact have to be single-valued *generically*, that is on sets which are "large" in a topological sense, specifically on a dense set which is a "G_δ" (a countable intersection of open sets)—see Exercise 17.

Returning to our main theme, the central result of this section extends Brouwer's theorem to the multifunction case.

Theorem 8.2.2 (Kakutani–Fan) *If the set $C \subset \mathbf{E}$ is nonempty, compact and convex, then any cusco $\Omega : C \to C$ has a fixed point.*

Before we prove this result, we outline a little more topology. A *cover* of a set $K \subset \mathbf{E}$ is a collection of sets in $\mathbf{E}$ whose union contains K. The cover is *open* if each set in the collection is open. A *subcover* is just a subcollection of the sets which is also a cover. The following result, which we state as a theorem, is in truth the definition of compactness in spaces more general than $\mathbf{E}$.

Theorem 8.2.3 (General definition of compactness) *Any open cover of a compact set in $\mathbf{E}$ has a finite subcover.*

Given a finite open cover $\{O_1, O_2, \ldots, O_m\}$ of a set $K \subset \mathbf{E}$, a *partition of unity subordinate to* this cover is a set of continuous functions $p_1, p_2, \ldots, p_m : K \to \mathbf{R}_+$ whose sum is identically equal to one and satisfying $p_i(x) = 0$ for all points x outside O_i (for each index i).. We outline the proof of the next result, a central topological tool, in the exercises.

Theorem 8.2.4 (Partition of unity) *There is a partition of unity subordinate to any finite open cover of a compact subset of $\mathbf{E}$.*

Besides fixed points, the other main theme of this section is the idea of a *continuous selection* of a multifunction Ω on a set $K \subset \mathbf{E}$, by which we mean a continuous map f on K satisfying $f(x) \in \Omega(x)$ for all points x in K. The central step in our proof of the Kakutani–Fan theorem is the following "approximate selection" theorem.

Theorem 8.2.5 (Cellina) *Given any compact set $K \subset \mathbf{E}$, suppose the multifunction $\Omega : K \to \mathbf{Y}$ is USC with nonempty convex images. Then for any real $\epsilon > 0$ there is a continuous map $f : K \to \mathbf{Y}$ which is an "approximate selection" of Ω :*

$$d_{G(\Omega)}(x, f(x)) < \epsilon \quad \textit{for all points } x \textit{ in } K. \tag{8.2.6}$$

Furthermore the range of f is contained in the convex hull of the range of Ω.

Proof. We can assume the norm on $\mathbf{E} \times \mathbf{Y}$ is given by

$$\|(x, y)\|_{\mathbf{E}\times\mathbf{Y}} = \|x\|_{\mathbf{E}} + \|y\|_{\mathbf{Y}} \quad \text{for all } x \in \mathbf{E} \text{ and } y \in \mathbf{Y}$$

(since all norms are equivalent—see Section 4.1, Exercise 2). Now, since Ω is USC, for each point x in K there is a real δ_x in the interval $(0, \epsilon/2)$ satisfying

$$\Omega(x + \delta_x B_{\mathbf{E}}) \subset \Omega(x) + \frac{\epsilon}{2} B_{\mathbf{Y}}.$$

Since the sets $x + (\delta_x/2)\mathrm{int}\, B_{\mathbf{E}}$ (as the point x ranges over K) comprise an open cover of the compact set K, there is a finite subset $\{x_1, x_2, \ldots, x_m\}$ of K with the sets $x_i + (\delta_i/2)\mathrm{int}\, B_{\mathbf{E}}$ comprising a finite subcover (where δ_i is shorthand for δ_{x_i} for each index i).

Theorem 8.2.4 shows there is a partition of unity $p_1, p_2, \ldots, p_m : K \to \mathbf{R}_+$ subordinate to this subcover. We now construct our desired approximate selection f by choosing a point y_i from $\Omega(x_i)$ for each i and defining

$$f(x) = \sum_{i=1}^{m} p_i(x) y_i \quad \text{for all points } x \text{ in } K. \tag{8.2.7}$$

Fix any point x in K and define the set $I = \{i \mid p_i(x) \neq 0\}$. By definition, x satisfies $\|x - x_i\| < \delta_i/2$ for each i in I. If we choose an index j in I maximizing δ_j, the triangle inequality shows $\|x_j - x_i\| < \delta_j$, whence we deduce the inclusions

$$y_i \in \Omega(x_i) \subset \Omega(x_j + \delta_j B_{\mathbf{E}}) \subset \Omega(x_j) + \frac{\epsilon}{2} B_{\mathbf{Y}}$$

for all i in I. In other words, for each i in I we know $d_{\Omega(x_j)}(y_i) \leq \epsilon/2$. Since the distance function is convex, equation (8.2.7) shows $d_{\Omega(x_j)}(f(x)) \leq \epsilon/2$. Since we also know $\|x - x_j\| < \epsilon/2$, this proves inequality (8.2.6). The final claim follows immediately from equation (8.2.7). □

Proof of the Kakutani–Fan theorem. With the assumption of the theorem, Cellina's result above shows for each positive integer r there is a continuous self map f_r of C satisfying

$$d_{G(\Omega)}(x, f_r(x)) < \frac{1}{r} \quad \text{for all points } x \text{ in } C.$$

By Brouwer's theorem (8.1.3), each f_r has a fixed point x^r in C, which therefore satisfies

$$d_{G(\Omega)}(x^r, x^r) < \frac{1}{r} \quad \text{for each } r.$$

Since C is compact, the sequence (x^r) has a convergent subsequence, and its limit must be a fixed point of Ω because Ω is closed by Exercise 3(c) (Closed versus USC). □

In the next section we describe some variational applications of the Kakutani–Fan theorem. But we end this section with an *exact* selection theorem parallel to Cellina's result but assuming an LSC rather than a USC multifunction.

Theorem 8.2.8 (Michael) *Given any closed set $K \subset \mathbf{E}$, suppose the multifunction $\Omega : K \to \mathbf{Y}$ is LSC with nonempty closed convex images. Then given any point $(\bar{x}, \bar{y})$ in $G(\Omega)$, there is a continuous selection f of Ω satisfying $f(\bar{x}) = \bar{y}$.*

We outline the proof in the exercises.

Exercises and Commentary

Many useful properties of cuscos are summarized in [27]. An excellent general reference on monotone operators is [153]. The topology we use in this section can be found in any standard text (see [67, 106], for example). The Kakutani–Fan theorem first appeared in [109] and was extended in [74]. Cellina's approximate selection theorem appears, for example, in [4, p. 84]. One example of the many uses of the Kakutani–Fan theorem is establishing equilibria in mathematical economics. The Michael selection theorem appeared in [137].

1. **(USC and continuity)** Consider a closed subset $K \subset \mathbf{E}$ and a multifunction $\Omega : K \to \mathbf{Y}$.

   (a) Prove the multifunction

   $$x \in \mathbf{E} \mapsto \begin{cases} \Omega(x) & \text{for } x \in K \\ \emptyset & \text{for } x \notin K \end{cases}$$

   is USC if and only if Ω is USC.

   (b) Prove a function $f : K \to \mathbf{Y}$ is continuous if and only if the multifunction $x \in K \mapsto \{f(x)\}$ is USC.

   (c) Prove a function $f : \mathbf{E} \to [-\infty, +\infty]$ is lower semicontinuous at a point x in $\mathbf{E}$ if and only if the multifunction whose graph is the epigraph of f is USC at x.

2. * **(Minimum norm)** If the set $U \subset \mathbf{E}$ is open and the multifunction $\Omega : U \to \mathbf{Y}$ is USC, prove the function $g : U \to \mathbf{Y}$ defined by

   $$g(x) = \inf\{\|y\| \mid y \in \Omega(x)\}$$

   is lower semicontinuous.

3. **(Closed versus USC)**

   (a) If the multifunction $\Phi : \mathbf{E} \to \mathbf{Y}$ is closed and the multifunction $\Omega : \mathbf{E} \to \mathbf{Y}$ is USC at the point x in $\mathbf{E}$ with $\Omega(x)$ compact, prove the multifunction

   $$z \in \mathbf{E} \mapsto \Omega(z) \cap \Phi(z)$$

   is USC at x.

(b) Hence prove that any closed multifunction with compact range is USC.

(c) Prove any USC multifunction with closed images is closed.

(d) If a USC multifunction has compact images, prove it is locally bounded.

4. **(Composition)** If the multifunctions Φ and Ω are USC prove their composition $x \mapsto \Phi(\Omega(x))$ is also.

5. * **(Clarke subdifferential)** If the set $U \subset \mathbf{E}$ is open and the function $f : U \to \mathbf{R}$ is locally Lipschitz, use Section 6.2, Exercise 12 (Closed subdifferentials) and Exercise 3 (Closed versus USC) to prove the Clarke subdifferential $x \in U \mapsto \partial_\circ f(x)$ is a cusco.

6. ** **(USC images of compact sets)** Consider a given multifunction $\Omega : K \to \mathbf{Y}$.

   (a) Prove Ω is USC if and only if for every open subset U of $\mathbf{Y}$ the set $\{x \in K \mid \Omega(x) \subset U\}$ is open in K.

   Now suppose K is compact and Ω is USC with compact images. Using the general definition of compactness (8.2.3), prove the range $\Omega(K)$ is compact by following the steps below.

   (b) Fix an open cover $\{U_\gamma \,|\, \gamma \in \Gamma\}$ of $\Omega(K)$. For each point x in K, prove there is a finite subset Γ_x of Γ with

$$\Omega(x) \subset \bigcup_{\gamma \in \Gamma_x} U_\gamma.$$

   (c) Construct an open cover of K by considering the sets

$$\left\{ z \in K \,\middle|\, \Omega(z) \subset \bigcup_{\gamma \in \Gamma_x} U_\gamma \right\}$$

   as the point x ranges over K.

   (d) Hence construct a finite subcover of the original cover of $\Omega(K)$.

7. * **(Partitions of unity)** Suppose the set $K \subset \mathbf{E}$ is compact with a finite open cover $\{O_1, O_2, \ldots, O_m\}$.

   (i) Show how to construct another open cover $\{V_1, V_2, \ldots, V_m\}$ of K satisfying $\mathrm{cl}\, V_i \subset O_i$ for each index i. (Hint: Each point x in K lies in some set O_i, so there is a real $\delta_x > 0$ with $x + \delta_x B \subset O_i$; now take a finite subcover of $\{x + \delta_x \mathrm{int}\, B \,|\, x \in K\}$ and build the sets V_i from it.)

(ii) For each index i, prove the function $q_i : K \to [0,1]$ given by

$$q_i = \frac{d_{K \setminus O_i}}{d_{K \setminus O_i} + d_{V_i}}$$

is well-defined and continuous, with q_i identically zero outside the set O_i.

(iii) Deduce that the set of functions $p_i : K \to \mathbf{R}_+$ defined by

$$p_i = \frac{q_i}{\sum_j q_j}$$

is a partition of unity subordinate to the cover $\{O_1, O_2, \ldots, O_m\}$.

8. Prove the Kakutani–Fan theorem is also valid under the weaker assumption that the images of the cusco $\Omega : C \to \mathbf{E}$ always intersect the set C using Exercise 3(a) (Closed versus USC).

9. ** **(Michael's theorem)** Suppose all the assumptions of Michael's theorem (8.2.8) hold. We consider first the case with K compact.

(a) Fix a real $\epsilon > 0$. By constructing a partition of unity subordinate to a finite subcover of the open cover of K consisting of the sets

$$O_y = \{x \in \mathbf{E} \mid d_{\Omega(x)}(y) < \epsilon\} \text{ for } y \text{ in } Y,$$

construct a continuous function $f : K \to Y$ satisfying

$$d_{\Omega(x)}(f(x)) < \epsilon \text{ for all points } x \text{ in } K.$$

(b) Construct a sequence of continuous functions $f_1, f_2, \ldots : K \to Y$ satisfying

$$\begin{aligned} d_{\Omega(x)}(f_i(x)) &< 2^{-i} \text{ for } i = 1, 2, \ldots \\ \|f_{i+1}(x) - f_i(x)\| &< 2^{1-i} \text{ for } i = 1, 2, \ldots \end{aligned}$$

for all points x in K. (Hint: Construct f_1 by applying part (a) with $\epsilon = 1/2$; then construct f_{i+1} inductively by applying part (a) to the multifunction

$$x \in K \mapsto \Omega(x) \cap (f_i(x) + 2^{-i} B_{\mathbf{Y}})$$

with $\epsilon = 2^{-i-1}$.

(c) The functions f_i of part (b) must converge uniformly to a continuous function f. Prove f is a continuous selection of Ω.

(d) Prove Michael's theorem by applying part (c) to the multifunction

$$\hat{\Omega}(x) = \begin{cases} \Omega(x) & \text{if } x \neq \bar{x} \\ \{\bar{y}\} & \text{if } x = \bar{x}. \end{cases}$$

(e) Now extend to the general case where K is possibly unbounded in the following steps. Define sets $K_n = K \cap nB_{\mathbf{E}}$ for each $n = 1, 2, \ldots$ and apply the compact case to the multifunction $\Omega_1 = \Omega|_{K_1}$ to obtain a continuous selection $g_1 : K_1 \to \mathbf{Y}$. Then inductively find a continuous selection $g_{n+1} : K_{n+1} \to \mathbf{Y}$ from the multifunction

$$\Omega_{n+1}(x) = \begin{cases} \{g_n(x)\} & \text{for } x \in K_n \\ \Omega(x) & \text{for } x \in K_{n+1} \setminus K_n \end{cases}$$

and prove the function defined by

$$f(x) = g_n(x) \ \text{ for } x \in K_n, \ n = 1, 2, \ldots$$

is the required selection.

10. **(Hahn–Katetov–Dowker sandwich theorem)** Suppose the set $K \subset \mathbf{E}$ is closed.

(a) For any two lower semicontinuous functions $f, g : K \to \mathbf{R}$ satisfying $f \geq -g$, prove there is a continuous function $h : K \to \mathbf{R}$ satisfying $f \geq h \geq -g$ by considering the multifunction $x \mapsto [-g(x), f(x)]$. Observe the result also holds for extended-real-valued f and g.

(b) **(Urysohn lemma)** Suppose the closed set V and the open set U satisfy $V \subset U \subset K$. By applying part (i) to suitable functions, prove there is a continuous function $f : K \to [0, 1]$ that is identically equal to one on V and to zero on U^c.

11. **(Continuous extension)** Consider a closed subset K of $\mathbf{E}$ and a continuous function $f : K \to \mathbf{Y}$. By considering the multifunction

$$\Omega(x) = \begin{cases} \{f(x)\} & \text{for } x \in K \\ \text{cl}\,(\text{conv}\, f(K)) & \text{for } x \notin K, \end{cases}$$

prove there is a continuous function $g : \mathbf{E} \to \mathbf{Y}$ satisfying $g|_K = f$ and $g(\mathbf{E}) \subset \text{cl}\,(\text{conv}\, f(K))$.

12. * **(Generated cuscos)** Suppose the multifunction $\Omega : K \to \mathbf{Y}$ is locally bounded with nonempty images.

(a) Among those cuscos containing Ω, prove there is a unique one with minimal graph, given by

$$\Phi(x) = \bigcap_{\epsilon>0} \mathrm{cl\,conv}\,(\Omega(x+\epsilon B)) \quad \text{for } x \in K.$$

(b) If K is nonempty, compact, and convex, $\mathbf{Y} = \mathbf{E}$, and Ω satisfies the conditions $\Omega(K) \subset K$ and

$$x \in \Phi(x) \Rightarrow x \in \Omega(x) \quad \text{for } x \in K,$$

prove Ω has a fixed point.

13. * **(Multifunctions containing cuscos)** Suppose the multifunction $\Omega : K \to \mathbf{Y}$ is closed with nonempty convex images, and the function $f : K \to \mathbf{Y}$ has the property that $f(x)$ is a point of minimum norm in $\Omega(x)$ for all points x in K. Prove Ω contains a cusco if and only if f is locally bounded. (Hint: Use Exercise 12 (Generated cuscos) to consider the cusco generated by f.)

14. * **(Singleton points)** For any subset D of $\mathbf{Y}$, define

$$s(D) = \inf\{r \in \mathbf{R} \mid D \subset y + rB_{\mathbf{Y}} \text{ for some } y \in \mathbf{Y}\}.$$

Consider an open subset U of $\mathbf{E}$.

(a) If the multifunction $\Omega : U \to \mathbf{Y}$ is USC with nonempty images, prove for any real $\epsilon > 0$ the set

$$S_\epsilon = \{x \in U \mid s(\Omega(x)) < \epsilon\}$$

is open. By considering the set $\cap_{n>1} S_{1/n}$, prove the set of points in U whose image is a singleton is a G_δ.

(b) Use Exercise 5 (Clarke subdifferential) to prove that the set of points where a locally Lipschitz function $f : U \to \mathbf{R}$ is strictly differentiable is a G_δ. If U and f are convex (or if f is regular throughout U), use Rademacher's theorem (in Section 6.2) to deduce f is generically differentiable.

15. **(Skew symmetry)** If the matrix $A \in \mathbf{M}^n$ satisfies $0 \neq A = -A^T$, prove the multifunction $x \in \mathbf{R}^n \mapsto x^T Ax$ is maximal monotone, yet is not the subdifferential of a convex function.

16. ** **(Monotonicity)** Consider a monotone multifunction $\Phi : \mathbf{E} \to \mathbf{E}$.

(a) **(Inverses)** Prove Φ^{-1} is monotone.

(b) Prove Φ^{-1} is maximal if and only if Φ is.

(c) **(Applying maximality)** Prove Φ is maximal if and only if it has the property

$$\langle u - v, x - y\rangle \geq 0 \text{ for all } (x, u) \in G(\Phi) \;\Rightarrow\; v \in \Phi(y).$$

(d) **(Maximality and closedness)** If Φ is maximal, prove it is closed with convex images.

(e) **(Continuity and maximality)** Supposing Φ is everywhere single-valued and *hemicontinuous* (that is, continuous on every line in $\mathbf{E}$), prove it is maximal. (Hint: Apply part (c) with $x = y + tw$ for w in $\mathbf{E}$ and $t \downarrow 0$ in $\mathbf{R}$.)

(f) We say Φ is *hypermaximal* if $\Phi + \lambda I$ is surjective for some real $\lambda > 0$. In this case, prove Φ is maximal. (Hint: Apply part (c) and use a solution $x \in \mathbf{E}$ to the inclusion $v + \lambda y \in (\Phi + \lambda I)(x)$.) What if just Φ is surjective?

(g) **(Subdifferentials)** If the function $f : \mathbf{E} \to (\infty, +\infty]$ is closed, convex, and proper, prove ∂f is maximal monotone. (Hint: For any element ϕ of $\mathbf{E}$, prove the function

$$x \in \mathbf{E} \mapsto f(x) + \|x\|^2 + \langle \phi, x\rangle$$

has a minimizer, and deduce ∂f is hypermaximal.)

(h) **(Local boundedness)** By completing the following steps, prove Φ is locally bounded at any point in the core of its domain.

(i) Assume $0 \in \Phi(0)$ and $0 \in \operatorname{core} D(\Phi)$, define a convex function $g : \mathbf{E} \to (\infty, +\infty]$ by

$$g(y) = \sup\{\langle u, y - x\rangle \mid x \in B,\ u \in \Phi(x)\}.$$

(ii) Prove $D(\Phi) \subset \operatorname{dom} g$.

(iii) Deduce g is continuous at zero.

(iv) Hence show $|g(y)| \leq 1$ for all small y, and deduce the result.

(j) **(Maximality and cuscos)** Use parts (d) and (h), and Exercise 3 (Closed versus USC) to conclude that any maximal monotone multifunction is a cusco on the interior of its domain.

(k) **(Surjectivity and growth)** If Φ is surjective, prove

$$\lim_{\|x\| \to \infty} \|\Phi(x)\| = +\infty.$$

(Hint: Assume the maximality of Φ, and hence of Φ^{-1}; deduce Φ^{-1} is a cusco on $\mathbf{E}$, and now apply Exercise 6 (USC images of compact sets).)

17. ** **(Single-valuedness and maximal monotonicity)** Consider a maximal monotone multifunction $\Omega : \mathbf{E} \to \mathbf{E}$ and an open subset U of its domain, and define the minimum norm function $g : U \to \mathbf{R}$ as in Exercise 2.

    (a) Prove g is lower semicontinuous. An application of the Baire category theorem now shows that any such function is generically continuous.

    (b) For any point x in U at which g is continuous, prove $\Omega(x)$ is a singleton. (Hint: Prove $\|\cdot\|$ is constant on $\Omega(x)$ by first assuming $y, z \in \Omega(x)$ and $\|y\| > \|z\|$, and then using the condition

    $$\langle w - y, x + ty - x \rangle \geq 0 \text{ for all small } t > 0 \text{ and } w \in \Omega(x + ty)$$

    to derive a contradiction.)

    (c) Conclude that any maximal monotone multifunction is generically single-valued.

    (d) Deduce that any convex function is generically differentiable on the interior of its domain.

8.3 Variational Inequalities

At the very beginning of this book we considered the problem of minimizing a differentiable function $f : \mathbf{E} \to \mathbf{R}$ over a convex set $C \subset \mathbf{E}$. A necessary optimality condition for a point x_0 in C to be a local minimizer is

$$\langle \nabla f(x_0), x - x_0 \rangle \geq 0 \text{ for all points } x \text{ in } C, \tag{8.3.1}$$

or equivalently

$$0 \in \nabla f(x_0) + N_C(x_0).$$

If the function f is convex instead of differentiable, the necessary and sufficient condition for optimality (assuming a constraint qualification) is

$$0 \in \partial f(x_0) + N_C(x_0),$$

and there are analogous nonsmooth necessary conditions.

We call problems like (8.3.1) "variational inequalities". Let us fix a multifunction $\Omega : C \to \mathbf{E}$. In this section we use the fixed point theory we have developed to study the *multivalued variational inequality*

$VI(\Omega, C)$: Find points x_0 in C and y_0 in $\Omega(x_0)$ satisfying $\langle y_0, x - x_0 \rangle \geq 0$ for all points x in C.

A more concise way to write the problem is this:

$$\text{Find a point } x_0 \text{ in } C \text{ satisfying } 0 \in \Omega(x_0) + N_C(x_0). \tag{8.3.2}$$

Suppose the set C is closed, convex, and nonempty. Recall that the projection $P_C : \mathbf{E} \to C$ is the (continuous) map that sends points in $\mathbf{E}$ to their unique nearest points in C (see Section 2.1, Exercise 8). Using this notation we can also write the variational inequality as a fixed point problem:

$$\text{Find a fixed point of } P_C \circ (I - \Omega) : C \to C. \tag{8.3.3}$$

This reformulation is useful if the multifunction Ω is single-valued, but less so in general because the composition will often not have convex images.

A more versatile approach is to define the (multivalued) *normal mapping* $\Omega_C = (\Omega \circ P_C) + I - P_C$, and repose the problem as follows:

$$\text{Find a point } \bar{x} \text{ in } \mathbf{E} \text{ satisfying } 0 \in \Omega_C(\bar{x}). \tag{8.3.4}$$

Then setting $x_0 = P_C(\bar{x})$ gives a solution to the original problem. Equivalently, we could phrase this as follows:

$$\text{Find a fixed point of } (I - \Omega) \circ P_C : \mathbf{E} \to \mathbf{E}. \tag{8.3.5}$$

As we shall see, this last formulation lets us immediately use the fixed point theory of the previous section.

The basic result guaranteeing the existence of solutions to variational inequalities is the following.

Theorem 8.3.6 (Solvability of variational inequalities) *If the subset C of $\mathbf{E}$ is compact, convex, and nonempty, then for any cusco $\Omega : C \to \mathbf{E}$ the variational inequality $VI(\Omega, C)$ has a solution.*

Proof. We in fact prove *Theorem 8.3.6 is equivalent to the Kakutani–Fan fixed point theorem (8.2.2).*

When Ω is a cusco its range $\Omega(C)$ is compact—we outline the proof in Section 8.2, Exercise 6. We can easily check that the multifunction $(I-\Omega)\circ P_C$ is also a cusco because the projection P_C is continuous. Since this multifunction maps the compact convex set $\operatorname{conv}(C-\Omega(C))$ into itself, the Kakutani–Fan theorem shows it has a fixed point, which, as we have already observed, implies the solvability of $VI(\Omega, C)$.

Conversely, suppose the set $C \subset \mathbf{E}$ is nonempty, compact, and convex. For any cusco $\Omega : C \to C$, the Solvability theorem (8.3.6) implies we can solve the variational inequality $VI(I-\Omega, C)$, so there are points x_0 in C and z_0 in $\Omega(x_0)$ satisfying

$$\langle x_0 - z_0, x - x_0 \rangle \geq 0 \quad \text{for all points } x \text{ in } C.$$

Setting $x = z_0$ shows $x_0 = z_0$, so x_0 is a fixed point. $\square$

An elegant application is von Neumann's minimax theorem, which we proved by a Fenchel duality argument in Section 4.2, Exercise 16. Consider Euclidean spaces $\mathbf{Y}$ and $\mathbf{Z}$, nonempty compact convex subsets $F \subset \mathbf{Y}$ and $G \subset \mathbf{Z}$, and a linear map $A : \mathbf{Y} \to \mathbf{Z}$. If we define a function $\Omega : F \times G \to \mathbf{Y} \times \mathbf{Z}$ by $\Omega(y,z) = (-A^*z, Ay)$, then it is easy to see that a point (y_0, z_0) in $F \times G$ solves the variational inequality $VI(\Omega, F \times G)$ if and only if it is a *saddlepoint*:

$$\langle z_0, Ay \rangle \leq \langle z_0, Ay_0 \rangle \leq \langle z, Ay_0 \rangle \quad \text{for all } y \in F,\ z \in G.$$

In particular, by the Solvability of variational inequalities theorem, there exists a saddlepoint, so

$$\min_{z \in G} \max_{y \in F} \langle z, Ay \rangle = \max_{y \in F} \min_{z \in G} \langle z, Ay \rangle.$$

Many interesting variational inequalities involve a noncompact set C. In such cases we need to impose a growth condition on the multifunction to guarantee solvability. The following result is an example.

Theorem 8.3.7 (Noncompact variational inequalities) *If the subset C of $\mathbf{E}$ is nonempty, closed, and convex, and the cusco $\Omega : C \to \mathbf{E}$ is* **coercive**, *that is, it satisfies the condition*

$$\liminf_{\|x\|\to\infty,\ x\in C} \inf \langle x, \Omega(x) + N_C(x)\rangle > 0, \tag{8.3.8}$$

then the variational inequality $VI(\Omega, C)$ has a solution.

Proof. For any large integer r, we can apply the solvability theorem (8.3.6) to the variational inequality $VI(\Omega, C \cap rB)$ to find a point x_r in $C \cap rB$ satisfying

$$\begin{aligned} 0 &\in \Omega(x_r) + N_{C\cap rB}(x_r) \\ &= \Omega(x_r) + N_C(x_r) + N_{rB}(x_r) \\ &\subset \Omega(x_r) + N_C(x_r) + \mathbf{R}_+ x_r \end{aligned}$$

(using Section 3.3, Exercise 10). Hence for all large r, the point x_r satisfies

$$\inf \langle x_r, \Omega(x_r) + N_C(x_r)\rangle \le 0.$$

This sequence of points (x_r) must therefore remain bounded, by the coercivity condition (8.3.8), and so x_r lies in $\operatorname{int} rB$ for large r and hence satisfies $0 \in \Omega(x_r) + N_C(x_r)$, as required. $\square$

A straightforward exercise shows in particular that the growth condition (8.3.8) holds whenever the cusco Ω is defined by $x \in \mathbf{R}^n \mapsto x^T A x$ for a matrix A in $\mathbf{S}^n_{++}$.

The most important example of a noncompact variational inequality is the case when the set C is a closed convex cone $S \subset \mathbf{E}$. In this case $VI(\Omega, S)$ becomes the *multivalued complementarity problem*:

$$\begin{array}{l} \text{Find points } x_0 \text{ in } S \text{ and } y_0 \text{ in } \Omega(x_0) \cap (-S^-) \\ \text{satisfying } \langle x_0, y_0\rangle = 0. \end{array} \tag{8.3.9}$$

As a particular example, we consider the dual pair of abstract linear programs (5.3.4) and (5.3.5):

$$\inf\{\langle c, z\rangle \mid Az - b \in H,\ z \in K\} \tag{8.3.10}$$

(where $\mathbf{Y}$ is a Euclidean space, the map $A : \mathbf{E} \to \mathbf{Y}$ is linear, the cones $H \subset \mathbf{Y}$ and $K \subset \mathbf{E}$ are closed and convex, and b and c are given elements of $\mathbf{Y}$ and $\mathbf{E}$ respectively), and

$$\sup\{\langle b, \phi\rangle \mid A^*\phi - c \in K^-,\ \phi \in -H^-\}. \tag{8.3.11}$$

As usual, we denote the corresponding primal and dual optimal values by p and d. We consider a corresponding variational inequality on the space $\mathbf{E} \times \mathbf{Y}$:

$$VI(\Omega, K \times (-H^-)) \text{ with } \Omega(z, \phi) = (c - A^*\phi, Ax - b). \tag{8.3.12}$$

Theorem 8.3.13 (Linear programming and variational inequalities) *Any solution of the above variational inequality (8.3.12) consists of a pair of optimal solutions for the linear programming dual pair (8.3.10) and (8.3.11). The converse is also true, providing there is no duality gap ($p = d$).*

We leave the proof as an exercise.

Notice that the linear map appearing in the above example, namely $M : \mathbf{E} \times \mathbf{Y} \to \mathbf{E} \times \mathbf{Y}$ defined by $M(z, \phi) = (-A^*\phi, Az)$, is monotone. We study monotone complementarity problems further in Exercise 7.

To end this section we return to the complementarity problem (8.3.9) in the special case where $\mathbf{E}$ is $\mathbf{R}^n$, the cone S is $\mathbf{R}^n_+$, and the multifunction Ω is single-valued: $\Omega(x) = \{F(x)\}$ for all points x in $\mathbf{R}^n_+$. In other words, we consider the following problem:

Find a point x_0 in $\mathbf{R}^n_+$ satisfying $F(x_0) \in \mathbf{R}^n_+$ and $\langle x_0, F(x_0)\rangle = 0$.

The lattice operation $\wedge$ is defined on $\mathbf{R}^n$ by $(x \wedge y)_i = \min\{x_i, y_i\}$ for points x and y in $\mathbf{R}^n$ and each index i. With this notation we can rewrite the above problem as the following *order complementarity problem.*

$OCP(F)$: Find a point x_0 in $\mathbf{R}^n_+$ satisfying $x_0 \wedge F(x_0) = 0$.

The map $x \in \mathbf{R}^n \mapsto x \wedge F(x) \in \mathbf{R}^n$ is sometimes amenable to fixed point methods.

As an example, let us fix a real $\alpha > 0$, a vector $q \in \mathbf{R}^n$, and an $n \times n$ matrix P with nonnegative entries, and define the map $F : \mathbf{R}^n \to \mathbf{R}^n$ by $F(x) = \alpha x - Px + q$. Then the complementarity problem $OCP(F)$ is equivalent to finding a fixed point of the map $\Phi : \mathbf{R}^n \to \mathbf{R}^n$ defined by

$$\Phi(x) = \frac{1}{\alpha}(0 \vee (Px - q)), \tag{8.3.14}$$

a problem that can be solved iteratively (see Exercise 9).

Exercises and commentary

A survey of variational inequalities and complementarity problems may be found in [93]. The normal mapping Ω_C is especially well studied when the multifunction Ω is single-valued with affine components and the set C is polyhedral. In this case the normal mapping is piecewise affine (see [164]). More generally, if we restrict the class of multifunctions Ω we wish to consider in the variational inequality, clearly we can correspondingly restrict the versions of the Kakutani–Fan theorem or normal mappings we study. Order complementarity problems are studied further in [26]. The Nash equilibrium theorem (Exercise 10(d)), which appeared in [147], asserts the existence of a Pareto efficient choice for n individuals consuming from n associated convex sets with n associated joint cost functions.

1. Prove the equivalence of the various formulations (8.3.2), (8.3.3), (8.3.4) and (8.3.5) with the original variational inequality $VI(\Omega, C)$.

2. Use Section 8.2, Exercise 4 (Composition) to prove the multifunction
$$(I - \Omega) \circ P_C$$
in the proof of Theorem 8.3.6 (Solvability of variational inequalities) is a cusco.

3. Consider Theorem 8.3.6 (Solvability of variational inequalities). Use the function
$$x \in [0,1] \mapsto \begin{cases} \dfrac{1}{x} & \text{if } x > 0 \\ -1 & \text{if } x = 0 \end{cases}$$
to prove the assumption in the theorem—that the multifunction Ω is USC—cannot be weakened to Ω closed.

4. * **(Variational inequalities containing cuscos)** Suppose the set $C \subset \mathbf{E}$ is nonempty, compact, and convex, and consider a multifunction $\Omega : C \to \mathbf{E}$.

   (a) If Ω contains a cusco, prove the variational inequality $VI(\Omega, C)$ has a solution.

   (b) Deduce from Michael's theorem (8.2.8) that if Ω is LSC with nonempty closed convex images then $VI(\Omega, C)$ has a solution.

5. Check the details of the proof of von Neumann's minimax theorem.

6. Prove Theorem 8.3.13 (Linear programming and variational inequalities).

7. **(Monotone complementarity problems)** Suppose the linear map $M : \mathbf{E} \to \mathbf{E}$ is monotone.

   (a) Prove the function $x \in \mathbf{E} \mapsto \langle Mx, x \rangle$ is convex.

   For a closed convex cone $S \subset \mathbf{E}$ and a point q in $\mathbf{E}$, consider the optimization problem
$$\inf\{\langle Mx + q, x \rangle \mid Mx + q \in -S^-,\ x \in S\}. \tag{8.3.15}$$

   (b) If the condition $-q \in \operatorname{core}(S^- + MS)$ holds, use the Fenchel duality theorem (3.3.5) to prove problem (8.3.15) has optimal value zero.

(c) If the cone S is polyhedral, problem (8.3.15) is a convex "quadratic program": when the optimal value is finite, it is known that there is no duality gap for such a problem and its (Fenchel) dual, and that both problems attain their optimal value. Deduce that when S is polyhedral and contains a point x with $Mx+q$ in $-S^-$, there is such a point satisfying the additional complementarity condition $\langle Mx+q, x\rangle = 0$.

8. * Consider a compact convex set $C \subset \mathbf{E}$ satisfying $C = -C$ and a continuous function $f : C \to \mathbf{E}$. If f has no zeroes, prove there is a point x on the boundary of C satisfying $\langle f(x), x\rangle < 0$. (Hint: For positive integers n, consider $VI(f + I/n, C)$.)

9. **(Iterative solution of OCP [26])** Consider the order complementarity problem $OCP(F)$ for the function F that we defined before equation (8.3.14). A point x^0 in $\mathbf{R}^n_+$ is *feasible* if it satisfies $F(x^0) \geq 0$.

   (a) Prove the map Φ in equation (8.3.14) is *isotone*: $x \geq y$ implies $\Phi(x) \geq \Phi(y)$ for points x and y in $\mathbf{R}^n$.

   (b) Suppose the point x^0 in $\mathbf{R}^n_+$ is feasible. Define a sequence (x^r) in $\mathbf{R}^n_+$ inductively by $x^{r+1} = \Phi(x^r)$. Prove this sequence decreases monotonically: $x_i^{r+1} \leq x_i^r$ for all r and i.

   (c) Prove the limit of the sequence in part (b) solves $OCP(F)$.

   (d) Define a sequence (y^r) in $\mathbf{R}^n_+$ inductively by $y^0 = 0$ and $y^{r+1} = \Phi(y^r)$. Prove this sequence increases monotonically.

   (e) If $OCP(F)$ has a feasible solution, prove the sequence in part (d) converges to a limit $\bar{y}$ which solves $OCP(F)$. What happens if $OCP(F)$ has no feasible solution?

   (f) Prove the limit $\bar{y}$ of part (e) is the *minimal* solution of $OCP(F)$: any other solution x satisfies $x \geq \bar{y}$.

10. * **(Fan minimax inequality [74])** We call a real function g on a convex set $C \subset \mathbf{E}$ *quasiconcave* if the set $\{x \in C \,|\, g(x) \geq \alpha\}$ is convex for all real α.

    Suppose the set $C \subset \mathbf{E}$ is nonempty, compact, and convex.

    (a) If the function $f : C \times C \to \mathbf{R}$ has the properties that the function $f(\cdot, y)$ is quasiconcave for all points y in C and the function $f(x, \cdot)$ is lower semicontinuous for all points x in C, prove *Fan's inequality*:

    $$\min_y \sup_x f(x,y) \leq \sup_x f(x,x).$$

(Hint: Apply the KKM theorem (Section 8.1, Exercise 15) to the family of sets

$$\{y \in C \mid f(x, y) \leq \beta\} \text{ for } x \in C,$$

where β denotes the right hand side of Fan's inequality.)

(b) If the function $F : C \to \mathbf{E}$ is continuous, apply Fan's inequality to the function $f(x, y) = \langle F(y), y - x \rangle$ to prove the variational inequality $VI(F, C)$ has a solution.

(c) Deduce Fan's inequality is equivalent to the Brouwer fixed point theorem.

(d) **(Nash equilibrium)** Define a set $C = C_1 \times C_2 \times \ldots \times C_n$, where each set $C_i \subset \mathbf{E}$ is nonempty, compact, and convex. For any continuous functions $f_1, f_2, \ldots, f_n : C \to \mathbf{R}$, if each function

$$x_i \in C_i \mapsto f_i(y_1, \ldots, x_i, \ldots, y_n)$$

is convex for all elements y of C, prove there is an element y of C satisfying the inequalities

$$f_i(y) \leq f_i(y_1, \ldots, x_i, \ldots, y_n) \text{ for all } x_i \in C_i,\ i = 1, 2, \ldots, n.$$

(Hint: Consider the function

$$f(x, y) = \sum_i (f_i(y) - f_i(y_1, \ldots, x_i, \ldots, y_n))$$

and apply Fan's inequality.)

(e) **(Minimax)** Apply the Nash equilibrium result from part (d) in the case $n = 2$ and $f_1 = -f_2$ to deduce the Kakutani minimax theorem (Section 4.3, Exercise 14).

11. **(Bolzano–Poincaré–Miranda intermediate value theorem)** Consider the box

$$J = \{x \in \mathbf{R}^n \mid 0 \leq x_i \leq 1 \text{ for all } i\}.$$

We call a continuous map $f : J \to \mathbf{R}^n$ *reversing* if it satisfies the condition

$$f_i(x) f_i(y) \leq 0 \text{ whenever } x_i = 0,\ y_i = 1, \text{ and } i = 1, 2, \ldots, n.$$

Prove any such map vanishes somewhere on J by completing the following steps:

(a) Observe the case $n = 1$ is just the classical intermediate value theorem.

(b) For all small real $\epsilon > 0$, prove the function $f^\epsilon = f + \epsilon I$ satisfies for all i

$$x_i = 0 \text{ and } y_i = 1 \Rightarrow \begin{cases} \text{either} & f_i^\epsilon(y) > 0 \text{ and } f_i^\epsilon(x) \leq 0 \\ \text{or} & f_i^\epsilon(y) < 0 \text{ and } f_i^\epsilon(x) \geq 0. \end{cases}$$

(c) ¿From part (b), deduce there is a function $\widetilde{f}^\epsilon$, defined coordinatewise by $\widetilde{f}_i^\epsilon = \pm f_i^\epsilon$, for some suitable choice of signs, satisfying the conditions (for each i)

$$\widetilde{f}_i^\epsilon(x) \leq 0 \text{ whenever } x_i = 0 \text{ and}$$
$$\widetilde{f}_i^\epsilon(x) > 0 \text{ whenever } x_i = 1.$$

(d) By considering the variational inequality $VI(\widetilde{f}^\epsilon, J)$, prove there is a point x^ϵ in J satisfying $\widetilde{f}^\epsilon(x^\epsilon) = 0$.

(e) Complete the proof by letting ϵ approach zero.

12. **(Coercive cuscos)** Consider a multifunction $\Omega : \mathbf{E} \to \mathbf{E}$ with nonempty images.

(a) If Ω is a coercive cusco, prove it is surjective.

(b) On the other hand, if Ω is monotone, use Section 8.2, Exercise 16 (Monotonicity) to deduce Ω is hypermaximal if and only if it is maximal. (We generalize this result in Exercise 13 (Monotone variational inequalities).)

13. ** **(Monotone variational inequalities)** Consider a continuous function $G : \mathbf{E} \to \mathbf{E}$ and a monotone multifunction $\Phi : \mathbf{E} \to \mathbf{E}$.

(a) Given a nonempty compact convex set $K \subset \mathbf{E}$, prove there is point x_0 in K satisfying

$$\langle x - x_0, y + G(x_0)\rangle \geq 0 \text{ for all } x \in K,\ y \in \Phi(x)$$

by completing the following steps:

(i) Assuming the result fails, show the collection of sets

$$\{x \in K \mid \langle z - x, w + G(x)\rangle < 0\} \text{ for } z \in K,\ w \in \Phi(z)$$

is an open cover of K.

(ii) For a partition of unity $p_1, p_2, \ldots, p_n$ subordinate to a finite subcover $K_1, K_2, \ldots K_n$ corresponding to points $z_i \in K$ and $w_i \in \Phi(z_i)$ (for $i = 1, 2, \ldots, n$), prove the function

$$f(x) = \sum_i p_i(x) z_i$$

is a continuous self map of K.

(iii) Prove the inequality

$$\begin{aligned}\langle f(x) - x, \textstyle\sum_i p_i(x)w_i + G(x)\rangle \\ &= \sum_{i,j} p_i(x)p_j(x)\langle z_j - x, w_i + G(x)\rangle \\ &< 0\end{aligned}$$

by considering the terms in the double sum where $i = j$ and sums of pairs where $i \neq j$ separately.

(iv) Deduce a contradiction with part (ii).

(b) Now assume G satisfies the growth condition

$$\lim_{\|x\| \to \infty} \|G(x)\| = +\infty \quad \text{and} \quad \liminf_{\|x\| \to \infty} \frac{\langle x, G(x)\rangle}{\|x\|\|G(x)\|} > 0.$$

(i) Prove there is a point x_0 in $\mathbf{E}$ satisfying

$$\langle x - x_0, y + G(x_0)\rangle \geq 0 \ \text{ whenever } y \in \Phi(x).$$

(Hint: Apply part (a) with $K = nB$ for $n = 1, 2, \ldots$.)

(ii) If Φ is maximal, deduce $-G(x_0) \in \Phi(x_0)$.

(c) Apply part (b) to prove that if Φ is maximal then for any real $\lambda > 0$, the multifunction $\Phi + \lambda I$ is surjective.

(d) **(Hypermaximal $\Leftrightarrow$ maximal)** Using Section 8.2, Exercise 16 (Monotonicity), deduce a monotone multifunction is maximal if and only if it is hypermaximal.

(e) **(Resolvent)** If Φ is maximal then for any real $\lambda > 0$ and any point y in $\mathbf{E}$ prove there is a unique point x satisfying the inclusion

$$y \in \Phi(x) + \lambda x.$$

(f) **(Maximality and surjectivity)** Prove a maximal Φ is surjective if and only if it satisfies the growth condition

$$\lim_{\|x\| \to \infty} \inf \|\Phi(x)\| = +\infty.$$

(Hint: The "only if" direction is Section 8.2, Exercise 16(k) (Monotonicity); for the "if" direction, apply part (e) with $\lambda = 1/n$ for $n = 1, 2, \ldots$, obtaining a sequence (x_n); if this sequence is unbounded, apply maximal monotonicity.)

14. * **(Semidefinite complementarity)** Define $F : \mathbf{S}^n \times \mathbf{S}^n \to \mathbf{S}^n$ by

$$F(U, V) = U + V - (U^2 + V^2)^{1/2}.$$

For any function $G : \mathbf{S}^n \to \mathbf{S}^n$, prove $U \in \mathbf{S}^n$ solves the variational inequality $VI(G, \mathbf{S}^n_+)$ if and only if $F(U, G(U)) = 0$. (Hint: See Section 5.2, Exercise 11.)

Monotonicity via convex analysis

Many important properties of monotone multifunctions can be derived using convex analysis, without using the Brouwer fixed point theorem (8.1.3). The following sequence of exercises illustrates the ideas. Throughout, we consider a monotone multifunction $\Phi : \mathbf{E} \to \mathbf{E}$. The point $(u, v) \in \mathbf{E} \times \mathbf{E}$ is *monotonically related* to Φ if $\langle x - u, y - v \rangle \geq 0$ whenever $y \in \Phi(x)$: in other words, appending this point to the graph of Φ does not destroy monotonicity. Our main aim is to prove a central case of the *Debrunner-Flor extension theorem* [59]. The full theorem states that if Φ has range contained in a nonempty compact convex set $C \subset \mathbf{E}$, and the function $f : C \to \mathbf{E}$ is continuous, then there is a point $c \in C$ such that the point $(f(c), c)$ is monotonically related to Φ. For an accessible derivation of this result from Brouwer's theorem, see [154]: the two results are in fact equivalent (see Exercise 19).

We call a convex function $\mathcal{H} : \mathbf{E} \times \mathbf{E} \to (\infty, +\infty]$ *representative* for Φ if all points $x, y \in \mathbf{E}$ satisfy $\mathcal{H}(x, y) \geq \langle x, y \rangle$, with equality if $y \in \Phi(x)$. Following [79], the *Fitzpatrick function* $\mathcal{F}_\Phi : \mathbf{E} \times \mathbf{E} \to [-\infty, +\infty]$ is defined by

$$\mathcal{F}_\Phi(x, y) = \sup\{\langle x, v \rangle + \langle u, y \rangle - \langle u, v \rangle \mid v \in \Phi(u)\},$$

while [171, 150] the *convexified representative* $\mathcal{P}_\Phi : \mathbf{E} \times \mathbf{E} \to [-\infty, +\infty]$ is defined by

$$\mathcal{P}_\Phi(x, y) = \inf\Big\{ \sum_{i=1}^{m} \lambda_i (x_i, y_i) \,\Big|\, m \in \mathbf{N},\ \lambda \in \mathbf{R}_+^m, \\ \sum_{i=1}^{m} \lambda_i (x_i, y_i, 1) = (x, y, 1),\ y_i \in \Phi(x_i)\ \forall i \Big\}.$$

These constructions are explored extensively in [30, 43, 172].

15. **(Fitzpatrick representatives)**

    (a) Prove the Fitzpatrick function $\mathcal{F}_\Phi$ is closed and convex.

    (b) Prove $\mathcal{F}_\Phi(x, y) = \langle x, y \rangle$ whenever $y \in \Phi(x)$.

    (c) Prove $\mathcal{F}_\Phi$ is representative providing Φ is maximal.

    (d) Find an example where $\mathcal{F}_\Phi$ is not representative.

16. **(Convexified representatives)** Consider points $x \in \mathbf{E}$ and $y \in \Phi(x)$.

    (a) Prove $\mathcal{P}_\Phi(x, y) \leq \langle x, y \rangle$.

    Now consider any points $u, v \in \mathbf{E}$.

(b) Prove $\mathcal{P}_\Phi(u,v) \geq \langle u,y\rangle + \langle x,v\rangle - \langle x,y\rangle$.

(c) Deduce $\mathcal{P}_\Phi(x,y) = \langle x,y\rangle$.

(d) Deduce $\mathcal{P}_\Phi(x,y) + \mathcal{P}_\Phi(u,v) \geq \langle u,y\rangle + \langle x,v\rangle$.

(e) Prove $\mathcal{P}_\Phi(u,v) \geq \langle u,v\rangle$ if $(u,v) \in \operatorname{conv} G(\Phi)$ and is $+\infty$ otherwise.

(f) Deduce that convexified representatives are indeed both convex and representative.

(g) Prove $\mathcal{P}_\Phi^* = \mathcal{F}_\Phi \leq \mathcal{F}_\Phi^*$.

17. * **(Monotone multifunctions with bounded range)** Suppose that the monotone multifunction $\Phi : \mathbf{E} \to \mathbf{E}$ has bounded range $R(\Phi)$, and let $C = \operatorname{cl}\operatorname{conv} R(\Phi)$. Apply Exercise 16 to prove the following properties.

(a) Prove the convexity of the function $f : \mathbf{E} \to [-\infty, +\infty]$ defined by

$$f(x) = \inf\{\mathcal{P}_\Phi(x,y) \mid y \in C\}.$$

(b) Prove that the function $g = \inf_{y \in C} \langle \cdot, y\rangle$ is a continuous concave minorant of f.

(c) Apply the Sandwich theorem (Exercise 13 in Section 3.3) to deduce the existence of an affine function α satisfying $f \geq \alpha \geq g$.

(d) Prove that the point $(0, \nabla\alpha)$ is monotonically related to Φ.

(e) Prove $\nabla\alpha \in C$.

(f) Given any point $x \in \mathbf{E}$, show that Φ is contained in a monotone multifunction Φ' with x in its domain and $R(\Phi') \subset C$.

(g) Give an alternative proof of part (f) using the Debrunner-Flor extension theorem.

(h) Extend part (f) to monotone multifunctions with unbounded ranges, by assuming that the point x lies in the set $\operatorname{int}\operatorname{dom} f - \operatorname{dom} \delta_C^*$. Express this condition explicitly in terms of C and the domain of Φ.

18. ** **(Maximal monotone extension)** Suppose the monotone multifunction $\Phi : \mathbf{E} \to \mathbf{E}$ has bounded range $R(\Phi)$.

(a) Use Exercise 17 and Zorn's lemma to prove that Φ is contained in a monotone multifunction Φ' with domain $\mathbf{E}$ and range contained in $\operatorname{cl}\operatorname{conv} R(\Phi)$.

(b) Deduce that if Φ is in fact maximal monotone, then its domain is $\mathbf{E}$.

(c) Using Exercise 16 (Local boundedness) in Section 8.2, prove that the multifunction $\Phi'' : \mathbf{E} \to \mathbf{E}$ defined by

$$\Phi''(x) = \bigcap_{\epsilon>0} \operatorname{cl} \operatorname{conv} \Phi'(x + \epsilon B)$$

is both monotone and a cusco.

(d) Prove that a monotone multifunction is a cusco on the interior of its domain if and only if it is maximal monotone.

(e) Deduce that Φ is contained in a maximal monotone multifunction with domain $\mathbf{E}$ and range contained in $\operatorname{cl} \operatorname{conv} R(\Phi)$.

(f) Apply part (e) to Φ^{-1} to deduce a parallel result.

19. ** **(Brouwer via Debrunner-Flor)** Consider a nonempty compact convex set $D \subset \operatorname{int} B$ and a continuous self map $g : D \to D$. By applying the Debrunner-Flor extension theorem in the case where $C = B$, the multifunction Φ is the identity map, and $f = g \circ P_D$ (where P_D is the nearest point projection), prove that g has a fixed point.

In similar fashion one may establish that the sum of two maximal monotone multifunctions S and T is maximal assuming the condition $0 \in \operatorname{core}(\operatorname{dom} T - \operatorname{dom} S)$. One commences with the *Fitzpatrick inequality* that

$$\mathcal{F}_T(x, x^*) + \mathcal{F}_S(x, -x^*) \geq 0,$$

for all x, x^* in $\mathbf{E}$. This and many other applications of representative functions are described in [30].

Chapter 9

More Nonsmooth Structure

9.1 Rademacher's Theorem

We mentioned Rademacher's fundamental theorem on the differentiability of Lipschitz functions in the context of the Intrinsic Clarke subdifferential formula (Theorem 6.2.5):

$$\partial_\circ f(x) = \operatorname{conv}\{\lim_r \nabla f(x^r) \mid x^r \to x,\ x^r \notin Q\}, \tag{9.1.1}$$

valid whenever the function $f : \mathbf{E} \to \mathbf{R}$ is locally Lipschitz around the point $x \in \mathbf{E}$ and the set $Q \subset \mathbf{E}$ has measure zero. We prove Rademacher's theorem in this section, taking a slight diversion into some basic measure theory.

Theorem 9.1.2 (Rademacher) *Any locally Lipschitz map between Euclidean spaces is Fréchet differentiable almost everywhere.*

Proof. Without loss of generality (Exercise 1), we can consider a locally Lipschitz function $f : \mathbf{R}^n \to \mathbf{R}$. In fact, we may as well further suppose that f has Lipschitz constant L throughout $\mathbf{R}^n$, by Exercise 2 in Section 7.1.

Fix a direction h in $\mathbf{R}^n$. For any $t \neq 0$, the function g_t defined on $\mathbf{R}^n$ by

$$g_t(x) = \frac{f(x + th) - f(x)}{t}$$

is continuous, and takes values in the interval $I = L\|h\|[-1, 1]$, by the Lipschitz property. Hence, for $k = 1, 2, \ldots$, the function $p_k : \mathbf{R}^n \to I$

defined by

$$p_k(x) = \sup_{0<|t|<1/k} g_t(x)$$

is lower semicontinuous and therefore Borel measurable. Consequently, the upper Dini derivative $D_h^+ f : \mathbf{R}^n \to I$ defined by

$$D_h^+ f(x) = \limsup_{t\to 0} g_t(x) = \inf_{k\in\mathbf{N}} p_k(x)$$

is measurable, being the infimum of a sequence of measurable functions. Similarly, the lower Dini derivative $D_h^- f : \mathbf{R}^n \to I$ defined by

$$D_h^- f(x) = \liminf_{t\to 0} g_t(x)$$

is also measurable.

The subset of $\mathbf{R}^n$ where f is not differentiable along the direction h, namely

$$A_h = \{x \in \mathbf{R}^n \mid D_h^- f(x) < D_h^+ f(x)\},$$

is therefore also measurable. Given any point $x \in \mathbf{R}^n$, the function $t \mapsto f(x+th)$ is absolutely continuous (being Lipschitz), so the fundamental theorem of calculus implies this function is differentiable (or equivalently, $x + th \notin A_h$) almost everywhere on $\mathbf{R}$.

Consider the nonnegative measurable function $\phi : \mathbf{R}^n \times \mathbf{R} \to \mathbf{R}$ defined by $\phi(x,t) = \delta_{A_h}(x+th)$. By our observation above, for any fixed $x \in \mathbf{R}^n$ we know $\int_{\mathbf{R}} \phi(x,t)\,dt = 0$. Denoting Lebesgue measure on $\mathbf{R}^n$ by μ, Fubini's theorem shows

$$0 = \int_{\mathbf{R}^n} \Big(\int_{\mathbf{R}} \phi(x,t)\,dt\Big)\,d\mu = \int_{\mathbf{R}} \Big(\int_{\mathbf{R}^n} \phi(x,t)\,d\mu\Big)\,dt = \int_{\mathbf{R}} \mu(A_h)\,dt$$

so the set A_h has measure zero. Consequently, we can define a measurable function $D_h f : \mathbf{R}^n \to \mathbf{R}$ having the property $D_h f = D_h^+ f = D_h^- f$ almost everywhere.

Denote the standard basis vectors in $\mathbf{R}^n$ by $e_1, e_2, \ldots, e_n$. The function $G : \mathbf{R}^n \to \mathbf{R}^n$ with components defined almost everywhere by

$$G_i = D_{e_i} f = \frac{\partial f}{\partial x_i} \tag{9.1.3}$$

for each $i = 1, 2, \ldots, n$ is the only possible candidate for the derivative of f. Indeed, if f (or $-f$) is regular at x, then it is easy to check that $G(x)$ is the Fréchet derivative of f at x (Exercise 2). The general case needs a little more work.

Consider any continuously differentiable function $\psi : \mathbf{R}^n \to \mathbf{R}$ that is zero except on a bounded set. For our fixed direction h, if $t \neq 0$ we have

$$\int_{\mathbf{R}^n} g_t(x)\,\psi(x)\,d\mu = \int_{\mathbf{R}^n} f(x)\frac{\psi(x-th) - \psi(x)}{t}\,d\mu.$$

As $t \to 0$, the bounded convergence theorem applies, since both f and ψ are Lipschitz, so

$$\int_{\mathbf{R}^n} D_h f(x)\, \psi(x)\, d\mu = -\int_{\mathbf{R}^n} f(x)\, \langle \nabla \psi(x), h \rangle\, d\mu.$$

Setting $h = e_i$ in the above equation, multiplying by h_i, and adding over $i = 1, 2, \ldots, n$, yields

$$\int_{\mathbf{R}^n} \langle h, G(x) \rangle\, \psi(x)\, d\mu = -\int_{\mathbf{R}^n} f(x)\, \langle \nabla \psi(x), h \rangle\, d\mu = \int_{\mathbf{R}^n} D_h f(x)\, \psi(x)\, d\mu.$$

Since ψ was arbitrary, we deduce $D_h f = \langle h, G \rangle$ almost everywhere.

Now extend the basis $e_1, e_2, \ldots, e_n$ to a dense sequence of unit vectors $\{h_k\}$ in the unit sphere $S_{n-1} \subset \mathbf{R}^n$. Define the set $A \subset \mathbf{R}^n$ to consist of those points where each function $D_{h_k} f$ is defined and equals $\langle h_k, G \rangle$. Our argument above shows A^c has measure zero. We aim to show, at each point $x \in A$, that f has Fréchet derivative $G(x)$.

Fix any $\epsilon > 0$. For any $t \neq 0$, define a function $r_t : \mathbf{R}^n \to \mathbf{R}$ by

$$r_t(h) = \frac{f(x + th) - f(x)}{t} - \langle G(x), h \rangle.$$

It is easy to check that r_t has Lipschitz constant $2L$. Furthermore, for each $k = 1, 2, \ldots$, there exists $\delta_k > 0$ such that

$$|r_t(h_k)| < \frac{\epsilon}{2} \quad \text{whenever } 0 < |t| < \delta_k.$$

Since the sphere S_{n-1} is compact, there is an integer M such that

$$S_{n-1} \subset \bigcup_{k=1}^{M} \Big(h_k + \frac{\epsilon}{4L} B \Big).$$

If we define $\delta = \min\{\delta_1, \delta_2, \ldots, \delta_M\} > 0$, we then have

$$|r_t(h_k)| < \frac{\epsilon}{2} \quad \text{whenever } 0 < |t| < \delta,\ k = 1, 2 \ldots, M.$$

Finally, consider any unit vector h. For some positive integer $k \leq M$ we know $\|h - h_k\| \leq \epsilon/4L$, so whenever $0 < |t| < \delta$ we have

$$|r_t(h)| \leq |r_t(h) - r_t(h_k)| + |r_t(h_k)| \leq 2L \frac{\epsilon}{4L} + \frac{\epsilon}{2} = \epsilon.$$

Hence $G(x)$ is the Fréchet derivative of f at x, as we claimed. □

An analogous argument using Fubini's theorem now proves the subdifferential formula (9.1.1)—see Exercise 3.

Exercises and Commentary

A basic reference for the measure theory and the version of the fundamental theorem of calculus we use in this section is [170]. Rademacher's theorem is also proved in [71]. Various implications of the insensitivity of Clarke's formula (9.1.1) to sets of measures zero are explored in [18]. In the same light, the generalized Jacobian of Exercise 4 is investigated in [72].

1. Assuming Rademacher's theorem with range $\mathbf{R}$, prove the general version.

2. * **(Rademacher's theorem for regular functions)** Suppose the function $f : \mathbf{R}^n \to \mathbf{R}$ is locally Lipschitz around the point $x \in \mathbf{R}^n$. Suppose the vector $G(x)$ is well-defined by equation (9.1.3). By observing

$$0 = f^-(x; e_i) + f^-(x; -e_i) = f^\circ(x; e_i) + f^\circ(x; -e_i)$$

and using the sublinearity of $f^\circ(x; \cdot)$, deduce $G(x)$ is the Fréchet derivative of f at x.

3. ** **(Intrinsic Clarke subdifferential formula)** Derive formula (9.1.1) as follows.

   (a) Using Rademacher's theorem (9.1.2), show we can assume that the function f is differentiable everywhere outside the set Q.

   (b) Recall the one-sided inclusion following from the fact that the Clarke subdifferential is a closed multifunction (Exercise 12 in Section 6.2)

   (c) For any vector $v \in \mathbf{E}$ and any point $z \in \mathbf{E}$, use Fubini's theorem to show that the set $\{t \in \mathbf{R} \mid z + tv \in Q\}$ has measure zero, and deduce

$$f(z + tv) - f(z) = \int_0^t \langle \nabla f(z + sv), v \rangle \, ds.$$

   (d) If formula (9.1.1) fails, show there exists $v \in \mathbf{E}$ such that

$$f^\circ(x; v) > \limsup_{w \to x,\ w \notin Q} \langle \nabla f(w), v \rangle.$$

   Use part (c) to deduce a contradiction.

4. ** **(Generalized Jacobian)** Consider a locally Lipschitz map between Euclidean spaces $h : \mathbf{E} \to \mathbf{Y}$ and a set $Q \subset \mathbf{E}$ of measure zero

outside of which h is everywhere Gâteaux differentiable. By analogy with formula (9.1.1) for the Clarke subdifferential, we call

$$\partial_Q h(x) = \text{conv}\,\{\lim_r \nabla h(x^r) \mid x^r \to x,\ x^r \notin Q\},$$

the *Clarke generalized Jacobian* of h at the point $x \in \mathbf{E}$.

(a) Prove that the set $J_h(x) = \partial_Q h(x)$ is independent of the choice of Q.

(b) **(Mean value theorem)** For any points $a, b \in \mathbf{E}$, prove

$$h(a) - h(b) \subset \text{conv}\, J_h[a, b](a - b).$$

(c) **(Chain rule)** If the function $g : \mathbf{Y} \to \mathbf{R}$ is locally Lipschitz, prove the formula

$$\partial_\circ (g \circ h)(x) \subset J_h(x)^* \partial_\circ g(h(x)).$$

(d) Propose a definition for the generalized Hessian of a continuously differentiable function $f : \mathbf{E} \to \mathbf{R}$.

9.2 Proximal Normals and Chebyshev Sets

We introduced the Clarke normal cone in Section 6.3 (Tangent Cones), via the Clarke subdifferential. An appealing alternative approach begins with a more geometric notion of a normal vector. We call a vector $y \in \mathbf{E}$ a *proximal normal* to a set $S \subset \mathbf{E}$ at a point $x \in S$ if, for some $t > 0$, the nearest point to $x + ty$ in S is x. The set of all such vectors is called the *proximal normal cone*, which we denote $N_S^p(x)$.

The proximal normal cone, which may not be convex, is contained in the Clarke normal cone (Exercise 3). The containment may be strict, but we can reconstruct the Clarke normal cone from proximal normals using the following result.

Theorem 9.2.1 (Proximal normal formula) *For any closed set $S \subset \mathbf{E}$ and any point $x \in S$, we have*

$$N_S(x) = \text{conv}\Big\{ \lim_r y_r \mid y_r \in N_S^p(x_r),\ x_r \in S,\ x_r \to x \Big\}.$$

One route to this result uses Rademacher's theorem (Exercise 7). In this section we take a more direct approach.

The Clarke normal cone to a set $S \subset \mathbf{E}$ at a point $x \in S$ is

$$N_S(x) = \text{cl}\,(\mathbf{R}_+ \partial_\circ d_S(x)),$$

by Theorem 6.3.8, where

$$d_S(x) = \inf_{z \in S} \|z - x\|$$

is the distance function. Notice the following elementary but important result that we use repeatedly in this section (Exercise 4(a) in Section 7.3).

Proposition 9.2.2 (Projections) *If $\bar{x}$ is a nearest point in the set $S \subset \mathbf{E}$ to the point $x \in \mathbf{E}$, then $\bar{x}$ is the unique nearest point in S to each point on the half-open line segment $[\bar{x}, x)$.*

To derive the proximal normal formula from the subdifferential formula (9.1.1), we can make use of some striking differentiability properties of distance functions, summarized in the next result.

Theorem 9.2.3 (Differentiability of distance functions) *Consider a nonempty closed set $S \subset \mathbf{E}$ and a point $x \notin S$. Then the following properties are equivalent:*

(i) *the Dini subdifferential $\partial_- d_S(x)$ is nonempty;*

(ii) *x has a unique nearest point $\bar{x}$ in S;*

(iii) *the distance function d_S is Fréchet differentiable at x.*

In this case,

$$\nabla d_S(x) = \frac{x - \bar{x}}{\|x - \bar{x}\|} \in N_S^p(\bar{x}) \subset N_S(\bar{x}).$$

The proof is outlined in Exercises 4 and 6.

For our alternate proof of the proximal normal formula without recourse to Rademacher's theorem, we return to an idea we introduced in Section 8.2. A cusco is a USC multifunction with nonempty compact convex images. In particular, the Clarke subdifferential of a locally Lipschitz function on an open set is a cusco (Exercise 5 in Section 8.2).

Suppose $U \subset \mathbf{E}$ is an open set, $\mathbf{Y}$ is a Euclidean space, and $\Phi : U \to \mathbf{Y}$ is a cusco. We call Φ *minimal* if its graph is minimal (with respect to set inclusion) among graphs of cuscos from U to Y. For example, the subdifferential of a continuous convex function is a minimal cusco (Exercise 8). We next use this fact to prove that Clarke subdifferentials of distance functions are also minimal cuscos.

Theorem 9.2.4 (Distance subdifferentials are minimal) *Outside a nonempty closed set $S \subset \mathbf{E}$, the distance function d_S can be expressed locally as the difference between a smooth convex function and a continuous convex function. Consequently, the Clarke subdifferential $\partial_\circ d_S : \mathbf{E} \to \mathbf{E}$ is a minimal cusco.*

Proof. Consider any closed ball T disjoint from S. For any point y in S, it is easy to check that the Fréchet derivative of the function $x \mapsto \|x - y\|$ is Lipschitz on T. Suppose the Lipschitz constant is $2L$. It follows that the function $x \mapsto L\|x\|^2 - \|x - y\|$ is convex on T (see Exercise 9). Since the function $h : T \to \mathbf{R}$ defined by

$$h(x) = L\|x\|^2 - d_S(x) = \sup_{y \in S}\{L\|x\|^2 - \|x - y\|\}$$

is convex, we obtain the desired expression $d_S = L\|\cdot\|^2 - h$.

To prove minimality, consider any cusco $\Phi : \mathbf{E} \to \mathbf{E}$ satisfying $\Phi(x) \subset \partial_\circ d_S(x)$ for all points x in $\mathbf{E}$. Notice that for any point $x \in \operatorname{int} T$ we have

$$\partial_\circ d_S(x) = -\partial_\circ(-d_S)(x) = \partial h(x) - Lx.$$

Since h is convex on $\operatorname{int} T$, the subdifferential ∂h is a minimal cusco on this set, and hence so is $\partial_\circ d_S$. Consequently, Φ must agree with $\partial_\circ d_S$ on $\operatorname{int} T$, and hence throughout S^c, since T was arbitrary.

On the set $\operatorname{int} S$, the function d_S is identically zero. Hence for all points x in $\operatorname{int} S$ we have $\partial_\circ d_S = \{0\}$ and therefore also $\Phi(x) = \{0\}$. We also deduce $0 \in \Phi(x)$ for all $x \in \operatorname{cl}(\operatorname{int} S)$.

Now consider a point $x \in \mathrm{bd}\, S$. The Mean value theorem (Exercise 9 in Section 6.1) shows

$$\begin{aligned}\partial_\circ d_S(x) &= \mathrm{conv}\Big\{0, \lim_r y^r \,\Big|\, y^r \in \partial_\circ d_S(x^r),\ x^r \to x,\ x^r \notin S\Big\} \\ &= \mathrm{conv}\Big\{0, \lim_r y^r \,\Big|\, y^r \in \Phi(x^r),\ x^r \to x,\ x^r \notin S\Big\},\end{aligned}$$

where 0 can be omitted from the convex hull unless $x \in \mathrm{cl}\,(\mathrm{int}\, S)$ (see Exercise 10). But the final set is contained in $\Phi(x)$, so the result now follows. □

The Proximal normal formula (Theorem 9.2.1), follows rather quickly from this result (and indeed can be strengthened), using the fact that Clarke subgradients of the distance function are proximal normals (Exercise 11).

We end this section with another elegant illustration of the geometry of nearest points. We call a set $S \subset \mathbf{E}$ a *Chebyshev set* if every point in E has a unique nearest point $P_S(x)$ in S. Any nonempty closed convex set is a Chebyshev set (Exercise 8 in Section 2.1). Much less obvious is the converse, stated in the following result.

Theorem 9.2.5 (Convexity of Chebyshev sets) *A subset of a Euclidean space is a Chebyshev set if and only if it is nonempty, closed and convex.*

Proof. Consider a Chebyshev set $S \subset \mathbf{E}$. Clearly S is nonempty and closed, and it is easy to verify that the projection $P_S : \mathbf{E} \to \mathbf{E}$ is continuous. To prove S is convex, we first introduce another new notion. We call S a *sun* if, for each point $x \in \mathbf{E}$, every point on the ray $P_S(x) + \mathbf{R}_+(x - P_S(x))$ has nearest point $P_S(x)$. We begin by proving that the following properties are equivalent (see Exercise 13):

(i) S is convex;

(ii) S is a sun;

(iii) P_S is nonexpansive.

So, we need to show that S is a sun.

Suppose S is not a sun, so there is a point $x \notin S$ with nearest point $P_S(x) = \bar{x}$ such that the ray $L = \bar{x} + \mathbf{R}_+(x - \bar{x})$ strictly contains

$$\{z \in L \mid P_S(z) = \bar{x}\}.$$

Hence by Proposition 9.2.2 (Projections) and the continuity of P_S, the above set is nontrivial closed line segment $[\bar{x}, x_0]$ containing x.

Choose a radius $\epsilon > 0$ so that the ball $x_0 + \epsilon B$ is disjoint from S. The continuous self map of this ball

$$z \mapsto x_0 + \epsilon \frac{x_0 - P_S(z)}{\|x_0 - P_S(z)\|}$$

has a fixed point by Brouwer's theorem (8.1.3). We then quickly derive a contradiction to the definition of the point x_0. □

Exercises and Commentary

Proximal normals provide an alternative comprehensive approach to nonsmooth analysis: a good reference is [56]. Our use of the minimality of distance subdifferentials here is modelled on [38]. Theorem 9.2.5 (Convexity of Chebyshev sets) is sometimes called the "Motzkin-Bunt theorem". Our discussion closely follows [62]. In the exercises, we outline three nonsmooth proofs. The first (Exercises 14, 15, 16) is a variational proof following [82]. The second (Exercises 17, 18, 19) follows [96], and uses Fenchel conjugacy. The third argument (Exercises 20 and 21) is due to Asplund [2]. It is the most purely geometric, first deriving an interesting dual result on furthest points, and then proceeding via inversion in the unit sphere. Asplund extended the argument to Hilbert space, where it remains unknown whether a norm-closed Chebyshev set must be convex. Asplund showed that, in seeking a nonconvex Chebyshev set, we can restrict attention to "Klee caverns": complements of closed bounded convex sets.

1. Consider a closed set $S \subset \mathbf{E}$ and a point $x \in S$.
   (a) Show that the proximal normal cone $N_S^p(x)$ may not be convex.
   (b) Prove $x \in \operatorname{int} S \Rightarrow N_S^p(x) = \{0\}$.
   (c) Is the converse to part (b) true?
   (d) Prove the set $\{z \in S \mid N_S^p(z) \neq \{0\}\}$ is dense in the boundary of S.

2. **(Projections)** Prove Proposition 9.2.2.

3. **(Proximal normals are normals)** Consider a set $S \subset \mathbf{E}$. Suppose the unit vector $y \in \mathbf{E}$ is a proximal normal to S at the point $x \in S$.
   (a) Use Proposition 9.2.2 (Projections) to prove $d_S'(x; y) = 1$.
   (b) Use the Lipschitz property of the distance function to prove $\partial_\circ d_S(x) \subset B$.
   (c) Deduce $y \in \partial_\circ d_S(x)$.
   (d) Deduce that any proximal normal lies in the Clarke normal cone.

4. * **(Unique nearest points)** Consider a closed set $S \subset \mathbf{E}$ and a point x outside S with unique nearest point $\bar{x}$ in S. Complete the following steps to prove

$$\frac{x - \bar{x}}{\|x - \bar{x}\|} \in \partial_- d_S(x).$$

(a) Assuming the result fails, prove there exists a direction $h \in \mathbf{E}$ such that

$$d_S^-(x; h) < \langle \|x - \bar{x}\|^{-1}(x - \bar{x}), h \rangle.$$

(b) Consider a sequence $t_r \downarrow 0$ such that

$$\frac{d_S(x + t_r h) - d_S(x)}{t_r} \to d_S^-(x; h)$$

and suppose each point $x + t_r h$ has a nearest point s_r in S. Prove $s_r \to \bar{x}$.

(c) Use the fact that the gradient of the norm at the point $x - s_r$ is a subgradient to deduce a contradiction.

5. **(Nearest points and Clarke subgradients)** Consider a closed set $S \subset \mathbf{E}$ and a point x outside S with a nearest point $\bar{x}$ in S. Use Exercise 4 to prove

$$\frac{x - \bar{x}}{\|x - \bar{x}\|} \in \partial_\circ d_S(x).$$

6. * **(Differentiability of distance functions)** Consider a nonempty closed set $S \subset \mathbf{E}$.

(a) For any points $x, z \in \mathbf{E}$, observe the identity

$$d_S^2(z) - d_S^2(x) = 2d_S(x)(d_S(z) - d_S(x)) + (d_S(z) - d_S(x))^2.$$

(b) Use the Lipschitz property of the distance function to deduce

$$2d_S(x)\partial_- d_S(x) \subset \partial_- d_S^2(x).$$

Now suppose $y \in \partial_- d_S(x)$.

(c) If $\bar{x}$ is any nearest point to x in S, use part (b) to prove $\bar{x} = x - d_S(x)y$, so $\bar{x}$ is in fact the unique nearest point.

(d) Prove $-2d_S(x)y \in \partial_-(-d_S^2)(x)$.

(e) Deduce d_S^2 is Fréchet differentiable at x.

Assume $x \notin S$.

(f) Deduce d_S is Fréchet differentiable at x.

(g) Use Exercises 3 and 4 to complete the proof of Theorem 9.2.3.

7. * **(Proximal normal formula via Rademacher)** Prove Theorem 9.2.1 using the subdifferential formula (9.1.1) and Theorem 9.2.3 (Differentiability of distance functions).

8. **(Minimality of convex subdifferentials)** If the open set $U \subset \mathbf{E}$ is convex and the function $f : U \to \mathbf{R}$ is convex, use the Max formula (Theorem 3.1.8) to prove that the subdifferential ∂f is a minimal cusco.

9. **(Smoothness and DC functions)** Suppose the set $C \subset \mathbf{E}$ is open and convex, and the Fréchet derivative of the function $g : C \to \mathbf{R}$ has Lipschitz constant $2L$ on C. Deduce that the function $L\|\cdot\|^2 - g$ is convex on C.

10. ** **(Subdifferentials at minimizers)** Consider a locally Lipschitz function $f : \mathbf{E} \to \mathbf{R}_+$, and a point x in $f^{-1}(0)$. Prove

$$\partial_\circ f(x) = \text{conv}\Big\{0, \lim_r y^r \,\Big|\, y^r \in \partial_\circ f(x^r),\ x^r \to x,\ f(x^r) > 0\Big\},$$

where 0 can be omitted from the convex hull if $\text{int}\, f^{-1}(0) = \emptyset$.

11. ** **(Proximal normals and the Clarke subdifferential)** Consider a closed set $S \subset \mathbf{E}$ and a point x in S Use Exercises 3 and 5 and the minimality of the subdifferential $\partial_\circ d_S : \mathbf{E} \to \mathbf{E}$ to prove

$$\partial_\circ d_S(x) = \text{conv}\Big\{0, \lim_r y^r \,\Big|\, y^r \in N_S^p(x^r),\ \|y^r\| = 1,\ x^r \to x,\ x^r \in S\Big\}.$$

Deduce the Proximal normal formula (Theorem 9.2.1). Assuming $x \in \text{bd}\, S$, prove the following stronger version. Consider any dense subset Q of S^c, and suppose $P : Q \to S$ maps each point in Q to a nearest point in S. Prove

$$\partial_\circ d_S(x) = \text{conv}\Big\{0, \lim_r \frac{x^r - P(x^r)}{\|x^r - P(x^r)\|} \,\Big|\, x^r \to x,\ x^r \in Q\Big\},$$

and derive a stronger version of the Proximal normal formula.

12. **(Continuity of the projection)** Consider a Chebyshev set S. Prove directly from the definition that the projection P_S is continuous.

13. * **(Suns)** Complete the details in the proof of Theorem 9.2.5 (Convexity of Chebyshev sets) as follows.

    (a) Prove (iii) $\Rightarrow$ (i).

    (b) Prove (i) $\Rightarrow$ (ii).

    (c) Denoting the line segment between points $y, z \in \mathbf{E}$ by $[y, z]$, prove property (ii) implies

$$P_S(x) = P_{[z, P_S(x)]}(x) \ \text{ for all } x \in \mathbf{E},\ z \in S. \tag{9.2.6}$$

(d) Prove (9.2.6) $\Rightarrow$ (iii).

(e) Fill in the remaining details of the proof.

14. ** **(Basic Ekeland variational principle [43])** Prove the following version of the Ekeland variation principle (Theorem 7.1.2). Suppose the function $f : \mathbf{E} \to (\infty, +\infty]$ is closed and the point $x \in \mathbf{E}$ satisfies $f(x) < \inf f + \epsilon$ for some real $\epsilon > 0$. Then for any real $\lambda > 0$ there is a point $v \in \mathbf{E}$ satisfying the conditions

    (a) $\|x - v\| \leq \lambda$,

    (b) $f(v) + (\epsilon/\lambda)\|x - v\| \leq f(x)$, and

    (c) v minimizes the function $f(\cdot) + (\epsilon/\lambda)\|\cdot - v\|$.

15. * **(Approximately convex sets)** Consider a closed set $C \subset \mathbf{E}$. We call C *approximately convex* if, for any closed ball $D \subset \mathbf{E}$ disjoint from C, there exists a closed ball $D' \supset D$ disjoint from C with arbitrarily large radius.

    (a) If C is convex, prove it is approximately convex.

    (b) Suppose C is approximately convex but not convex.

    (i) Prove there exist points $a, b \in C$ and a closed ball D centered at the point $c = (a + b)/2$ and disjoint from C.

    (ii) Prove there exists a sequence of points $x_1, x_2, \ldots \in \mathbf{E}$ such that the balls $B_r = x_r + rB$ are disjoint from C and satisfy $D \subset B_r \subset B_{r+1}$ for all $r = 1, 2, \ldots$.

    (iii) Prove the set $H = \mathrm{cl} \bigcup_r B_r$ is closed and convex, and its interior is disjoint from C but contains c.

    (iv) Suppose the unit vector u lies in the polar set H°. By considering the quantity $\langle u, \|x_r - x\|^{-1}(x_r - x)\rangle$ as $r \to \infty$, prove H° must be a ray.

    (v) Deduce a contradiction.

    (c) Conclude that a closed set is convex if and only if it is approximately convex.

16. ** **(Chebyshev sets and approximate convexity)** Consider a Chebyshev set $C \subset \mathbf{E}$, and a ball $x + \beta B$ disjoint from C.

    (a) Use Theorem 9.2.3 (Differentiability of distance functions) to prove

$$\limsup_{v \to x} \frac{d_C(v) - d_C(x)}{\|v - x\|} = 1.$$

(b) Consider any real $\alpha > d_C(x)$. Fix reals $\sigma \in (0,1)$ and ρ satisfying

$$\frac{\alpha - d_C(x)}{\sigma} < \rho < \alpha - \beta.$$

By applying the Basic Ekeland variational principle (Exercise 14) to the function $-d_C + \delta_{x+\rho B}$, prove there exists a point $v \in \mathbf{E}$ satisfying the conditions

$$\begin{aligned} d_C(x) + \sigma\|x - v\| &\leq d_C(v) \\ d_C(z) - \sigma\|z - v\| &\leq d_C(v) \text{ for all } z \in x + \rho B. \end{aligned}$$

Use part (a) to deduce $\|x - v\| = \rho$, and hence $x + \beta B \subset v + \alpha B$.

(c) Conclude that C is approximately convex, and hence convex by Exercise 15.

(d) Extend this argument to an arbitrary norm on $\mathbf{E}$.

17. ** **(Smoothness and biconjugacy)** Consider a function $f : \mathbf{E} \to (\infty, +\infty]$ that is closed and bounded below and satisfies the condition

$$\lim_{\|x\| \to \infty} \frac{f(x)}{\|x\|} = +\infty.$$

Consider also a point $x \in \operatorname{dom} f$.

(a) Using Carathéodory's theorem (Section 2.2, Exercise 5), prove there exist points $x_1, x_2, \ldots, x_m \in \mathbf{E}$ and real $\lambda_1, \lambda_2, \ldots, \lambda_m > 0$ satisfying

$$\sum_i \lambda_i = 1, \quad \sum_i \lambda_i x_i = x, \quad \sum_i \lambda_i f(x_i) = f^{**}(x).$$

(b) Use the Fenchel-Young inequality (Proposition 3.3.4) to prove

$$\partial(f^{**})(x) = \bigcap_i \partial f(x_i).$$

Suppose furthermore that the conjugate f^* is everywhere differentiable.

(c) If $x \in \operatorname{ri}(\operatorname{dom}(f^{**}))$, prove $x_i = x$ for each i.

(d) Deduce $\operatorname{ri}(\operatorname{epi}(f^{**})) \subset \operatorname{epi}(f)$.

(e) Use the fact that f is closed to deduce $f = f^{**}$, so f is convex.

18. * **(Chebyshev sets and differentiability)** Use Theorem 9.2.3 (Differentiability of distance functions) to prove that a closed set $S \subset \mathbf{E}$ is a Chebyshev set if and only if the function d_S^2 is Fréchet differentiable throughout $\mathbf{E}$.

19. * **(Chebyshev convexity via conjugacy)** For any nonempty closed set $S \subset \mathbf{E}$, prove

$$\left(\frac{\|\cdot\|^2 + \delta_S}{2}\right)^* = \frac{\|\cdot\|^2 - d_S^2}{2}$$

Deduce, using Exercises 17 and 18, that Chebyshev sets are convex.

20. ** **(Unique furthest points)** Consider a set $S \subset \mathbf{E}$, and define a function $r_S : \mathbf{E} \to [-\infty, +\infty]$ by

$$r_S(x) = \sup_{y \in S} \|x - y\|.$$

Any point y attaining the above supremum is called a *furthest point* in S to the point $x \in \mathbf{E}$.

(a) Prove that the function $(r_S^2 - \|\cdot\|^2)/2$ is the conjugate of the function

$$g_S = \frac{\delta_{-S} - \|\cdot\|^2}{2}.$$

(b) Prove that the function r_S^2 is strictly convex on its domain.

Now suppose each point $x \in \mathbf{E}$ has a unique nearest point $q_S(x)$ in S.

(c) Prove that the function q_S is continuous.

We consider two alternative proofs that a set has the unique furthest point property if and only if it is a singleton.

(d) (i) Use Section 6.1 , Exercise 10 (Max-functions) to show that the function $r_S^2/2$ has Clarke subdifferential the singleton $\{x - q_S(x)\}$ at any point $x \in \mathbf{E}$, and hence is everywhere differentiable.

(ii) Use Exercise 17 (Smoothness and biconjugacy) to deduce that the function g_S is convex, and hence that S is a singleton.

(e) Alternatively, suppose S is not a singleton. Denote the unique minimizer of the function r_S by y. By investigating the continuity of the function q_S on the line segment $[y, q_S(y)]$, derive a contradiction without using part (d).

21. ** **(Chebyshev convexity via inversion)** The map $\iota : \mathbf{E}\backslash\{0\} \to \mathbf{E}$ defined by $\iota(x) = \|x\|^{-2}x$ is called the *inversion in the unit sphere*.

(a) If $D \subset \mathbf{E}$ is a ball with $0 \in \operatorname{bd} D$, prove $\iota(D \setminus \{0\})$ is a halfspace disjoint from 0.

(b) For any point $x \in \mathbf{E}$ and radius $\delta > \|x\|$, prove

$$\iota((x + \delta B) \setminus \{0\}) = \frac{1}{\delta^2 - \|x\|^2} \{y \in \mathbf{E} : \|y + x\| \geq \delta\}.$$

Prove that any Chebyshev set $C \subset \mathbf{E}$ must be convex as follows. Without loss of generality, suppose $0 \notin C$ but $0 \in \operatorname{cl}(\operatorname{conv} C)$. Consider any point $x \in \mathbf{E}$.

(c) Prove the quantity

$$\rho = \inf\{\delta > 0 \mid \iota C \subset x + \delta B\}$$

satisfies $\rho > \|x\|$.

(d) Let z denote the unique nearest point in C to the point

$$\frac{-x}{\rho^2 - \|x\|^2}.$$

Use part (b) to prove that ιz is the unique furthest point in ιC to x.

(e) Use Exercise 20 to derive a contradiction.

9.3 Amenable Sets and Prox-Regularity

In the previous section we saw that nonempty closed convex subsets S of the Euclidean space $\mathbf{E}$ are characterized by the attractive global property that every point in $\mathbf{E}$ has a unique nearest point in S. The corresponding local property is also a useful tool: we begin with a condition guaranteeing this property.

We call the closed set S *prox-regular* at a point $\bar{x}$ in S if there exists a constant $\rho > 0$ such that all distinct points $x, x' \in S$ near $\bar{x}$ and small vectors $v \in N_S(x)$ satisfy the inequality

$$\langle v, x' - x \rangle < \rho \|x' - x\|^2. \tag{9.3.1}$$

Geometrically, this condition states that the ball centered at the point $x + \frac{1}{2\rho}v$ containing the point x on its boundary has interior disjoint from S.

Proposition 9.3.2 (Prox-regularity and projections) *If a closed set $S \subset \mathbf{E}$ is prox-regular at a point $\bar{x} \in S$, then each point in $\mathbf{E}$ close to $\bar{x}$ has a unique nearest point in S.*

Proof. If the result fails, then there exist sequences of points $u_r \to \bar{x}$ in $\mathbf{E}$ and $x_r \neq x'_r$ in S such that both x_r and x'_r are nearest points in S to u_r. Clearly we have $x_r \to \bar{x}$ and $x'_r \to \bar{x}$, and Exercise 3 in Section 9.1 implies $0 \neq u_r - x_r \in N_S(x_r)$. Applying inequality (9.3.1), there exist constants $\epsilon, \rho > 0$ such that

$$\left\langle \epsilon \frac{u_r - x_r}{\|u_r - x_r\|}, x'_r - x_r \right\rangle < \rho \|x'_r - x_r\|^2$$

for all large r. However, the fact that $\|u_r - x_r\| = \|u_r - x'_r\|$ easily implies

$$\langle u_r - x_r, x'_r - x_r \rangle = \frac{1}{2}\|x'_r - x_r\|^2,$$

contradicting the preceding inequality. □

In this section we study an important class of structured prox-regular sets. Our key tool is the chain rule we outlined in Section 7.1 (Exercise 6 (Transversality)).

We proceed by filling in the details of the chain rule. Throughout this section we consider a Euclidean space $\mathbf{Y}$, open sets $U \subset \mathbf{E}$ and $V \subset \mathbf{Y}$, closed sets $S \subset U$ and $R \subset V$, and a continuous map $h : U \to Y$. Our aim is to calculate the tangent cone to the set $S \cap h^{-1}(R)$: the first step is an easy inclusion for the contingent cone, generalizing Proposition 7.1.1.

Proposition 9.3.3 *Suppose the function h is Fréchet differentiable at the point $x \in S \cap h^{-1}(R)$. Then*

$$K_{S\cap h^{-1}(R)}(x) \subset K_S(x) \cap (\nabla h(x))^{-1} K_R(h(x)).$$

To obtain an inclusion in the opposite direction, we need the transversality condition

$$\nabla h(x)(T_S(x)) - T_R(h(x)) = \mathbf{Y}. \tag{9.3.4}$$

Theorem 9.3.5 (Chain rule for sets) *If the function h is strictly differentiable at the point $x \in S \cap h^{-1}(R)$, and the transversality condition (9.3.4) holds, then*

$$T_{S\cap h^{-1}(R)}(x) \supset T_S(x) \cap (\nabla h(x))^{-1} T_R(h(x)) \tag{9.3.6}$$

$$N_{S\cap h^{-1}(R)}(x) \subset N_S(x) + (\nabla h(x))^* N_R(h(x)). \tag{9.3.7}$$

If furthermore the sets S and R are tangentially regular at the points x and $h(x)$ respectively then the set $S \cap h^{-1}(R)$ is tangentially regular at x, and

$$\begin{aligned} T_{S\cap h^{-1}(R)}(x) &= T_S(x) \cap (\nabla h(x))^{-1} T_R(h(x)) \\ N_{S\cap h^{-1}(R)}(x) &= N_S(x) + (\nabla h(x))^* N_R(h(x)). \end{aligned}$$

Proof. The function $g : U \times V \to \mathbf{Y}$, defined by $g(z, y) = h(z) - y$, is strictly differentiable at the point $(x, h(x))$, with derivative $(\nabla h(x), -I)$ (where I denotes the identity map). Section 6.3, Exercise 9 (Products) shows $T_{S\times R}(x, h(x)) = T_S(x) \times T_R(h(x))$, so the transversality condition says $\nabla g(x, h(x)) T_{S\times R}(x, h(x)) = \mathbf{Y}$.

We can now apply Theorem 7.1.5 (Surjectivity and metric regularity) to deduce that the function g is weakly metrically regular on the set $S \times R$ at the point $(x, h(x))$: in other words, there is a constant k' such that

$$d_{(S\times R)\cap g^{-1}(0)}(z, y) \le k' \|h(z) - y\|$$

for all points $(z, y) \in S \times R$ close to $(x, h(x))$. Thus the locally Lipschitz function

$$(z, y) \mapsto k' \|h(z) - y\| - d_{(S\times R)\cap g^{-1}(0)}(z, y)$$

has a local minimizer on $S \times R$ at $(x, h(x))$, so by Proposition 6.3.2 (Exact penalization), there is a constant $L > 0$ such that $(x, h(x))$ is an unconstrained local minimizer of the function

$$(z, y) \mapsto k' \|h(z) - y\| - d_{(S\times R)\cap g^{-1}(0)}(z, y) + L d_{S\times R}(z, y).$$

Since $d_{S\times R}(z, y) \le d_S(z) + d_R(y)$, if we set $k = \max\{k', L\}$, we obtain the inequalities

$$d_{S\cap h^{-1}(R)}(z) \le d_{(S\times R)\cap g^{-1}(0)}(z, h(z)) \le k(d_S(z) + d_R(h(z))), \tag{9.3.8}$$

for all $z \in U$ close to x.

Now consider a vector $p \in T_S(x) \cap (\nabla h(x))^{-1} T_R(h(x))$, and sequences $x^r \to x$ in $S \cap h^{-1}(R)$ and $t_r \downarrow 0$. According to Theorem 6.3.6 (Tangent cones), to prove inclusion (9.3.6) we need to find vectors $p^r \to p$ satisfying $x^r + t_r p^r \in S \cap h^{-1}(R)$ for all r. To this end, note that the inequalities (9.3.8) show

$$d_{S\cap h^{-1}(R)}(x^r + t_r p) \leq k(d_S(x^r + t_r p) + d_R(h(x^r + t_r p))),$$

so there exist points $z^r \in S \cap h^{-1}(R)$ such that

$$\|x^r + t_r p - z_r\| \leq k(d_S(x^r + t_r p) + d_R(h(x^r + t_r p))).$$

We now claim the vectors $p_r = t_r^{-1}(z^r - x^r)$ satisfy our desired properties. Clearly, $x^r + t_r p^r \in S \cap h^{-1}(R)$, by definition, and

$$\|p - p^r\| = \frac{1}{t_r}\|x^r + t_r p - z_r\| \leq k\Big(\frac{1}{t_r} d_S(x^r + t_r p) + \frac{1}{t_r} d_R(h(x^r + t_r p))\Big).$$

Since $p \in T_S(x)$, we know $t_r^{-1} d_S(x^r + t_r p) \to 0$. On the other hand, by strict differentiability,

$$h(x_r + t_r p) = h(x_r) + t_r \nabla h(x) p + o(t_r)$$

as $r \to \infty$, so

$$\frac{1}{t_r} d_R(h(x^r + t_r p)) = \frac{1}{t_r} d_R(h(x_r) + t_r \nabla h(x) p) + \frac{o(t_r)}{t_r}.$$

The first term on the right approaches zero, since $\nabla h(x) p \in T_R(h(x))$ by assumption, so we have proved $p_r \to p$ as desired.

We have thus proved inclusion (9.3.6). The Krein–Rutman polar cone calculus (3.3.13) and the transversality condition imply

$$\Big(T_S(x) \cap (\nabla h(x))^{-1} T_R(h(x))\Big)^- = N_S(x) + (\nabla h(x))^* N_R(h(x)),$$

so we immediately obtain inclusion (9.3.7). With the extra tangential regularity assumptions, Proposition 9.3.3 implies

$$\begin{aligned} T_{S\cap h^{-1}(R)}(x) &\subset K_{S\cap h^{-1}(R)}(x) \subset K_S(x) \cap (\nabla h(x))^{-1} K_R(h(x)) \\ &= T_S(x) \cap (\nabla h(x))^{-1} T_R(h(x)) \subset T_{S\cap h^{-1}(R)}(x), \end{aligned}$$

so the final claims now follow. □

Inverse images of convex sets under smooth mappings are particularly common examples of nonsmooth nonconvex sets. We call a set $Q \subset \mathbf{E}$

amenable at a point $\bar{x} \in Q$ if there exists an open neighbourhood U of $\bar{x}$, a Euclidean space $\mathbf{Y}$, a closed convex set $R \subset \mathbf{Y}$, and a continuously differentiable map $h : U \to Y$, such that

$$Q \cap U = h^{-1}(R), \tag{9.3.9}$$

and furthermore, the transversality condition

$$N_R(h(\bar{x})) \cap N((\nabla h(\bar{x}))^*) = \{0\} \tag{9.3.10}$$

holds. If furthermore we can choose h to be *twice* continuously differentiable at $\bar{x}$, then we call Q *fully amenable* at $\bar{x}$. It is easy to check that if the condition (9.3.10) holds, then it also holds with the point $\bar{x}$ replaced by any nearby point in Q.

The following straightforward exercise using the preceding chain rule shows that amenable sets are well-behaved.

Corollary 9.3.11 (Amenable sets) *If the set $Q \subset \mathbf{E}$ is amenable at the point $\bar{x} \in Q$, then Q is tangentially regular at $\bar{x}$. Furthermore, given the representation (9.3.9), we have*

$$\begin{aligned} T_Q(\bar{x}) &= \nabla h(\bar{x})^{-1} T_R(h(\bar{x})) \\ N_Q(\bar{x}) &= \nabla h(\bar{x})^* N_R(h(\bar{x})). \end{aligned}$$

With the extra assumption of *full* amenability, we arrive, as promised, at a broad class of prox-regular sets.

Theorem 9.3.12 (Amenability and prox-regularity) *If a set $Q \subset \mathbf{E}$ is fully amenable at a point $\bar{x} \in Q$, then Q is prox-regular at $\bar{x}$.*

Proof. Suppose we have the representation (9.3.9), where the function h is twice continuously differentiable, and suppose the transversality condition (9.3.10) holds. If prox-regularity fails, then there exist sequences of points $x_r \neq x'_r$ approaching $\bar{x}$ in Q, and vectors $v_r \in N_Q(x_r)$ approaching 0, such that

$$\langle v_r, x'_r - x_r \rangle > r\|x'_r - x_r\|^2. \tag{9.3.13}$$

As we observed above, the condition (9.3.10) implies

$$N_R(h(x_r)) \cap N((\nabla h(x_r))^*) = \{0\}$$

for all large r. Hence Corollary 9.3.11 (Amenable sets) implies there exist vectors $y_r \in N_R(h(x_r))$ such that $v_r = (\nabla h(x_r))^* y_r$, for each large r.

We next observe that the vectors y_r approach 0. Indeed, if this were not the case, we could find a subsequence for which $\|y_r\| \geq \epsilon$ for some $\epsilon > 0$ and with $\|y_r\|^{-1} y_r \to u$ for some unit vector u. Since the normal cone $N_R(\cdot)$ is a closed multifunction (Section 4.2, Exercise 8), we deduce $u \in N_R(h(\bar{x}))$. On the other hand, as $r \to \infty$ in the subsequence,

$$(\nabla h(x_r))^*\Big(\frac{1}{\|y_r\|}y_r\Big) = \Big(\frac{1}{\|y_r\|}\Big)v_r \to 0,$$

so $u \in N(\nabla h(\bar{x})^*)$, contradicting the transversality condition (9.3.10).

Returning to our proof, for all large r we have by Taylor's theorem,

$$\begin{aligned} 0 &\geq \langle y_r, h(x'_r) - h(x_r)\rangle \\ &= \langle y_r, \nabla h(x_r)(x'_r - x_r)\rangle + \Big\langle y_r\,,\, \frac{1}{2}\nabla^2 h(z_r)(x'_r - x_r, x'_r - x_r)\Big\rangle, \end{aligned}$$

for some point z_r between x_r and x'_r. For large r, inequality (9.3.13) shows that the first term on the right hand side is bigger than $r\|x'_r - x_r\|^2$, which is eventually larger than minus the second term. This contradiction completes the proof. □

Exercises and Commentary

Prox-regularity as a tool for nonsmooth analysis was introduced in [156]. Its relationship with the differentiability of the distance function is studied in [157]. Amenability is surveyed in [155].

1. Prove Proposition 9.3.3.

2. **(Persistence of amenability)** Prove that if a set Q is amenable at a point $\bar{x} \in Q$, then it is amenable at every nearby point in Q.

3. * Use the chain rule for sets, Theorem 9.3.5, and Section 3.3, Exercise 16 (Sums of closed cones), to prove Corollary 9.3.11 (Amenable sets).

4. **(Amenability and Mangasarian-Fromowitz)** Compare Corollary 9.3.11 (Amenable sets) with the formula for the contingent cone to a feasible region satisfying the Mangasarian-Fromowitz constraint qualification (Theorem 7.2.6).

9.4 Partly Smooth Sets

Having solved an optimization problem, we often wish to study the sensitivity of the solution to small changes in the problem. Such "sensitivity analysis" often depends on a judicious blend of smooth and nonsmooth analysis. In this section we consider a class of sets particularly well structured for such analysis.

We begin by formalizing the idea of a smooth surface in $\mathbf{E}$. We call a set $M \subset \mathbf{E}$ a *manifold of codimension m around* a point $\bar{x} \in M$ if there is a neighbourhood V of $\bar{x}$ and a twice continuously differentiable ($C^{(2)}$) map $F : V \to \mathbf{R}^m$ with surjective derivative $\nabla F(\bar{x})$ such that points $x \in V$ lie in M if and only if $F(x) = 0$. A set M is simply a *manifold (of codimension m)* if this condition holds for all points $\bar{x} \in M$.

The set in $\mathbf{R}^3$ defined by the inequality $z \geq |x| + y^2$ has a sharp "ridge" described by the manifold M of points satisfying the equations $x = 0$ and $z = y^2$. Minimizing the linear function $(x, y, z) \mapsto z$ over this set gives the optimal solution $(0, 0, 0)$, and minimizing any nearby linear function gives a nearby optimal solution lying on M. We isolate this kind of stable structure of the solution in the following definition.

We call a closed set $S \subset \mathbf{E}$ *partly smooth* relative to a manifold $M \subset S$ if, for all points $x \in M$, the set S is tangentially regular at x with $N_M(x) = N_S(x) - N_S(x)$, and furthermore, for any normal $y \in N_S(x)$ and sequence of points $x_r \in M$ approaching x, there exists a sequence of normals $y_r \in N_S(x_r)$ approaching y. A simple example is the positive orthant.

Proposition 9.4.1 (Partly smooth orthant) *The positive orthant $\mathbf{R}^n_+$ is partly smooth relative to each manifold*

$$\left\{ x \in \mathbf{R}^n_+ \,\middle|\, |\{i \,|\, x_i = 0\}| = k \right\}$$

(for $k = 0, 1, 2, \ldots, n$).

A *face* of a polyhedron is its intersection with a supporting hyperplane. The result above shows that the positive orthant is partly smooth relative to the relative interior of any face: not surprisingly, this property holds for *any* polyhedron.

An analogous, less obvious result concerns the semidefinite cone: we state it without proof.

Theorem 9.4.2 (Partial smoothness of $\mathbf{S}^n_+$) *The semidefinite cone $\mathbf{S}^n_+$ is partly smooth relative to each manifold $\{X \in \mathbf{S}^n \mid \operatorname{rank} X = k\}$ (for $k = 0, 1, 2, \ldots, n$).*

The following easy result describes another basic example.

Proposition 9.4.3 (Partial smoothness and cones) *Any closed convex cone $K \subset \mathbf{E}$ is partly smooth relative to $K \cap (-K)$.*

By building on the chain rule for sets (Theorem 9.3.5), more complex examples follow from the building blocks above. We leave the following result as an exercise.

Theorem 9.4.4 (Partly smooth chain rule) *Given a Euclidean space $\mathbf{Y}$ and a map $h : \mathbf{E} \to \mathbf{Y}$ that is continuously differentiable around a point $x \in \mathbf{E}$, suppose the closed set $R \subset \mathbf{Y}$ is partly smooth relative to a manifold M containing $h(x)$. Assuming the transversality condition*

$$\nabla h(x)\mathbf{E} - T_M(h(x)) = \mathbf{Y},$$

then for some open neighbourhood U of x, the set $h^{-1}(M) \cap U$ is a manifold, relative to which the set $h^{-1}(R)$ is partly smooth.

An easy consequence is the partial smoothness of inequality-constrained sets, assuming a typical constraint qualification.

Corollary 9.4.5 (Inequalities and partial smoothness) *Given maps $g_i : \mathbf{E} \to \mathbf{R}$ (for i in some finite index set I) that are continuously differentiable around a point*

$$\bar{x} \in S = \{x \in \mathbf{E} \mid g_i(x) \leq 0 \ (i \in I)\},$$

define the active index set at $x \in \mathbf{E}$ by

$$I(x) = \{i \in I \mid g_i(x) = 0\},$$

and suppose the set of active gradients $\{\nabla g_i(\bar{x}) \mid i \in I(\bar{x})\}$ is linearly independent. Then for some open neighbourhood U of $\bar{x}$, the set

$$\{x \in U \mid I(x) = I(\bar{x})\}$$

is a manifold, relative to which S is partly smooth.

Our aim is sensitivity analysis and sufficient optimality conditions for problems with partly smooth feasible regions. To accomplish this, we combine a variety of conditions familiar in optimization theory: a smooth second order condition like that of Theorem 2.1.5; the partial smoothness condition we introduced above; a kind of "strict complementarity condition"; the prox-regularity condition we discussed in the previous section.

Given a set $S \subset \mathbf{E}$ that is partly smooth relative to a manifold $M \subset S$, we call a point $\bar{x} \in M$ *strongly critical* for the minimization problem $\inf_S \langle \bar{c}, \cdot \rangle$ if

$$-\bar{c} \in \operatorname{ri} N_S(\bar{x}) \tag{9.4.6}$$

and

$$\liminf_{x \to \bar{x} \text{ in } M} \frac{\langle \bar{c}, x - \bar{x} \rangle}{\|x - \bar{x}\|^2} > 0. \tag{9.4.7}$$

We can write this latter condition rather more constructively as follows. According to our definition, there exists a neighbourhood $V \subset \mathbf{E}$ of $\bar{x}$ and $C^{(2)}$ functions $h_i : V \to \mathbf{R}$ (for $i = 1, 2, \ldots, m$) such that the gradients $\nabla h_i(\bar{x})$ are linear independent and

$$M \cap V = \{x \in V \mid h_i(x) = 0 \ \forall i\}.$$

The condition (9.4.6) together with partial smoothness of S implies the existence of scalars $\bar{\lambda}_i$ satisfying $-\bar{c} = \sum_i \bar{\lambda}_i \nabla h_i(\bar{x})$. In fact, by linear independence, these $\bar{\lambda}_i$ are unique. Now an exercise shows that the second order condition (9.4.7) is equivalent to the condition

$$0 \neq d \in \{\nabla h_i(\bar{x}) \mid i = 1, 2, \ldots, m\}^{\perp} \ \Rightarrow \ \sum_{i=1}^{m} \bar{\lambda}_i \nabla^2 h_i(\bar{x})(d, d) > 0. \tag{9.4.8}$$

Theorem 9.4.9 (Strong critical points) *Suppose the closed set $S \subset \mathbf{E}$ is partly smooth relative to the manifold $M \subset \mathbf{E}$. If the point $\bar{x} \in M$ is strongly critical for the problem $\inf_S \langle \bar{c}, \cdot \rangle$, then for all vectors $c \in \mathbf{E}$ close to $\bar{c}$ the problem $\inf_S \langle c, \cdot \rangle$ has a strong critical point $x(c) \in M$ whose dependence on c is continuously differentiable, and satisfying $x(\bar{c}) = \bar{x}$. If S is also prox-regular at $\bar{x}$, then for all c close to $\bar{c}$ the point $x(c)$ is a strict local minimizer for $\inf_S \langle c, \cdot \rangle$.*

Proof. Describe the manifold M as in the previous paragraph, and consider the system of equations in variables $x \in \mathbf{E}$ and $\lambda \in \mathbf{R}^m$,

$$\begin{aligned} h_i(x) &= 0 \quad (i = 1, 2, \ldots, m) \\ \sum_{i=1}^{m} \lambda_i \nabla h_i(x) &= c. \end{aligned}$$

Using the positive definiteness condition (9.4.8), the inverse function theorem shows the existence of a solution $(x(c), \lambda(c))$ for c close to $\bar{c}$, whose dependence on c is continuously differentiable, and satisfying $x(\bar{c}) = \bar{x}$. An exercise shows that, providing c is close to $\bar{c}$, any nonzero vector d orthogonal to each vector $\nabla h_i(x(c))$ satisfies

$$\sum_{i=1}^{m} \lambda_i(c) \nabla^2 h_i(x(c))(d, d) > 0.$$

To complete the proof that the point $x(c)$ is strongly critical, we just need to check $-c \in \operatorname{ri} N_S(x(c))$, for all vectors c close to $\bar{c}$. Recall that the subspace spanned by the cone $N_S(x(c))$ is $N_M(x(c))$, and c lies in this subspace by the definition of $x(c)$.

By way of contradiction, suppose there is a sequence of vectors $c_r \to \bar{c}$ in $\mathbf{E}$ satisfying $-c_r \notin \mathrm{ri}\, N_S(x(c_r))$ for each r. Hence we can separate $-c_r$ from $N_S(x(c_r))$ in the subspace $N_M(x(c_r))$, so some unit vector $z_r \in N_M(x(c_r))$ satisfies $\inf \langle z_r, N_S(x(c_r)) + c_r \rangle \geq 0$. By taking a subsequence, we can suppose z_r approaches a unit vector $\bar{z}$, which must lie in $N_M(\bar{x})$.

Now consider any vector $\bar{y} \in N_S(\bar{x})$. By partial smoothness, there are vectors $y_r \in N_S(x(c_r))$ approaching $\bar{y}$. Since $\langle z_r, y_r + c_r \rangle \geq 0$, we deduce $\langle \bar{z}, \bar{y} + \bar{c} \rangle \geq 0$. We have thus shown that the unit vector $\bar{z}$ separates the vector $-\bar{c}$ from the cone $N_S(\bar{x})$ in its span, $N_M(\bar{x})$, contradicting the fact that $-\bar{c} \in \mathrm{ri}\, N_S(\bar{x})$.

Now suppose the set S is prox-regular at $\bar{x}$ (and hence also at any nearby point in S). Clearly it suffices to prove the strict local minimizer property just for the point $\bar{x}$. By prox-regularity, there is a constant $\rho > 0$ such that all distinct points $x, x' \in S$ near $\bar{x}$ and small vectors $v \in N_S(x)$ satisfy

$$\langle v, x' - x \rangle < \rho \|x' - x\|^2.$$

On the other hand, by the second order condition (9.4.7), there is a constant $\delta > 0$ such that all points $x'' \in M$ near $\bar{x}$ satisfy the inequality

$$\langle \bar{c}, x'' - \bar{x} \rangle \geq \delta \|x'' - \bar{x}\|^2. \tag{9.4.10}$$

We claim that this inequality in fact holds for all $x'' \in S$ near $\bar{x}$.

If the claim fails, there is a sequence $x'_r \to \bar{x}$ in S satisfying the inequality

$$\langle \bar{c}, x'_r - \bar{x} \rangle < \delta \|x'_r - \bar{x}\|^2,$$

for all r. Since manifolds are prox-regular (Exercise 9), for all large r the point x'_r has a unique nearest point x_r in M. Inequality (9.4.10) implies $x_r \neq x'_r$, so after taking a subsequence, we can suppose the unit vectors

$$z_r = \frac{x'_r - x_r}{\|x'_r - x_r\|} \in N_M(x_r)$$

approach a unit vector $\bar{z}$. Notice

$$\langle \bar{c}, z_r \rangle = \frac{\langle \bar{c}, x'_r - \bar{x} \rangle - \langle \bar{c}, x_r - \bar{x} \rangle}{\|x'_r - x_r\|} < \delta \frac{\|x'_r - \bar{x}\|^2 - \|x_r - \bar{x}\|^2}{\|x'_r - x_r\|}.$$

Letting $r \to \infty$ shows the inequality

$$\langle \bar{c}, \bar{z} \rangle \leq 0. \tag{9.4.11}$$

Since $x_r \to \bar{x}$, we know $\bar{z}$ lies in the subspace $N_M(\bar{x})$, which is the span of the cone $N_S(\bar{x})$, so by condition (9.4.6), there exists a scalar $\lambda > 0$ such that $-\bar{c}+\lambda\bar{z} \in N_S(\bar{x})$. Hence, by partial smoothness, there exist vectors $v_r \in N_S(x_r)$ approaching $-\bar{c}+\lambda\bar{z}$. By prox-regularity, there exists a scalar $\kappa > 0$ such that

$$\langle \kappa v_r, x'_r - x_r \rangle < \rho \|x'_r - x_r\|^2$$

for all r, so $\langle v_r, z_r \rangle < \rho\|x'_r - x_r\|/\kappa$. Letting $r \to \infty$ shows the inequality $\langle -\bar{c}+\lambda\bar{z}, \bar{z}\rangle \leq 0$, contradicting inequality (9.4.11). □

Exercises and Commentary

The material in this section is taken from [126, 92].

1. Prove Proposition 9.4.1 (Partly smooth orthant).

2. * Prove that any polyhedron is partly smooth relative to the relative interior of any face.

3. Prove Proposition 9.4.3 (Partly smooth cones).

4. Identify all the manifolds relative to which the *second-order cone* epi $(\|\cdot\|)$ is partly smooth.

5. * Prove Theorem 9.4.4 (Partly smooth chain rule).

6. * **(Strict complementarity)** Prove Corollary 9.4.5 (Inequalities and partial smoothness). With the assumptions of this result, prove that the strict complementarity condition (9.4.6) holds if and only if there exist Lagrange multipliers $\lambda_i > 0$ (for $i \in I(\bar{x})$) such that $\bar{x}$ is a critical point of the Lagrangian defined by

$$L(x) = \langle \bar{c}, x\rangle + \sum_{i \in I(\bar{x})} \lambda_i g_i(x).$$

7. * **(Constructive second order condition)** Verify the claim before Theorem 9.4.9 that the two second order conditions (9.4.7) and (9.4.8) are equivalent.

8. * Complete the details of the proof of Theorem 9.4.9 (Strong critical points).

9. * **(Prox-regularity of manifolds)** If the set $M \subset \mathbf{E}$ is a manifold around the point $\bar{x} \in M$, prove M is prox-regular at $\bar{x}$.

10. * **(Necessity of prox-regularity)** Partition $\mathbf{R}^2$ into four disjoint sets,

$$\begin{aligned} S_1 &= \{(x,y) \mid y \leq 0\} \\ S_2 &= \{(x,y) \mid 0 < y < 2x^2\} \\ S_3 &= \{(x,y) \mid 0 < 2x^2 \leq y \leq 4x^2\} \\ S_4 &= \{(x,y) \mid 4x^2 < y\}, \end{aligned}$$

and define a function $f : \mathbf{R}^2 \to \mathbf{R}$ by

$$f(x,y) = \begin{cases} x^2 - y & \text{on } S_1 \\ \sqrt{x^4 + 2x^2y - y^2} & \text{on } S_2 \\ 3x^2 - y & \text{on } S_3 \\ y - 5x^2 & \text{on } S_4. \end{cases}$$

(i) Prove that f is everywhere locally Lipschitz.

(ii) Prove that f is everywhere regular.

(iii) Prove that the set $\operatorname{epi} f$ is partly smooth at zero relative to each of the manifolds

$$\begin{aligned} M_1 &= \{(x,y,z) \mid y = 0,\ z = x^2\}, \\ M_2 &= \{(x,y,z) \mid y = 4x^2,\ z = -x^2\}. \end{aligned}$$

(iv) Prove that zero is a strong critical point relative to M_1 for the problem of minimizing the function $(x,y,z) \mapsto z$ over $\operatorname{epi} f$, but is not a local minimizer.

(v) Is $\operatorname{epi} f$ prox-regular at zero?

Chapter 10

Postscript: Infinite Versus Finite Dimensions

10.1 Introduction

We have chosen to finish this book by indicating many of the ways in which finite dimensionality has played a critical role in the previous chapters. While our list is far from complete it should help illuminate the places in which care is appropriate when "generalizing". Many of our main results (on subgradients, variational principles, open mappings, Fenchel duality, metric regularity) immediately generalize to at least reflexive Banach spaces. When they do not, it is principally because the compactness properties and support properties of convex sets have become significantly more subtle. There are also significantly many properties that characterize Hilbert space. The most striking is perhaps the deep result that a Banach space X is (isomorphic to) Hilbert space if and only if every closed vector subspace is complemented in X. Especially with respect to best approximation properties, it is Hilbert space that best captures the properties of Euclidean space.

Since this chapter will be primarily helpful to those with some knowledge of Banach space functional analysis, we make use of a fair amount of standard terminology without giving details. In the exercises more specific cases are considered.

Throughout, X is a real Banach space with continuous dual space X^* and $f : X \to (\infty, +\infty]$ is usually convex and proper (somewhere finite). If f is everywhere finite and lower semicontinuous then f is continuous—since a Banach space is *barreled*, as it is a *Baire space* (see Exercise 1). This is one of the few significant analytic properties which hold in a large class of incomplete normed spaces. By contrast, it is known that completeness is characterized by the nonemptiness or maximality of subdifferentials on a normed space. For example, on every incomplete normed space there is a closed convex function with an empty subdifferential, and a closed convex set with no support points.

The *convex subdifferential* at a point x where f is finite is defined by

$$\partial f(x) = \{x^* \in X^* \mid \langle x^*, h\rangle \leq f(x+h) - f(x) \ \text{ for all } h \in X\}.$$

In what follows, sets are usually closed and convex and $B(X)$ denotes the closed unit ball: $B(X) = \{x \mid \|x\| \leq 1\}$. In general our notation and terminology are consistent with the Banach space literature. We will interchangeably write $\langle x^*, h\rangle$ or $x^*(h)$ depending whether functional or vectorial ideas are first in our minds.

A point x^* of a convex set C is a *(proper) support point* of C if there exists a linear continuous functional ϕ with

$$\phi(x^*) = \sigma = \sup_C \phi > \inf_C \phi.$$

Then ϕ is said to be a (nontrivial) *supporting functional* and $H = \phi^{-1}(\sigma)$ is a *supporting hyperplane*. In the case when $C = B(X)$, ϕ is said to be *norm-attaining*.

We complete the preliminaries by recalling some derivative notions. Let β denote a *bornology*, that is, a family of bounded and centrally symmetric subsets of X, closed under positive scalar multiplication and finite unions, and whose union is X. We write $x^* \in \partial^\beta f(x)$ if for all sets B in β and real $\epsilon > 0$, there exists real $\delta > 0$ such that

$$\langle x^*, h\rangle \leq \frac{f(x+th) - f(x)}{t} + \epsilon \ \text{ for all } t \in (0, \delta) \text{ and } h \in B.$$

It is useful to identify the following bornologies:

$$\begin{aligned} \text{points} &\leftrightarrow \text{Gâteaux } (G) \\ \text{(norm) compacts} &\leftrightarrow \text{Hadamard } (H) \\ \text{weak compacts} &\leftrightarrow \text{weak Hadamard } (WH) \\ \text{bounded} &\leftrightarrow \text{Fréchet } (F). \end{aligned}$$

Then $\partial^H f(x) = \partial^G f(x)$ for any locally Lipschitz f, while $\partial^F f(x) = \partial^{WH} f(x)$ when X is a reflexive space. With this language we may define the *β-derivative of f at x* by

$$\{\nabla^\beta f(x)\} = \partial^\beta f(x) \cap -\partial^\beta(-f)(x)$$

so that

$$\{\nabla^\beta f(x)\} = \partial^\beta f(x) \ \text{ for concave } f.$$

For convex functions there is a subtle interplay between these notions. For example, a convex function that is weak Hadamard differentiable at a point of X is Fréchet differentiable at that point if $\ell_1(\mathbf{N}) \not\subset X$. For general Lipschitz mappings the situation is much simpler. For example, on every nonreflexive but smooth Banach space there is a distance function that is everywhere weak Hadamard differentiable but not Fréchet differentiable at some point. Hence the situation on $c_0(\mathbf{N})$ differs entirely for convex and distance functions.

10.2 Finite Dimensionality

We begin with a compendium of standard and relatively easy results whose proofs may be pieced together from many sources. Sometimes the separable version of these results is simpler.

Theorem 10.2.1 (Closure, continuity, and compactness) *The following statements are equivalent:*

(*i*) *X is finite-dimensional.*

(*ii*) *Every vector subspace of X is closed.*

(*iii*) *Every linear map taking values in X has closed range.*

(*iv*) *Every linear functional on X is continuous.*

(*v*) *Every convex function $f : X \to \mathbf{R}$ is continuous.*

(*vi*) *The closed unit ball in X is (pre-)compact.*

(*vii*) *For each nonempty closed set C in X and for each x in X, the distance*

$$d_C(x) = \inf\{\|x - y\| \mid y \in C\}$$

is attained.

(*viii*) *The weak and norm topologies coincide on X.*

(*ix*) *The weak-star and norm topologies coincide on X^*.*

Turning from continuity to tangency properties of convex sets we have the following result.

Theorem 10.2.2 (Support and separation) *The following statements are equivalent:*

(*i*) *X is finite-dimensional.*

(*ii*) *Whenever a lower semicontinuous convex* $f : X \to (\infty, +\infty]$ *has a unique subgradient at* x *then* f *is Gâteaux differentiable at* x.

(*iii*) X *is separable and every (closed) convex set in* X *has a supporting hyperplane at each boundary point.*

(*iv*) *Every (closed) convex set in* X *has nonempty relative interior.*

(*v*) $A \cap R = \emptyset$, *A closed and convex,* R *a ray (or line)* $\Rightarrow$ A *and* R *are separated by a closed hyperplane.*

It is conjectured but not proven that the property described in part (iii) of the above result holds in all nonseparable Banach spaces X.

In essence these two results say "don't trust finite-dimensionally derived intuitions". In Exercise 6 we present a nonconvex tangency characterization.

By comparison, the following is a much harder and less well known set of results.

Theorem 10.2.3 *The following statements are equivalent:*

(*i*) X *is finite-dimensional.*

(*ii*) *Weak-star and norm convergence agree for sequences in* X^*.

(*iii*) *Every continuous convex* $f : X \to \mathbf{R}$ *is bounded on bounded sets.*

(*iv*) *For every continuous convex* $f : X \to \mathbf{R}$, *the subdifferential* ∂f *is bounded on bounded sets.*

(*v*) *For every continuous convex* $f : X \to \mathbf{R}$, *any point of Gâteaux differentiability is a point of Fréchet differentiability.*

Proof sketch. (i) $\Rightarrow$ (iii) or (v) is clear; (iii) $\Rightarrow$ (iv) is easy.

To see (v) $\Rightarrow$ (ii) and (iii) $\Rightarrow$ (ii) we proceed as follows. Consider sequences (x_n^*) in X^* and (α_n) in $\mathbf{R}$ satisfying $\|x_n^*\| = 1$ and $0 < \alpha_n \downarrow 0$. Define

$$f(x) = \sup_{n \in \mathbf{N}} \{\langle x_n^*, x \rangle - \alpha_n\}.$$

Then f is convex and continuous and satisfies

$$\text{Gâteaux differentiable at } 0 \;\Leftrightarrow\; x_n^* \xrightarrow{w^*} 0$$

and

$$\text{Fréchet differentiable at } 0 \;\Leftrightarrow\; \|x_n^*\|_* \to 0.$$

Thus (v) $\Rightarrow$ (ii).

Now consider the function

$$f(x) = \sum_n \varphi_n(\langle x_n^*, x\rangle),$$

where $\varphi_n(t) = n(|t| - (1/2))^+$. Then f is

$$\text{finite (continuous)} \;\Leftrightarrow\; x_n^* \xrightarrow{w^*} 0,$$

and is

$$\text{bounded on bounded sets} \;\Leftrightarrow\; \|x_n^*\|_* \to 0.$$

Thus (iii) $\Rightarrow$ (ii). □

Note that the sequential coincidence of weak and norm topologies characterizes the so called *Schur spaces* (such as $\ell_1(\mathbf{N})$), while the sequential coincidence of weak and weak-star topologies characterizes the *Grothendieck spaces* (reflexive spaces and nonreflexive spaces such as $\ell_\infty(\mathbf{N})$).

The last four statements of the previous theorem are equivalent in the strong sense that they are easily interderived while no "easy proof" is known of (ii) $\Rightarrow$ (i). (This is the Josephson–Nissenzweig theorem, first established in 1975.) For example, (ii) $\Rightarrow$ (iii) follows from the next result.

Proposition 10.2.4 *Suppose that $f : X \to \mathbf{R}$ is continuous and convex and that (x_n) is bounded while $f(x_n) \to \infty$. Then*

$$x_n^* \in \partial f(x_n) \;\Rightarrow\; \frac{x_n^*}{\|x_n^*\|} \xrightarrow{w^*} 0.$$

Thus each such function yields a *Josephson–Nissenzweig sequence* of unit vectors w*-convergent to zero.

Theorem 10.2.3 highlights the somewhat disconcerting fact that even innocent-seeming examples of convex functions inevitably involve deeper questions about the structure of Banach spaces. The following are some examples.

(i) In $c_0(\mathbf{N})$ with the supremum norm $\|\cdot\|_\infty$, one may find an equivalent norm ball $B_0(X)$ so that the sum $B_\infty(X) + B_0(X)$ is open. This is certainly not possible in a reflexive space, where closed bounded convex sets are weakly compact.

(ii) A Banach space X is reflexive if and only if each continuous linear functional is norm-attaining, that is, achieves its norm on the unit ball in X. (This is the celebrated theorem of James.) In consequence, in each nonreflexive space there is a closed hyperplane H such that for no point x outside H is $d_H(x)$ attained.

(iii) In most nonseparable spaces there exist closed convex sets C each of whose points is a proper support point. This is certainly not possible in a separable space, wherein *quasi relative interior points* must exist.

10.3 Counterexamples and Exercises

1. **(Absorbing sets)** A convex set C satisfying $X = \cup\{tC \mid t \geq 0\}$ is said to be *absorbing* (and zero is said to be in the *core* of C).

   (a) A normed space is said to be *barreled* if every closed convex absorbing subset C has zero in its interior. Use the Baire category theorem to show that Banach spaces are barreled. (There are normed spaces which are barreled but in which the Baire category theorem fails, and there are Baire normed spaces which are not complete: appropriate dense hyperplanes and countable codimension subspaces will do the job.)

   (b) Let f be proper lower semicontinuous and convex. Suppose that zero lies in the core of the domain of f. By considering the set

   $$C = \{x \in X \mid f(x) \leq 1\},$$

   deduce that f is continuous at zero.

   (c) Show that an infinite-dimensional Banach space cannot be written as a countable union of finite-dimensional subspaces, and so cannot have a countable but infinite vector space basis.

   (d) Let $X = \ell_2(\mathbf{N})$ and let $C = \{x \in X \mid |x_n| \leq 2^{-n}\}$. Show

   $$X \neq \bigcup\{tC \mid t \geq 0\} \quad \text{but} \quad X = \operatorname{cl} \bigcup\{tC \mid t \geq 0\}.$$

   (e) Let $X = \ell_p(\mathbf{N})$ for $1 \leq p < \infty$. Let

   $$C = \{x \in X \mid |x_n| \leq 4^{-n}\},$$

   and let

   $$D = \{x \in X \mid x_n = 2^{-n}t, \ t \geq 0\}.$$

   Show $C \cap D = \{0\}$, and so

   $$T_{C \cap D}(0) = \{0\}$$

   but

   $$T_C(0) \cap T_D(0) = D.$$

   (In general, we need to require something like $0 \in \operatorname{core}(C - D)$, which fails in this example—see also Section 7.1, Exercise 6(h).)

(f) Show that every (separable) infinite-dimensional Banach space X contains a proper vector subspace Y with $\mathrm{cl}\,(Y) = X$. Thus show that in every such space there is a nonclosed convex set with empty interior whose closure has interior.

2. **(Unique subgradients)**

(a) Show that in any Banach space, a lower semicontinuous convex function is continuous at any point of Gâteaux differentiability.

(b) Let f be the indicator function of the nonnegative cone in $\ell_p(\mathbf{N})$ for $1 \le p < \infty$. Let x^* have strictly positive coordinates. Then prove zero is the unique element of $\partial f(x^*)$ but f is not continuous at x^*.

(c) Let $X = L_1[0,1]$ with Lebesgue measure. Consider the negative *Boltzmann–Shannon entropy*:

$$B(x) = \int_0^1 x(t) \log x(t)\, dt$$

for $x(t) \ge 0$ almost everywhere and $B(x) = +\infty$ otherwise. Show B is convex, nowhere continuous (but lower semicontinuous), and has a unique subgradient when $x > 0$ almost everywhere, namely $1 + \log x(t)$.

3. **(Norm-attaining functionals)**

(a) Find a non-norm-attaining functional in $c_0(\mathbf{N})$, in $\ell_\infty(\mathbf{N})$, and in $\ell_1(\mathbf{N})$.

(b) Consider the unit ball of $\ell_1(\mathbf{N})$ as a set C in $\ell_2(\mathbf{N})$. Show that C is closed and bounded and has empty interior. Determine the support points of C.

4. **(Support points)**

(a) Let X be separable and let $C \subset X$ be closed, bounded, and convex. Let $\{x_n \,|\, n \in \mathbf{N}\}$ be dense in C. Let $x^* = \sum_{n=1}^{\infty} 2^{-n} x_n$. Then any linear continuous functional f with $f(x^*) = \sup_C f$ must be constant on C and so x^* is not a proper support point of C.

(b) Show that every point of the nonnegative cone in the space $\ell_1(\mathbf{R})$ is a support point.

5. **(Sums of closed cones)**

(a) Let $X = \ell_2(\mathbf{N})$. Construct two closed convex cones (subspaces) S and T such that $S \cap T = \{0\}$ while $S^- + T^- \neq \ell_2(\mathbf{N})$. Deduce that the sum of closed subspaces may be dense.

(b) Let $X = \ell_2(\mathbf{N})$. Construct two continuous linear operators mapping X to itself such that each has dense range but their ranges intersect only at zero. (This is easier if one uses the Fourier identification of L_2 with ℓ_2.)

6. **(Epigraphical and tangential regularity)**

(a) Let C be a closed subset of a finite-dimensional space. Show that

$$d_C^-(0;h) = d_{K_C(0)}(h)$$

for all $h \in X$. Show also that d_C is regular at $x \in C$ if and only if C is regular at x.

(b) In every infinite-dimensional space X there is necessarily a sequence of unit vectors (u_n) such that $\inf\{\|u_n - u_m\| \,|\, n \neq m\} > 0$. Consider the set

$$C = \left\{4^{-n}\left(u_0 + \frac{1}{4}u_n\right) \,\Big|\, n = 0,1,2,\ldots\right\} \cup \{0\}.$$

Show the following results:

(i) $T_C(0) = K_C(0) = 0$.

(ii) For all $h \in X$,

$$\|h\| = d_C^\circ(0;h) = d_{K_C(0)}(h) \geq d_C^-(0;h) \geq -(-d)_C^\circ(0;h) = -\|h\|.$$

(iii) $d_C^\circ(0;u_0) = d_{K_C(0)}(u_0) > d_C^-(0;u_0)$.

(iv) $(-d)_C^\circ(0;u_0) > (-d)_C^-(0;u_0)$.

Conclude that C is regular at zero, but that neither d_C nor $-d_C$ is regular at zero.

(c) Establish that X is finite-dimensional if and only if regularity of sets coincides with regularity defined via distance functions.

7. **(Polyhedrality)** There is one particularly striking example where finite-dimensional results "lift" well to the infinite-dimensional setting. A set in a Banach space is a *polyhedron* if it is the intersection of a finite number of halfspaces. The definition of a *polytope* is unchanged since its span is finite-dimensional.

(a) Observe that polyhedra and polytopes coincide if and only if X is finite-dimensional.

(b) Show that a set is a polyhedron if and only if it is the sum of a finite-dimensional polyhedron and a closed finite-codimensional subspace of X.

So each polyhedron really "lives" in a finite-dimensional quotient space. In essence, this is why convex problems subject to a finite number of linear inequality constraints are so tractable. By contrast, note that Theorem 10.2.2(v) shows that even a ray may cause difficulties when the other set is not polyhedral.

8. **(Semicontinuity of separable functions on ℓ_p)** Let functions $\varphi_i : \mathbf{R} \to [0, +\infty]$ be given for $i \in \mathbf{N}$. Let the function F be defined on $X = \ell_p$ for $1 \leq p < \infty$ by

$$F(x) = \sum_i \varphi_i(x_i).$$

Relatedly, suppose the function $\varphi : \mathbf{R} \to (\infty, +\infty]$ is given, and consider the function

$$F_\varphi(x) = \sum_i \varphi(x_i).$$

(a) Show that F is convex and lower semicontinuous on X if and only if each φ_i is convex and lower semicontinuous on $\mathbf{R}$.

(b) Suppose $0 \in \operatorname{dom} F_\varphi$. Show that F_φ is convex and lower semicontinuous on X if and only if

(i) φ is convex and lower semicontinuous on $\mathbf{R}$, and

(ii) $\inf_{\mathbf{R}} \varphi = 0 = \varphi(0)$.

Thus, for $\varphi = \exp^*$ we have F_φ is a natural convex function which is not lower semicontinuous.

9. **(Sums of subspaces)**

(a) Let M and N be closed subspaces of X. Show that $M + N$ is closed when N is finite-dimensional. (Hint: First consider the case when $M \cap N = \{0\}$.)

(b) Let $X = \ell_p$ for $1 \leq p < \infty$. Define closed subspaces M and N by

$$M = \{x \mid x_{2n} = 0\} \text{ and } N = \{x \mid x_{2n} = 2^{-n} x_{2n-1}\}.$$

Show that $M + N$ is not closed. Observe that the same result obtains if M is replaced by the cone

$$K = \{x \mid x_{2n} = 0,\ x_{2n-1} \geq 0\}.$$

(Hint: Denote the unit vectors by (u_n). Let

$$x^n = \sum_{k<n} u_{2k-1} \text{ and } y^n = x^n + \sum_{k<n} 2^{-k} u_{2k}..$$

Then $x^n \in M$, $y^n \in N$, but $x^n - y^n \in M + N$ converges to $\sum_{k<\infty} 2^k u_{2k} \notin M + N$.)

(c) Relatedly, let $X := \ell_2$ and denote the unit vectors by (u_n). Suppose (α_n) is a sequence of positive real numbers with $1 > \alpha_n > 0$ and $\alpha_n \to 1$ sufficiently fast. Set

$$e_n = u_{2n-1}, \quad f_n = \alpha_n u_{2n-1} + \sqrt{1-\alpha_n^2}\, u_{2n}.$$

Consider the subspaces

$$M_1 = \operatorname{cl}\operatorname{span}\{e_1, e_2, \ldots\} \quad \text{and} \quad M_2 = \operatorname{cl}\operatorname{span}\{f_1, f_2, \ldots\}.$$

(i) Show $M_1 \cap M_2 = \{0\}$ and that the sum $M_1^\perp + M_2^\perp$ is dense in X but not closed.

(ii) Dually, show that $M_1^\perp \cap M_2^\perp = \{0\}$ and that the sum $M_1 + M_2$ is dense in X but not closed.

(iii) Find two continuous linear operators on X, T_1, and T_2 such that both have dense range but $R(T_1) \cap R(T_2) = \{0\}$. (Such subspaces are called *disjoint operator ranges.*)

10.4 Notes on Previous Chapters

Chapter 1: Background

In infinite-dimensional spaces, the separation theorem is known as the geometric version of the Hahn–Banach theorem and is one of the basic principles of functional analysis (for example, see [179] or [169]).

The Bolzano–Weierstrass theorem requires some assumption on the space to hold. One of its main applications can be stated as follows: *any lower semicontinuous real-valued function on a countably compact space (a space for which every countable open cover has a finite subcover) is bounded below and assumes its minimum* [169].

Exercise 13 in Section 1.1 (The relative interior) does not extend to the infinite-dimensional setting. As a simple counterexample, consider the nullspace H of a discontinuous linear functional. It is dense (and so not closed), convex, and nonempty but has empty relative interior. To overcome that difficulty, new definitions were given to classify sets that are big enough in some sense (compactly epi-Lipschitz sets, epi-Lipschitz-like sets, ...). All these definitions agree in finite dimensions. Another approach considers the "quasi relative interior" (see [34]).

Chapter 2: Inequality Constraints

First order necessary conditions hold in general spaces [105, 131]. However, one has to be careful about nearest point properties (Section 2.1, Exercise 8). We have existence and unicity of the nearest point to a closed convex set in a Hilbert space or for weakly compact convex sets in a strictly convex norm, but no longer without any assumptions. Often it is possible to deal with approximations by using density results such as the Bishop–Phelps theorem, which states: *the set of continuous linear functionals that attain their norm on the unit ball in a Banach space is norm dense in the dual* [153, 82].

Chapter 3: Fenchel Duality

The main results (Fenchel duality, Lagrange multiplier theorem) still hold in a very general setting [105, 131]. Properties of convex functions defined on Banach spaces are investigated in [153, 82]. Note that many properties of cones coincide in finite dimensions, while one has to be more careful in the infinite-dimensional setting (see [29]).

Chapter 4: Convex Analysis

Convexity in general linear spaces is studied in [98].

In infinite dimensions, Minkowski's theorem requires some assumption on the space since there may be bounded closed convex sets that do not have supporting hyperplanes (indeed, James' theorem states that a *Banach space is reflexive if and only if every continuous linear functional achieves its maximum on the closed unit ball*). Here is a generalization of Minkowski's theorem: Any weakly compact (respectively, closed bounded) convex subset of a Banach space (respectively, Banach space with the Radon–Nikodým property) is the closed convex hull of its strongly exposed points [63].

The Open mapping theorem extends to general Banach spaces (for example, see [179]). Similarly, the Moreau–Rockafellar theorem holds in general spaces [146, 165]. Furthermore, Lagrangian duality, which is equivalent to Fenchel duality, can be established in great generality [131, 105].

Chapter 5: Special Cases

The theory of linear operators is well-developed in infinite dimensions. See [149] for spectral theory in Banach algebras and [188] on compact operators. Many of the eigenvalue results have extensions for compact selfadjoint operators [37].

As we saw, closed convex processes are natural generalizations of linear mappings; in Banach space they admit open mapping, closed graph, and uniform boundedness theorems (see [5], and also [3] for applications to differential inclusions).

Chapter 6: Nonsmooth Optimization

All the calculus rules and the mean value theorem extend. Note however that Hadamard and Fréchet derivatives are no longer equal (see [55] and also this chapter). Density theorems extend (see [153]).

Various subdifferentials have been defined in infinite dimensions. See the recent survey [42] for how calculus rules and main properties are proved, as well as for some applications.

Chapter 7: Karush–Kuhn–Tucker Theory

Ekeland's variational principle holds in complete metric spaces (see [3]). It has numerous applications. For example, it is used in [153] to obtain the Brønsted–Rockafellar theorem, which in turn implies the Bishop–Phelps theorem (see also [82]).

The idea of a variational principle is to consider a point where the function is almost minimized and show it is the minimum of a slightly perturbed function. In Ekeland's variational principle, the perturbed function is obtained by adding a Lipschitz function to the original function. On the other hand, the Borwein–Preiss variational principle adds a smooth convex function. This latter principle is used in [42] to obtain several results on subdifferentials.

There are several other such principles. Examples include Stella's variational principle [56] (which adds a linear function), and the Deville–Godefroy–Zizler variational principle (see [153, §4]).

Metric regularity results extend to Banach space (see [145], for example). Constraint qualifications take various forms in infinite dimensions (see [105, 131] for some examples).

Chapter 8: Fixed Points

The Banach contraction principle holds in complete metric spaces. Moreover, in the Banach space setting, fixed point theorems hold not only for contractions but also for certain nonexpansive maps; see [63] for more precise formulations. See also [189] for a more extensive reference on fixed point theorems and applications.

Brouwer's theorem holds in Banach spaces for continuous self maps on a compact convex set [189]. Michael's selection theorem extends to appropriate multifunctions from a paracompact space into a Banach space [3], as does the Cellina selection theorem.

Chapter 9: More Nonsmooth Structure

A complete analogue of the Intrinsic Clarke subdifferential formula exists in separable Banach spaces, using Gâteaux derivatives and the notion of a "Haar null set", and due originally to Thibault: see [43]. The proximal normal formula has satisfactory extensions to Hilbert space, and, with appropriate modifications, to all reflexive spaces [43].

Not every Clarke subdifferential is minimal. In particular, there is a nonexpansive real function whose Clarke subdifferential is $[-1, 1]$ throughout **R** . An example is the integral of the characteristic function of a "ubiquitous" subset of **R** (one which intersects each interval (a, b) in a set of measure strictly between 0 and $b - a$): for the existence of such sets, see [177, 43]. A much deeper result [43, Cor 5.2.34] is that in every Banach space there is a Lipschitz function whose Clarke subdifferential is the dual ball throughout the unit ball. In such cases, every point is Clarke-critical, and the Clarke subdifferential is maximal and provides no information. In Banach space, the minimality of various Clarke subdifferentials (as in Theorem 9.2.4) is a rich subject that may be followed up in [43].

A Banach space is finite-dimensional if and only if every point has a nearest point in every norm-closed bounded set. To see this, in any infinite-dimensional space, take a set of unit vectors u_n with no norm-cluster point, and consider the sequence $\{(1 + n^{-1})u_n\}$. The existence or non-existence of a non-convex Chebyshev set in Hilbert space remains one of the most challenging problems in the field. While many partial results have been established, the overall state of play remains little different than when Asplund wrote his paper [2]. A recent and detailed discussion is to be found in [62].

Chapter 11

List of Results and Notation

11.1 Named Results

Section 1.1: Euclidean Spaces

Theorem 1.1.1 (Basic separation)
Theorem 1.1.2 (Bolzano–Weierstrass)
Proposition 1.1.3 (Weierstrass)
Exercise 4 (Radstrom cancellation)
Exercise 5 (Strong separation)
Exercise 6 (Recession cones)
Exercise 9 (Composing convex functions)
Exercise 10 (Convex growth conditions)
Exercise 11 (Accessibility lemma)
Exercise 12 (Affine sets)
Exercise 13 (The relative interior)

Section 1.2: Symmetric Matrices

Theorem 1.2.1 (Fan)
Proposition 1.2.4 (Hardy–Littlewood–Pólya)
Theorem 1.2.5 (Birkhoff)
Exercise 3 ($\mathbf{S}^3_+$ is not strictly convex)
Exercise 4 (A nonlattice ordering)
Exercise 5 (Order preservation)
Exercise 6 (Square-root iteration)
Exercise 7 (The Fan and Cauchy–Schwarz inequalities)
Exercise 12 (Fan's inequality)

Exercise 13 (A lower bound)
Exercise 14 (Level sets of perturbed log barriers)
Exercise 15 (Theobald's condition)
Exercise 16 (Generalizing Theobald's condition)
Exercise 17 (Singular values and von Neumann's lemma)

Section 2.1: Optimality Conditions

Proposition 2.1.1 (First order necessary condition)
Proposition 2.1.2 (First order sufficient condition)
Corollary 2.1.3 (First order conditions, linear constraints)
Theorem 2.1.5 (Second order conditions)
Theorem 2.1.6 (Basic separation)
Exercise 2 (Examples of normal cones)
Exercise 3 (Self-dual cones)
Exercise 4 (Normals to affine sets)
Exercise 6 (The Rayleigh quotient)
Exercise 8 (Nearest points)
Exercise 8(e) (Projection on $\mathbf{R}^n_+$ and $\mathbf{S}^n_+$)
Exercise 9 (Coercivity)
Exercise 11 (Kirchhoff's law)
Exercise 12 (Matrix completion)
Exercise 13 (BFGS update)
Exercise 15 (Nearest polynomial with a given root)

Section 2.2: Theorems of the Alternative

Theorem 2.2.1 (Gordan)
Lemma 2.2.7 (Farkas)
Exercise 5 (Carathéodory's theorem)
Exercise 7 (Ville's theorem)
Exercise 8 (Stiemke's theorem)
Exercise 9 (Schur-convexity)

Section 2.3: Max-functions

Proposition 2.3.2 (Directional derivatives of max-functions)
Theorem 2.3.6 (Fritz John conditions)
Assumption 2.3.7 (The Mangasarian–Fromovitz constraint qualification)
Theorem 2.3.8 (Karush–Kuhn–Tucker conditions)
Exercise 2 (Failure of Karush–Kuhn–Tucker)
Exercise 3 (Linear independence implies Mangasarian–Fromovitz)

Exercise 5 (Cauchy–Schwarz and steepest descent)
Exercise 6 (Hölder's inequality)
Exercise 8 (Minimum volume ellipsoid)

Section 3.1: Subgradients and Convex Functions

Proposition 3.1.1 (Sublinearity)
Proposition 3.1.2 (Sublinearity of the directional derivative)
Proposition 3.1.5 (Subgradients at optimality)
Proposition 3.1.6 (Subgradients and the directional derivative)
Theorem 3.1.8 (Max formula)
Corollary 3.1.10 (Differentiability of convex functions)
Theorem 3.1.11 (Hessian characterization of convexity)
Exercise 2 (Core versus interior)
Exercise 4 (Subgradients and normal cones)
Exercise 8 (Subgradients of norm)
Exercise 9 (Subgradients of maximum eigenvalue)
Exercise 12 (Recognizing convex functions)
Exercise 13 (Local convexity)
Exercise 14 (Examples of convex functions)
Exercise 15 (Examples of convex functions)
Exercise 16 (Bregman distances)
Exercise 17 (Convex functions on $\mathbf{R}^2$)
Exercise 19 (Domain of subdifferential)
Exercise 20 (Monotonicity of gradients)
Exercise 21 (The log barrier)
Exercise 24 (Minimizers of essentially smooth functions)
Exercise 25 (Convex matrix functions)
Exercise 26 (Log-convexity)
Exercise 27 (Maximum entropy)
Exercise 28 (DAD problems)
Exercise 29 (Relativizing the Max formula)

Section 3.2: The Value Function

Proposition 3.2.3 (Lagrangian sufficient conditions)
Theorem 3.2.8 (Lagrangian necessary conditions)
Exercise 5 (Mixed constraints)
Exercise 6 (Extended convex functions)
Exercise 7 (Nonexistence of multiplier)
Exercise 8 (Duffin's duality gap)
Exercise 9 (Karush–Kuhn–Tucker vectors)
Exercise 11 (Normals to epigraphs)
Exercise 12 (Normals to level sets)

Exercise 13 (Subdifferential of max-function)
Exercise 14 (Minimum volume ellipsoid)

Section 3.3: The Fenchel Conjugate

Proposition 3.3.3 (Log barriers)
Proposition 3.3.4 (Fenchel–Young inequality)
Theorem 3.3.5 (Fenchel duality and convex calculus)
Corollary 3.3.11 (Fenchel duality for linear constraints)
Proposition 3.3.12 (Self-dual cones)
Corollary 3.3.13 (Krein–Rutman polar cone calculus)
Theorem 3.3.14 (Bipolar cone)
Theorem 3.3.15 (Pointed cones)
Exercise 2 (Quadratics)
Exercise 4 (Self-conjugacy)
Exercise 5 (Support functions)
Exercise 7 (Maximum entropy example)
Exercise 9 (Fenchel duality and convex calculus)
Exercise 10 (Normals to an intersection)
Exercise 11 (Failure of convex calculus)
Exercise 12 (Infimal convolution)
Exercise 13 (Applications of Fenchel duality)
Exercise 13(a) (Sandwich theorem)
Exercise 13(c) (Pshenichnii–Rockafellar conditions)
Exercise 13(e) (Hahn–Banach extension)
Exercise 15 (Bipolar theorem)
Exercise 16 (Sums of closed cones)
Exercise 17 (Subdifferential of a max-function)
Exercise 18 (Order convexity)
Exercise 19 (Order convexity of inversion)
Exercise 20 (Pointed cones and bases)
Exercise 21 (Order-subgradients)
Exercise 22 (Linearly constrained examples)
Exercise 22(a) Separable problems
Exercise 22(a)(i) (Nearest points in polyhedrons)
Exercise 22(a)(ii) (Analytic centre)
Exercise 22(a)(iii) (Maximum entropy)
Exercise 22(b) (BFGS update)
Exercise 22(c) (DAD problem)
Exercise 23 (Linear inequalities)
Exercise 24 (Symmetric Fenchel duality)
Exercise 25 (Divergence bounds)

Section 4.1: Continuity of Convex Functions

Theorem 4.1.1 (Local boundedness)
Theorem 4.1.3 (Convexity and continuity)
Theorem 4.1.4 (Core and interior)
Theorem 4.1.5 (Bipolar set)
Theorem 4.1.6 (Supporting hyperplane)
Theorem 4.1.8 (Minkowski)
Exercise 1 (Points of continuity)
Exercise 2 (Equivalent norms)
Exercise 3 (Examples of polars)
Exercise 4 (Polar sets and cones)
Exercise 5 (Polar sets)
Exercise 6 (Polar sets and strict separation)
Exercise 7 (Polar calculus)
Exercise 8 (Polar calculus)
Exercise 9 (Open mapping theorem)
Exercise 10 (Conical absorption)
Exercise 11 (Hölder's inequality)
Exercise 12 (Pareto minimization)
Exercise 12(d) (Scalarization)
Exercise 13 (Existence of extreme points)
Exercise 16 (A converse of Minkowski's theorem)
Exercise 17 (Extreme points)
Exercise 18 (Exposed points)
Exercise 19 (Tangency conditions)
Exercise 20 (Properties of the relative interior)
Exercise 22 (Birkhoff's theorem)

Section 4.2: Fenchel Biconjugation

Theorem 4.2.1 (Fenchel biconjugation)
Corollary 4.2.3 (Support functions)
Theorem 4.2.4 (Moreau–Rockafellar)
Theorem 4.2.5 (Strict-smooth duality)
Proposition 4.2.7 (Lower semicontinuity and closure)
Exercise 2 (Lower semicontinuity and closedness)
Exercise 3 (Pointwise maxima)
Exercise 5 (Midpoint convexity)
Exercise 7 (Inverse of subdifferential)
Exercise 8 (Closed subdifferential)
Exercise 9 (Support functions)
Exercise 10 (Almost homogeneous functions)
Exercise 12 (Compact bases for cones)

Exercise 14 (Lower semicontinuity and closure)
Exercise 16 (Von Neumann's minimax theorem)
Exercise 17 (Recovering primal solutions)
Exercise 19 (Strict-smooth duality)
Exercise 20 (Logarithmic homogeneity)
Exercise 21 (Cofiniteness)
Exercise 22 (Computing closures)
Exercise 23 (Recession functions)
Exercise 24 (Fisher information function)
Exercise 25 (Finiteness of biconjugate)
Exercise 26 (Self-dual cones)
Exercise 27 (Conical open mapping)
Exercise 28 (Expected surprise)

Section 4.3: Lagrangian Duality

Proposition 4.3.5 (Dual optimal value)
Corollary 4.3.6 (Zero duality gap)
Theorem 4.3.7 (Dual attainment)
Theorem 4.3.8 (Primal attainment)
Exercise 1 (Weak duality)
Exercise 3 (Slater and compactness)
Exercise 4 (Examples of duals)
Exercise 5 (Duffin's duality gap, continued)
Exercise 8 (Mixed constraints)
Exercise 9 (Fenchel and Lagrangian duality)
Exercise 10 (Trust region subproblem duality)
Exercise 12 (Conjugates of compositions)
Exercise 13 (A symmetric pair)
Exercise 14 (Convex minimax theory)

Section 5.1: Polyhedral Convex Sets and Functions

Proposition 5.1.1 (Polyhedral functions)
Proposition 5.1.3 (Finitely generated functions)
Theorem 5.1.7 (Polyhedrality)
Proposition 5.1.8 (Polyhedral algebra)
Corollary 5.1.9 (Polyhedral Fenchel duality)
Corollary 5.1.10 (Mixed Fenchel duality)
Exercise 6 (Tangents to polyhedra)
Exercise 7 (Polyhedral algebra)
Exercise 9 (Polyhedral cones)
Exercise 11 (Generalized Fenchel duality)
Exercise 12 (Relativizing Mixed Fenchel duality)

Exercise 13 (Geometric programming)

Section 5.2: Functions of Eigenvalues

Theorem 5.2.2 (Spectral conjugacy)
Corollary 5.2.3 (Davis)
Corollary 5.2.4 (Spectral subgradients)
Corollary 5.2.5 (Spectral differentiability)
Exercise 4 (Examples of convex spectral functions)
Exercise 8 (Orthogonal invariance)
Exercise 10 (Fillmore–Williams)
Exercise 11 (Semidefinite complementarity)
Exercise 12 (Eigenvalue sums)
Exercise 13 (Davis' theorem)
Exercise 14 (DC functions)

Section 5.3: Duality for Linear and Semidefinite Programming

Corollary 5.3.6 (Cone programming duality)
Corollary 5.3.7 (Linear programming duality)
Corollary 5.3.10 (Semidefinite programming duality)
Exercise 3 (Linear programming duality gap)
Exercise 7 (Complementary slackness)
Exercise 8 (Semidefinite programming duality)
Exercise 9 (Semidefinite programming duality gap)
Exercise 10 (Central path)
Exercise 11 (Semidefinite central path)
Exercise 12 (Relativizing cone programming duality)

Section 5.4: Convex Process Duality

Proposition 5.4.1 (Openness and lower semicontinuity)
Theorem 5.4.8 (Adjoint process duality)
Theorem 5.4.10 (Norm duality)
Theorem 5.4.12 (Open mapping)
Theorem 5.4.13 (Closed graph)
Exercise 1 (Inverse multifunctions)
Exercise 2 (Convex images)
Exercise 5 (LSC and lower semicontinuity)
Exercise 7 (Biconjugation)
Exercise 14 (Linear maps)
Exercise 15 (Normal cones)
Exercise 15(c) (Krein–Grossberg)

Exercise 16 (Inverse boundedness)
Exercise 17 (Localization)
Exercise 18 (Cone duality)
Exercise 19 (Order epigraphs)
Exercise 20 (Condition number)
Exercise 21 (Distance to inconsistency)

Section 6.1: Generalized Derivatives

Corollary 6.1.2 (Nonsmooth max formulae)
Theorem 6.1.5 (Nonsmooth calculus)
Theorem 6.1.8 (Nonsmooth necessary condition)
Exercise 1 (Examples of nonsmooth derivatives)
Exercise 2 (Continuity of Dini derivative)
Exercise 4 (Surjective Dini subdifferential)
Exercise 6 (Failure of Dini calculus)
Exercise 9 (Mean value theorem)
Exercise 9(b) (Monotonicity and convexity)
Exercise 10 (Max-functions)
Exercise 11 (Order statistics)

Section 6.2: Regularity and Strict Differentiability

Proposition 6.2.1 (Unique Michel–Penot subgradient)
Theorem 6.2.2 (Regularity of convex functions)
Theorem 6.2.3 (Strict differentiability)
Theorem 6.2.4 (Unique Clarke subgradient)
Theorem 6.2.5 (Intrinsic Clarke subdifferential)
Exercise 2 (Regularity and nonsmooth calculus)
Exercise 9 (Mixed sum rules)
Exercise 10 (Types of differentiability)
Exercise 12 (Closed subdifferentials)
Exercise 13 (Dense Dini subgradients)
Exercise 14 (Regularity of order statistics)
Exercise 15 (Subdifferentials of eigenvalues)
Exercise 15(h) (Isotonicity of λ)

Section 6.3: Tangent Cones

Proposition 6.3.2 (Exact penalization)
Theorem 6.3.6 (Tangent cones)
Corollary 6.3.7 (Convex tangent cone)
Corollary 6.3.9 (Nonsmooth necessary conditions)
Proposition 6.3.10 (Contingent necessary condition)

Exercise 1 (Exact penalization)
Exercise 2 (Distance function)
Exercise 3 (Examples of tangent cones)
Exercise 4 (Topology of contingent cone)
Exercise 5 (Topology of Clarke cone)
Exercise 6 (Intrinsic tangent cones)
Exercise 8 (Isotonicity)
Exercise 9 (Products)
Exercise 10 (Tangents to graphs)
Exercise 11 (Graphs of Lipschitz functions)
Exercise 12 (Proper Pareto minimization)
Exercise 12(c) (Scalarization)
Exercise 13 (Boundary properties)
Exercise 13(f) (Nonconvex separation)
Exercise 14 (Pseudoconvexity and sufficiency)
Exercise 15 (No ideal tangent cone exists)
Exercise 16 (Distance function)

Section 6.4: The Limiting Subdifferential

Theorem 6.4.1 (Fuzzy sum rule)
Theorem 6.4.4 (Limiting subdifferential sum rule)
Exercise 3 (Local minimizers)
Exercise 4 (Failure of sum rule)
Exercise 7 (Limiting and Clarke subdifferentials)
Exercise 8 (Topology of limiting subdifferential)
Exercise 9 (Tangents to graphs)
Exercise 10 (Composition)
Exercise 10(e) (Composition rule)
Exercise 10(f) (Mean value theorem)
Exercise 10(g) (Max rule)
Exercise 11 (Viscosity subderivatives)
Exercise 12 (Order statistic)

Section 7.1: An Introduction to Metric Regularity

Theorem 7.1.2 (Ekeland variational principle)
Theorem 7.1.5 (Surjectivity and metric regularity)
Theorem 7.1.6 (Liusternik)
Exercise 2 (Lipschitz extension)
Exercise 3 (Closure and the Ekeland principle)
Exercise 6 (Transversality)
Exercise 6(g) (Guignard)
Exercise 7 (Liusternik via inverse functions)

Section 7.2: The Karush–Kuhn–Tucker Theorem

Assumption 7.2.3 (The Mangasarian–Fromovitz constraint qualification)
Theorem 7.2.9 (Karush–Kuhn–Tucker conditions)
Exercise 1 (Linear independence implies Mangasarian–Fromovitz)
Exercise 3 (Linear constraints)
Exercise 4 (Bounded multipliers)
Exercise 5 (Slater condition)
Exercise 6 (Largest eigenvalue)
Exercise 7 (Largest singular value)
Exercise 7(f) (Jordan)
Exercise 8 (Hadamard's inequality)
Exercise 9 (Nonexistence of multipliers)
Exercise 10 (Guignard optimality conditions)
Exercise 11 (Quadratic penalties)

Section 7.3: Metric Regularity and the Limiting Subdifferential

Theorem 7.3.3 (Limiting subdifferential and regularity)
Corollary 7.3.4 (Surjectivity and metric regularity)
Corollary 7.3.6 (Distance to level sets)
Exercise 3 (Metric regularity and openness)
Exercise 4 (Limiting normals and distance functions)
Exercise 5 (Normals to products)
Exercise 8 (Limiting versus Clarke conditions)
Exercise 9 (Normals to level sets)

Section 7.4: Second Order Conditions

Theorem 7.4.2 (Second order necessary conditions)
Theorem 7.4.8 (Second order sufficient condition)
Exercise 1 (Higher order conditions)
Exercise 2 (Uniform multipliers)
Exercise 3 (Standard second order necessary conditions)
Exercise 4 (Narrow and broad critical cones are needed)
Exercise 5 (Standard second order sufficient conditions)
Exercise 6 (Guignard-type conditions)

Section 8.1: The Brouwer Fixed Point Theorem

Theorem 8.1.2 (Banach contraction)

Theorem 8.1.3 (Brouwer)
Theorem 8.1.4 (Stone–Weierstrass)
Theorem 8.1.5 (Change of variable)
Theorem 8.1.6 (Retraction)
Exercise 1 (Banach iterates)
Exercise 2 (Nonexpansive maps)
Exercise 2(c) (Browder–Kirk)
Exercise 3 (Non-uniform contractions)
Exercise 11 (Convex sets homeomorphic to the ball)
Exercise 12 (A nonclosed nonconvex set with the fixed point property)
Exercise 13 (Change of variable and Brouwer)
Exercise 14 (Brouwer and inversion)
Exercise 15 (Knaster–Kuratowski–Mazurkiewicz principle)
Exercise 15(b) (KKM implies Brouwer)
Exercise 15(c) (Brouwer implies KKM)
Exercise 16 (Hairy ball theorem)
Exercise 16(h) (Hedgehog theorem)
Exercise 17 (Borsuk–Ulam theorem)
Exercise 17(d) (Borsuk–Ulam implies Brouwer)
Exercise 18 (Generalized Riesz lemma)
Exercise 19 (Riesz implies Borsuk)

Section 8.2: Selection and the Kakutani–Fan Fixed Point Theorem

Theorem 8.2.1 (Maximal monotonicity)
Theorem 8.2.2 (Kakutani–Fan)
Theorem 8.2.3 (General definition of compactness)
Theorem 8.2.4 (Partition of unity)
Theorem 8.2.5 (Cellina)
Theorem 8.2.8 (Michael)
Exercise 1 (USC and continuity)
Exercise 2 (Minimum norm)
Exercise 3 (Closed versus USC)
Exercise 4 (Composition)
Exercise 5 (Clarke subdifferential)
Exercise 6 (USC images of compact sets)
Exercise 7 (Partitions of unity)
Exercise 9 (Michael's theorem)
Exercise 10 (Hahn–Katetov–Dowker sandwich theorem)
Exercise 10(b) (Urysohn lemma)
Exercise 11 (Continuous extension)
Exercise 12 (Generated cuscos)

Exercise 13 (Multifunctions containing cuscos)
Exercise 14 (Singleton points)
Exercise 15 (Skew symmetry)
Exercise 16 (Monotonicity)
Exercise 16(a) (Inverses)
Exercise 16(c) (Applying maximality)
Exercise 16(d) (Maximality and closedness)
Exercise 16(e) (Continuity and maximality)
Exercise 16(g) (Subdifferentials)
Exercise 16(h) (Local boundedness)
Exercise 16(j) (Maximality and cuscos)
Exercise 16(k) (Surjectivity and growth)
Exercise 17 (Single-valuedness and maximal monotonicity)

Section 8.3: Variational Inequalities

Theorem 8.3.6 (Solvability of variational inequalities)
Theorem 8.3.7 (Noncompact variational inequalities)
Theorem 8.3.13 (Linear programming and variational inequalities)
Exercise 4 (Variational inequalities containing cuscos)
Exercise 7 (Monotone complementarity problems)
Exercise 9 (Iterative solution of OCP)
Exercise 10 (Fan minimax inequality)
Exercise 10(d) (Nash equilibrium)
Exercise 10(e) (Minimax)
Exercise 11 (Bolzano–Poincaré–Miranda intermediate value theorem)
Exercise 12 (Coercive cuscos)
Exercise 13 (Monotone variational inequalities)
Exercise 13(d) (Hypermaximal $\Leftrightarrow$ maximal)
Exercise 13(e) (Resolvent)
Exercise 13(f) (Maximality and surjectivity)
Exercise 14 (Semidefinite complementarity)
Exercise 15 (Fitzpatrick representatives)
Exercise 16 (Convexified representatives)
Exercise 17 (Monotone multifunctions with bounded range)
Exercise 18 (Maximal monotone extension)
Exercise 19 (Brouwer via Debrunner-Flor)

Section 9.1: Rademacher's Theorem

Theorem 9.1.2 (Rademacher)
Exercise 2 (Rademacher's theorem for regular functions)

Exercise 3 (Intrinsic Clarke subdifferential formula)
Exercise 4 (Generalized Jacobian)
Exercise 4(b) (Mean value theorem)
Exercise 4(c) (Chain rule)

Section 9.2: Proximal Normals and Chebyshev Sets

Theorem 9.2.1 (Proximal normal formula)
Proposition 9.2.2 (Projections)
Theorem 9.2.3 (Differentiability of distance functions)
Theorem 9.2.4 (Distance subdifferentials are minimal)
Theorem 9.2.5 (Convexity of Chebyshev sets)
Exercise 2 (Projections)
Exercise 3 (Proximal normals are normals)
Exercise 4 (Unique nearest points)
Exercise 5 (Nearest points and Clarke subgradients)
Exercise 6 (Differentiability of distance functions)
Exercise 7 (Proximal normal formula via Rademacher)
Exercise 8 (Minimality of convex subdifferentials)
Exercise 9 (Smoothness and DC functions)
Exercise 10 (Subdifferentials at minimizers)
Exercise 11 (Proximal normals and the Clarke subdifferential)
Exercise 12 (Continuity of the projection)
Exercise 13 (Suns)
Exercise 14 (Basic Ekeland variational principle)
Exercise 15 (Approximately convex sets)
Exercise 16 (Chebyshev sets and approximate convexity)
Exercise 17 (Smoothness and biconjugacy)
Exercise 18 (Chebyshev sets and differentiability)
Exercise 19 (Chebyshev convexity via conjugacy)
Exercise 20 (Unique furthest points)
Exercise 21 (Chebyshev convexity via inversion)

Section 9.3: Amenable Sets and Prox-Regularity

Proposition 9.3.2 (Prox-regularity and projections)
Theorem 9.3.5 (Chain rule for sets)
Corollary 9.3.11 (Amenable sets)
Theorem 9.3.12 (Amenability and prox-regularity)
Exercise 2 (Persistence of amenability)
Exercise 4 (Amenability and Mangasarian-Fromowitz)

Section 9.4: Partly Smooth Sets

Proposition 9.4.1 (Partly smooth orthant)
Theorem 9.4.2 (Partial smoothness of $\mathbf{S}^n_+$)
Proposition 9.4.3 (Partial smoothness and cones)
Theorem 9.4.4 (Partly smooth chain rule)
Corollary 9.4.5 (Inequalities and partial smoothness)
Theorem 9.4.9 (Strong critical points)
Exercise 6 (Strict complementarity)
Exercise 7 (Constructive second order condition)
Exercise 9 (Prox-regularity of manifolds)
Exercise 10 (Necessity of prox-regularity)

Section 10.2: Finite Dimensionality

Theorem 10.2.1 (Closure, continuity, and compactness)
Theorem 10.2.2 (Support and separation)

Section 10.3: Counterexamples and Exercises

Exercise 1 (Absorbing sets)
Exercise 2 (Unique subgradients)
Exercise 3 (Norm attaining functionals)
Exercise 4 (Support points)
Exercise 5 (Sums of closed cones)
Exercise 6 (Epigraphical and tangential regularity)
Exercise 7 (Polyhedrality)
Exercise 8 (Semicontinuity of separable functions on ℓ_p)
Exercise 9 (Sums of subspaces)

11.2 Notation

Section 1.1: Euclidean Spaces

$\mathbf{E}$: a Euclidean space

$\mathbf{R}$: the reals

$\langle \cdot, \cdot \rangle$: inner product

$\mathbf{R}^n$: the real n-vectors

$\| \cdot \|$: the norm

B: the unit ball

$C + D$, $C - D$, ΛC: set sum, difference, and scalar product

$\times$: Cartesian product

$\mathbf{R}_+$: the nonnegative reals

$\mathbf{R}^n_+$: the nonnegative orthant

$\mathbf{R}^n_{\geq}$: the vectors with nonincreasing components

span: linear span

conv: convex hull

int: interior

$\mathbf{R}^n_{++}$: the interior of the nonnegative orthant

$\to$, lim: (vector) limit

cl: closure

bd: boundary

D^c: set complement

A^*: adjoint map

$N(\cdot)$: null space

$G^{\perp}$: orthogonal complement

inf: infimum

sup: supremum

$\circ$: composition of functions

$0^+(\cdot)$: recession cone

aff: affine hull

ri: relative interior

Section 1.2: Symmetric Matrices

$\mathbf{S}^n$: the $n \times n$ real symmetric matrices

$\mathbf{S}^n_+$: the positive semidefinite matrices

$\leq, \ <, \ \geq, \ >$: componentwise ordering

$\preceq, \ \prec, \ \succeq, \ \succ$: semidefinite ordering

$\mathbf{S}^n_{++}$: the positive definite matrices

I: identity matrix

tr: trace

$\lambda_i(\cdot)$: ith largest eigenvalue

Diag $(\cdot)$: diagonal matrix

det: determinant

$\mathbf{O}^n$: the orthogonal matrices

$X^{1/2}$: matrix square-root

$[\cdot]$: nonincreasing rearrangement

$\mathbf{P}^n$: the permutation matrices

$\mathbf{\Gamma}^n$: the doubly stochastic matrices

$\mathbf{M}^n$: the $n \times n$ real matrices

$\sigma_i(\cdot)$: ith largest singular value

Section 2.1: Optimality Conditions

$f'(\cdot;\cdot)$: directional derivative

∇: Gâteaux derivative

$N_C(\cdot)$: normal cone

∇^2: Hessian

y^+: positive part of vector

P_C: projection on C

Section 2.2: Theorems of the Alternative

$P_{\mathbf{Y}}$: orthogonal projection

Section 2.3: Max-functions

$I(\cdot)$: active set

$\mathbf{N}$: the natural numbers

$L(\cdot;\cdot)$: Lagrangian

Section 3.1: Subgradients and Convex Functions

δ_C: indicator function

dom: domain

lin: lineality space

core: core

∂: subdifferential

$\mathrm{dom}\,\partial f$: domain of subdifferential

$\Gamma(\cdot)$: gamma function

Section 3.2: The Value Function

$L(\cdot;\cdot)$: Lagrangian

$v(\cdot)$: value function

epi: epigraph

dom: domain

Section 3.3: The Fenchel Conjugate

h^*: conjugate

lb: log barrier on $\mathbf{R}^n_{++}$

ld: log det on $\mathbf{S}^n_{++}$

cont: points of continuity

K^-: polar cone

$T_C(\cdot)$: (convex) tangent cone

$\odot$: infimal convolution

d_C: distance function

g_*: concave conjugate

Section 4.1: Continuity of Convex Functions

Δ: the simplex

γ_C: gauge function

C°: polar set

$\text{ext}\,(\cdot)$: extreme points

Section 4.2: Fenchel Biconjugation

$\liminf h(x^r)$: $\liminf$ of sequence

$\text{cl}\, h$: closure of function

$0^+ f$: recession function

$h_\circ$: concave polar

Section 4.3: Lagrangian Duality

Φ: dual function

Section 5.4: Convex Process Duality

$D(\cdot)$: domain of multifunction

$\Phi(C)$: image under a multifunction

$R(\cdot)$: range of multifunction

$G(\cdot)$: graph of multifunction

$B_{\mathbf{E}}$: unit ball in $\mathbf{E}$

Φ^{-1}: inverse multifunction

Φ^*: adjoint multifunction

$\|\cdot\|_l$: lower norm

$\|\cdot\|_u$: upper norm

Section 6.1: Generalized Derivatives

$f^-(\cdot;\cdot)$: Dini directional derivative

$f^\circ(\cdot;\cdot)$: Clarke directional derivative

$f^\diamond(\cdot;\cdot)$: Michel–Penot directional derivative

$\partial_\circ$: Clarke subdifferential

∂_-: Dini subdifferential

$\partial_\diamond$: Michel–Penot subdifferential

$f \vee g$: pointwise maximum of functions

Section 6.3: Tangent Cones

d_S: distance function

$T_S(\cdot)$: Clarke tangent cone

$K_S(\cdot)$: contingent cone

$N_S(\cdot)$: Clarke normal cone

$[x, y]$, (x, y): line segments

star: star of a set

$P_S(\cdot)$: pseudotangent cone

Section 6.4: The Limiting Subdifferential

$f^-(\cdot;\cdot)$: Dini directional derivative

∂_-: Dini subdifferential

∂_a: limiting subdifferential

$N_S^a(\cdot)$: limiting normal cone

$U(f; x; \delta)$: f-neighbourhood of x.

Section 7.1: An Introduction to Metric Regularity

$h|_S$: h restricted to S

Section 7.2: The Karush–Kuhn–Tucker Theorem

sgn: sign function

Section 7.4: Second Order Conditions

$L(\mathbf{E}, \mathbf{Y})$: the linear maps from $\mathbf{E}$ to $\mathbf{Y}$

$\nabla^2 h(\bar{x})$: second derivative

$\nabla^2 h(\bar{x})(v, v)$: evaluated second derivative

$C(\bar{x})$: narrow critical cone

$L(\cdot), \overline{L}(\cdot)$: Lagrangians

$\overline{C}(\bar{x})$: broad critical cone

Section 8.1: The Brouwer Fixed Point Theorem

γ_f: contraction constant

$C^{(1)}$: continuously differentiable

S: unit sphere

S_n: unit sphere in $\mathbf{R}^{n+1}$

$S(U)$: unit sphere in U

Section 8.2: Selection and the Kakutani–Fan Fixed Point Theorem

G_δ: countable intersection of open sets

Section 8.3: Variational Inequalities

$VI(\Omega, C)$: variational inequality

$\mathcal{F}_\Phi$: Fitzpatrick function

$\mathcal{P}_\Phi$: convexified representative

Section 9.1: Rademacher's Theorem

$D_h^+ f(\cdot)$: upper Dini derivative

$D_h^- f(\cdot)$: lower Dini derivative

$\partial_Q h(\cdot)$: Clarke generalized Jacobian

$J_h(\cdot)$: Clarke generalized Jacobian

Section 9.2: Proximal Normals and Chebyshev Sets

$N_S^p(\cdot)$: proximal normal cone

ι: inversion in the unit sphere

Section 10.1: Euclidean Space

X: a real Banach space

X^*: continuous dual space

Section 10.2: Finite Dimensionality

x^*: a continuous linear functional

$B(X)$: closed unit ball

β, G, H, WH, F: a bornology, Gâteaux, Hadamard, weak Hadamard, Fréchet

∂^β: bornological subdifferential

∇^β: bornological derivative

$\ell_p(\mathbf{N}), c_0(\mathbf{N})$: classical sequence spaces

$\|\cdot\|_*$: dual norm

Bibliography

[1] T.M. Apostol. *Linear Algebra: A First Course, with Applications to Differential Equations.* Wiley, New York, 1997.

[2] E. Asplund. Cebysev sets in Hilbert space. *Transactions of the American Mathematical Society*, 144:235–240, 1969.

[3] J.-P. Aubin. *Viability Theory.* Birkhäuser, Boston, 1991.

[4] J.-P. Aubin and A. Cellina. *Differential Inclusions.* Springer-Verlag, Berlin, 1984.

[5] J.-P. Aubin and H. Frankowska. *Set-Valued Analysis.* Birkhäuser, Boston, 1990.

[6] M. Avriel. *Nonlinear Programming.* Prentice-Hall, Englewood Cliffs, N.J., 1976.

[7] S. Banach. Sur les opérations dans les ensembles abstraits et leur application aux équations intégrales. *Fundamenta Mathematicae*, 3:133–181, 1922.

[8] H.H. Bauschke, J.M. Borwein, and P. Tseng. Bounded linear regularity, strong CHIP, and CHIP are distinct properties. *Journal of Convex Analysis*, 7:395–412, 2000.

[9] M.S. Bazaraa and C.M. Shetty. *Nonlinear Programming.* Wiley, New York, 1979.

[10] R. Bellman. *Introduction to Matrix Analysis.* SIAM, Philadelphia, 1997.

[11] A. Ben-Tal and J. Zowe. Necessary and sufficient conditions for a class of nonsmooth minimization problems. *Mathematical Programming*, 24:70–91, 1982.

[12] A. Ben-Tal and J. Zowe. A unified theory of first-order and second-order conditions for extremum problems. *Mathematical Programming Study*, 19:39–76, 1982.

[13] C. Berge. *Espaces Topologiques et Fonctions Multivoques*. Dunod, Paris, 1959.

[14] D.N. Bessis and F.H. Clarke. Partial subdifferentials, derivates and Rademacher's Theorem. *Transactions of the American Mathematical Society*, 351:2899–2926, 1999.

[15] G. Birkhoff. Tres observaciones sobre el algebra lineal. *Universidad Nacionale Tucamán Revista*, 5:147–151, 1946.

[16] B. Bollobás. *Linear Analysis*. Cambridge University Press, Cambridge, U.K., 1999.

[17] K. Borsuk. Drei Sätze über die n-dimensionale Euklidische Sphäre. *Fundamenta Mathematicae*, 21:177–190, 1933.

[18] D. Borwein, J.M. Borwein, and P. Maréchal. Surprise maximization. *American Mathematical Monthly*, 107:517–527, 2000.

[19] J.M. Borwein. The direct method in semi-infinite programming. *Mathematical Programming*, 21:301–318, 1981.

[20] J.M. Borwein. Continuity and differentiability properties of convex operators. *Proceedings of the London Mathematical Society*, 44:420–444, 1982.

[21] J.M. Borwein. Necessary and sufficient conditions for quadratic minimality. *Numerical Functional Analysis and Optimization*, 5:127–140, 1982.

[22] J.M. Borwein. A note on the existence of subgradients. *Mathematical Programming*, 24:225–228, 1982.

[23] J.M. Borwein. Adjoint process duality. *Mathematics of Operations Research*, 8:403–434, 1983.

[24] J.M. Borwein. Norm duality for convex processes and applications. *Journal of Optimization Theory and Applications*, 48:53–64, 1986.

[25] J.M. Borwein. Stability and regular points of inequality systems. *Journal of Optimization Theory and Applications*, 48:9–52, 1986.

[26] J.M. Borwein. The linear order-complementarity problem. *Mathematics of Operations Research*, 14:534–558, 1989.

[27] J.M. Borwein. Minimal cuscos and subgradients of Lipschitz functions. In J.-B. Baillon and M. Thera, editors, *Fixed Point Theory and its Applications*, Pitman Lecture Notes in Mathematics, pages 57–82, Essex, U.K., 1991. Longman.

[28] J.M. Borwein. A generalization of Young's l^p inequality. *Mathematical Inequalities and Applications*, 1:131–136, 1997.

[29] J.M. Borwein. Cones and orderings. Technical report, CECM, Simon Fraser University, 1998.

[30] J.M. Borwein. Maximal monotonicity via convex analysis. *Journal of Convex Analysis*, 2005. To appear.

[31] J.M. Borwein and S. Fitzpatrick. Characterization of Clarke subgradients among one-dimensional multifunctions. In *Proceedings of the Optimization Miniconference II*, pages 61–73. University of Balarat Press, 1995.

[32] J.M. Borwein, S.P. Fitzpatrick, and J.R. Giles. The differentiability of real functions on normed linear spaces using generalized gradients. *Journal of Optimization Theory and Applications*, 128:512–534, 1987.

[33] J.M. Borwein and J.R. Giles. The proximal normal formula in Banach spaces. *Transactions of the American Mathematical Society*, 302:371–381, 1987.

[34] J.M. Borwein and A.S. Lewis. Partially finite convex programming, Part I, Duality theory. *Mathematical Programming B*, 57:15–48, 1992.

[35] J.M. Borwein, A.S. Lewis, and D. Noll. Maximum entropy spectral analysis using first order information. Part I: Fisher information and convex duality. *Mathematics of Operations Research*, 21:442–468, 1996.

[36] J.M. Borwein, A.S. Lewis, and R. Nussbaum. Entropy minimization, DAD problems and doubly-stochastic kernels. *Journal of Functional Analysis*, 123:264–307, 1994.

[37] J.M. Borwein, A.S. Lewis, J. Read, and Q. Zhu. Convex spectral functions of compact operators. *International Journal of Convex and Nonlinear Analysis*, 1:17–35, 1999.

[38] J.M. Borwein and W.B. Moors. Null sets and essentially smooth Lipschitz functions. *SIAM Journal on Optimization*, 8:309–323, 1998.

[39] J.M. Borwein and H.M. Strojwas. Tangential approximations. *Nonlinear Analysis: Theory, Methods and Applications*, 9:1347–1366, 1985.

[40] J.M. Borwein and H.M. Strojwas. Proximal analysis and boundaries of closed sets in Banach space, Part I: theory. *Canadian Journal of Mathematics*, 38:431 452, 1986.

[41] J.M. Borwein and H.M. Strojwas. Proximal analysis and boundaries of closed sets in Banach space, Part II. *Canadian Journal of Mathematics*, 39:428–472, 1987.

[42] J.M. Borwein and Q. Zhu. A survey of smooth subdifferential calculus with applications. *Nonlinear Analysis: Theory, Methods and Applications*, 38:687–773, 1999.

[43] J.M. Borwein and Q. Zhu. *Techniques of Variational Analysis*. Springer-Verlag, New York, 2005.

[44] J.M. Borwein and D. Zhuang. Super-efficient points in vector optimization. *Transactions of the American Mathematical Society*, 338:105–122, 1993.

[45] G. Bouligand. Sur les surfaces dépourvues de points hyperlimites. *Annales de la Societé Polonaise de Mathématique*, 9:32–41, 1930.

[46] S. Boyd, L. El Ghaoui, E. Feron, and V. Balikrishnan. *Linear Matrix Inequalities in System and Control Theory*. SIAM, Philadelphia, 1994.

[47] S. Boyd and L. Vandenberghe. *Convex Optimization*. Cambridge University Press, Cambridge, U.K, 2004.

[48] L.M. Bregman. The method of successive projection for finding a common point of convex sets. *Soviet Mathematics Doklady*, 6:688–692, 1965.

[49] L.E.J. Brouwer. On continuous one-to-one transformations of surfaces into themselves. *Proc. Kon. Ned. Ak. V. Wet. Ser. A*, 11:788–798, 1909.

[50] L.E.J. Brouwer. Uber Abbildungen vom Mannigfaltigkeiten. *Mathematische Annalen*, 71:97–115, 1912.

[51] F.E. Browder. Nonexpansive nonlinear operators in a Banach space. *Proc. Nat. Acad. Sci. U.S.A.*, 54:1041–1044, 1965.

[52] C. Carathéodory. Uber den Variabiletätsbereich der Fourier'schen Konstanten von positiven harmonischen Funktionen. *Rendiconti del Circolo Matematico de Palermo*, 32:193–217, 1911.

[53] V. Chvátal. *Linear Programming*. Freeman, New York, 1983.

[54] F.H. Clarke. Generalized gradients and applications. *Transactions of the American Mathematical Society*, 205:247–262, 1975.

[55] F.H. Clarke. *Optimization and Nonsmooth Analysis.* Wiley, New York, 1983.

[56] F.H. Clarke, Y.S. Ledyaev, R.J. Stern, and P.R. Wolenski. *Nonsmooth Analysis and Control Theory.* Springer-Verlag, New York, 1998.

[57] G.B. Dantzig. *Linear Programming and Its Extensions.* Princeton University Press, Princeton, N.J., 1963.

[58] C. Davis. All convex invariant functions of hermitian matrices. *Archiv der Mathematik*, 8:276–278, 1957.

[59] H. Debrunner and P. Flor. Ein Erweiterungssatz für monotone Mengen. *Archiv der Mathematik*, 15:445–447, 1964.

[60] V.F. Dem'yanov and V.M. Malozemov. *Introduction to Minimax.* Dover, New York, 1990.

[61] J.E. Dennis and R.B. Schnabel. *Numerical Methods for Unconstrained Optimization and Nonlinear Equations.* Prentice-Hall, Englewood Cliffs, N.J., 1983.

[62] F.R. Deutsch. *Best Approximation in Inner Product Spaces.* Springer, New York, 2001.

[63] J. Diestel. *Geometry of Banach Spaces—Selected Topics*, volume 485 of *Lecture Notes in Mathematics.* Springer-Verlag, New York, 1975.

[64] U. Dini. Fondamenti per la teoria delle funzioni di variabili reali. Pisa, 1878.

[65] A. Dontchev. The Graves theorem revisited. *Journal of Convex Analysis*, 3:45–54, 1996.

[66] A.L. Dontchev, A.S. Lewis, and R.T. Rockafellar. The radius of metric regularity. *Transactions of the American Mathematical Society*, 355:493–517, 2003.

[67] J. Dugundji. *Topology.* Allyn and Bacon, Boston, 1965.

[68] J. Dugundji and A. Granas. *Fixed Point Theory.* Polish Scientific Publishers, Warsaw, 1982.

[69] I. Ekeland. On the variational principle. *Journal of Mathematical Analysis and Applications*, 47:324–353, 1974.

[70] I. Ekeland and R. Temam. *Convex Analysis and Variational Problems.* North-Holland, Amsterdam, 1976.

[71] L.C. Evans and R.F. Gariepy. *Measure Theory and Fine Properties of Functions.* CRC Press, Boca Raton, Florida, 1992.

[72] M. Fabian and D. Preiss. On the Clarke's generalized Jacobian. *Rend. Circ. Mat. Palermo*, 14:305–307, 1987.

[73] K. Fan. On a theorem of Weyl concerning eigenvalues of linear transformations. *Proceedings of the National Academy of Sciences of U.S.A.*, 35:652–655, 1949.

[74] K. Fan. Fixed point and minimax theorems in locally convex topological linear spaces. *Proceedings of the National Academy of Sciences of U.S.A.*, 38:431–437, 1952.

[75] J. Farkas. Theorie der einfachen Ungleichungen. *Journal für die reine und angewandte Mathematik*, 124:1–27, 1902.

[76] W. Fenchel. On conjugate convex functions. *Canadian Journal of Mathematics*, 1:73–77, 1949.

[77] L.A. Fernández. On the limits of the Lagrange multiplier rule. *SIAM Review*, 39:292–297, 1997.

[78] P.A. Fillmore and J.P. Williams. Some convexity theorems for matrices. *Glasgow Mathematical Journal*, 12:110–117, 1971.

[79] S. Fitzpatrick. Representing monotone operators by convex functions. *Proceedings of the Centre for Mathematical Analysis, Australian National University*, 20:59–65, 1988.

[80] R. Fletcher. A new variational result for quasi-Newton formulae. *SIAM Journal on Optimization*, 1:18–21, 1991.

[81] D. Gale. A geometric duality theorem with economic applications. *Review of Economic Studies*, 34:19–24, 1967.

[82] J.R. Giles. *Convex Analysis with Application in the Differentiation of Convex Functions.* Pitman, Boston, 1982.

[83] K. Goebel and W.A. Kirk. *Topics in Metric Fixed Point Theory.* Cambridge University Press, Cambridge, U.K., 1990.

[84] P. Gordan. Uber die Auflösung linearer Gleichungen mit reelen Coefficienten. *Mathematische Annalen*, 6:23–28, 1873.

[85] L.M. Graves. Some mapping theorems. *Duke Mathematical Journal*, 17:111–114, 1950.

[86] B. Grone, C.R. Johnson, E. Marques de Sá, and H. Wolkowicz. Positive definite completions of partial Hermitian matrices. *Linear Algebra and Its Applications*, 58:109–124, 1984.

[87] M. Guignard. Generalized Kuhn–Tucker conditions for mathematical programming in Banach space. *SIAM Journal on Control and Optimization*, 7:232–241, 1969.

[88] J. Hadamard. Résolution d'une question relative aux déterminants. *Bull. Sci. Math.*, 2:240–248, 1893.

[89] J. Hadamard. Sur quelques applications de l'indice de Kronecker. In J. Tannery, *Introduction à la Théorie des Fonctions d'une Variable*, volume II. Hermann, Paris, second edition, 1910.

[90] P.R. Halmos. *Finite-Dimensional Vector Spaces.* Van Nostrand, Princeton, N.J., 1958.

[91] G.H. Hardy, J.E. Littlewood, and G. Pólya. *Inequalities.* Cambridge University Press, Cambridge, U.K., 1952.

[92] W.L. Hare and A.S. Lewis. Identifying active constraints via partial smoothness and prox-regularity. *Journal of Convex Analysis*, 11:251–266, 2004.

[93] P.T. Harker and J.-S. Pang. Finite-dimensional variational inequality and nonlinear complementarity problems: a survey of theory, algorithms and applications. *Mathematical Programming*, 48:161–220, 1990.

[94] J.-B. Hiriart-Urruty. A short proof of the variational principle for approximate solutions of a minimization problem. *American Mathematical Monthly*, 90:206–207, 1983.

[95] J.-B. Hiriart-Urruty. What conditions are satisfied at points minimizing the maximum of a finite number of differentiable functions. In *Nonsmooth Optimization: Methods and Applications.* Gordan and Breach, New York, 1992.

[96] J.-B. Hiriart-Urruty. Ensemble de Tchebychev vs ensemble convexe: l'état de la situation vu via l'analyse convexe nonlisse. *Annales Scientifiques et Mathematiques du Québec*, 22:47–62, 1998.

[97] J.-B. Hiriart-Urruty and C. Lemaréchal. *Convex Analysis and Minimization Algorithms.* Springer-Verlag, Berlin, 1993.

[98] R.B. Holmes. *Geometric Functional Analysis and Its Applications.* Springer-Verlag, New York, 1975.

[99] R.A. Horn and C. Johnson. *Matrix Analysis.* Cambridge University Press, Cambridge, U.K., 1985.

[100] R.A. Horn and C.R. Johnson. *Topics in Matrix Analysis.* Cambridge University Press, Cambridge, U.K., 1991.

[101] A.D. Ioffe. Regular points of Lipschitz functions. *Transactions of the American Mathematical Society,* 251:61–69, 1979.

[102] A.D. Ioffe. Sous-différentielles approchées de fonctions numériques. *Comptes Rendus de l'Académie des Sciences de Paris,* 292:675–678, 1981.

[103] A.D. Ioffe. Approximate subdifferentials and applications. I: The finite dimensional theory. *Transactions of the American Mathematical Society,* 281:389–416, 1984.

[104] J. Jahn. Scalarization in multi objective optimization. In P. Serafini, editor, *Mathematics of Multi Objective Optimization,* pages 45–88. Springer-Verlag, Vienna, 1985.

[105] J. Jahn. *An Introduction to the Theory of Nonlinear Optimization.* Springer-Verlag, Berlin, 1996.

[106] G.J.O. Jameson. *Topology and Normed Spaces.* Chapman and Hall, New York, 1974.

[107] Fritz John. Extremum problems with inequalities as subsidiary conditions. In *Studies and Essays, Courant Anniversary Volume,* pages 187–204. Interscience, New York, 1948.

[108] C. Jordan. Mémoire sur les formes bilinéaires. *Journal de Mathematiques Pures et Appliquées,* 2:35–54, 1874.

[109] S. Kakutani. A generalization of Brouwer's fixed point theorem. *Duke Mathematical Journal,* 8:457–459, 1941.

[110] S. Karlin. *Mathematical Methods and Theory in Games, Programming and Economics.* McGraw-Hill, New York, 1960.

[111] W. Karush. Minima of functions of several variables with inequalities as side conditions. Master's thesis, University of Chicago, 1939.

[112] W.A. Kirk. A fixed point theorem for nonexpansive mappings which do not increase distance. *American Mathematical Monthly,* 72:1004–1006, 1965.

[113] E. Klein and A.C. Thompson. *Theory of Correspondences, Including Applications to Mathematical Economics.* Wiley, New York, 1984.

[114] B. Knaster, C. Kuratowski, and S. Mazurkiewicz. Ein Beweis des Fixpunktsatzes für n-dimesionale Simplexe. *Fundamenta Mathematicae*, 14:132–137, 1929.

[115] D. König. *Theorie der Endlichen und Unendlichen Graphen*. Akademische Verlagsgesellschaft, Leipzig, 1936.

[116] A.Y. Kruger and B.S. Mordukhovich. Extremal points and the Euler equation in nonsmooth optimization. *Doklady Akademia Nauk BSSR (Belorussian Academy of Sciences)*, 24:684–687, 1980.

[117] H.W. Kuhn and A.W. Tucker. Nonlinear programming. In *Proceedings of the Second Berkeley Symposium on Mathematical Statistics and Probability*. University of California Press, Berkeley, 1951.

[118] P. Lax. Change of variables in multiple integrals. *American Mathematical Monthly*, 106:497–501, 1999.

[119] G. Lebourg. Valeur moyenne pour gradient généralisé. *Comptes Rendus de l'Académie des Sciences de Paris*, 281:795–797, 1975.

[120] A.S. Lewis. Convex analysis on the Hermitian matrices. *SIAM Journal on Optimization*, 6:164–177, 1996.

[121] A.S. Lewis. Derivatives of spectral functions. *Mathematics of Operations Research*, 6:576–588, 1996.

[122] A.S. Lewis. Group invariance and convex matrix analysis. *SIAM Journal on Matrix Analysis and Applications*, 17:927–949, 1996.

[123] A.S. Lewis. Ill-conditioned convex processes and conic linear systems. *Mathematics of Operations Research*, 24:829–834, 1999.

[124] A.S. Lewis. Lidskii's theorem via nonsmooth analysis. *SIAM Journal on Matrix Analysis*, 21:379–381, 1999.

[125] A.S. Lewis. Nonsmooth analysis of eigenvalues. *Mathematical Programming*, 84:1–24, 1999.

[126] A.S. Lewis. Active sets, nonsmoothness and sensitivity. *SIAM Journal on Optimization*, 13:702–725, 2003.

[127] A.S. Lewis. The mathematics of eigenvalue optimization. *Mathematical Programming*, 97:155–176, 2003.

[128] A.S. Lewis and M.L. Overton. Eigenvalue optimization. *Acta Numerica*, 5:149–190, 1996.

[129] R. Lipschitz. *Lehrbuch der Analysis*. Cohen und Sohn, Bonn, 1877.

[130] L.A. Liusternik. On the conditional extrema of functionals. *Matematicheskii Sbornik*, 41:390–401, 1934.

[131] D.G. Luenberger. *Optimization by Vector Space Methods.* Wiley, New York, 1969.

[132] D.G. Luenberger. *Linear and Nonlinear Programming.* Addison-Wesley, Reading, Mass., 1984.

[133] O.L. Mangasarian and S. Fromovitz. The Fritz John necessary optimality conditions in the presence of equality and inequality constraints. *Journal of Mathematical Analysis and Applications*, 17:37–47, 1967.

[134] A.W. Marshall and I. Olkin. *Inequalities: Theory of Majorization and Its Applications.* Academic Press, New York, 1979.

[135] M. Matić, C.E.M. Pearce, and J. Pečarić. Improvements on some bounds on entropy measures in information theory. *Mathematical Inequalities and Applications*, 1:295–304, 1998.

[136] E.J. McShane. The Lagrange multiplier rule. *American Mathematical Monthly*, 80:922–924, 1973.

[137] E. Michael. Continuous selections I. *Annals of Mathematics*, 63:361–382, 1956.

[138] P. Michel and J.-P. Penot. Calcul sous-différentiel pour les fonctions lipschitziennes et non lipschitziennes. *Comptes Rendus de l'Académie des Sciences de Paris*, 298:269–272, 1984.

[139] P. Michel and J.-P. Penot. A generalized derivative for calm and stable functions. *Differential and Integral Equations*, 5:433–454, 1992.

[140] J. Milnor. Analytic proofs of the Hairy ball theorem and the Brouwer fixed point theorem. *American Mathematical Monthly*, 85:521–524, 1978.

[141] H. Minkowski. *Geometrie der Zahlen.* Teubner, Leipzig, 1910.

[142] H. Minkowski. Theorie der konvexen Körper, insbesondere Begründung ihres Oberflächenbegriffs. In *Gesammelte Abhandlungen II.* Chelsea, New York, 1967.

[143] B.S. Mordukhovich. Maximum principle in the problem of time optimal response with nonsmooth constraints. *Journal of Applied Mathematics and Mechanics*, 40:960–969, 1976.

[144] B.S. Mordukhovich. Nonsmooth analysis with nonconvex generalized differentials and adjoint mappings. *Doklady Akademia Nauk BSSR*, 28:976–979, 1984.

[145] B.S. Mordukhovich. Complete characterization of openness, metric regularity, and Lipschitzian properties of multifunctions. *Transactions of the American Mathematical Society*, 340:1–35, 1993.

[146] J.-J. Moreau. Sur la fonction polaire d'une fonction semi-continue supérieurement. *Comptes Rendus de l'Académie des Sciences de Paris*, 258:1128–1130, 1964.

[147] J. Nash. Non-cooperative games. *Annals of Mathematics*, 54:286–295, 1951.

[148] Y. Nesterov and A. Nemirovskii. *Interior-Point Polynomial Algorithms in Convex Programming.* SIAM Publications, Philadelphia, 1994.

[149] G.K. Pedersen. *Analysis Now.* Springer-Verlag, New York, 1989.

[150] J.-P. Penot. Is convexity useful for the study of monotonicity. In R.P. Agarwal and D. O'Regan, eds: *Nonlinear Analysis and Applications*, pages 807–822. Kluwer, Dordrecht, 2003.

[151] A.L. Peressini. *Ordered Topological Vector Spaces.* Harper and Row, New York, 1967.

[152] A.L. Peressini, F.E. Sullivan, and J.J. Uhl. *The Mathematics of Nonlinear Programming.* Springer-Verlag, New York, 1988.

[153] R.R. Phelps. *Convex Functions, Monotone Operators, and Differentiability*, volume 1364 of *Lecture Notes in Mathematics.* Springer-Verlag, New York, 1989.

[154] R.R. Phelps. Lectures on maximal monotone operators. `arXiv:math.FA/9302209`, 1993.

[155] R.A. Poliquin and R.T. Rockafellar. Amenable functions in optimization. In F. Giannessi, editor, *Nonsmooth Optimization Methods and Applications*, pages 338–353. Gordan and Breach, Philadelphia, 1992.

[156] R.A. Poliquin and R.T. Rockafellar. Prox-regular functions in variational analysis. *Transactions of the American Mathematical Society*, 348:1805–1838, 1996.

[157] R.A. Poliquin, R.T. Rockafellar, and L. Thibault. Local differentiability of distance functions. *Transactions of the American Mathematical Society*, 352:5231–5249, 2000.

[158] B.H. Pourciau. Modern multiplier rules. *American Mathematical Monthly*, 87:433–452, 1980.

[159] B. Pshenichnii. *Necessary Conditions for an Extremum.* Marcel Dekker, New York, 1971.

[160] J. Renegar. Linear programming, complexity theory and elementary functional analysis. *Mathematical Programming*, 70:279–351, 1995.

[161] S.M. Robinson. Normed convex processes. *Transactions of the American Mathematical Society*, 174:127–140, 1972.

[162] S.M. Robinson. Regularity and stability for convex multivalued functions. *Mathematics of Operations Research*, 1:130–143, 1976.

[163] S.M. Robinson. Stability theory for systems of inequalities, part II: differentiable nonlinear systems. *SIAM Journal on Numerical Analysis*, 13:497–513, 1976.

[164] S.M. Robinson. Normal maps induced by linear transformations. *Mathematics of Operations Research*, 17:691–714, 1992.

[165] R.T. Rockafellar. Level sets and continuity of conjugate convex functions. *Transactions of the American Mathematical Society*, 123:46–63, 1966.

[166] R.T. Rockafellar. *Monotone Processes of Convex and Concave Type.* Memoirs of the American Mathematical Society, 1967. No. 77.

[167] R.T. Rockafellar. *Convex Analysis.* Princeton University Press, Princeton, N.J., 1970.

[168] R.T. Rockafellar and R.J.-B. Wets. *Variational Analysis.* Springer, Berlin, 1998.

[169] H.L. Royden. *Real Analysis.* Macmillan, New York, 1988.

[170] W. Rudin. *Real and Complex Analysis.* McGraw-Hill, New York, 1966.

[171] S. Simons. *Minimax and Monotonicity.* Springer-Verlag, Berlin, 1998.

[172] S. Simons. Dualized and scaled Fitzpatrick functions. *Proceedings of the American Mathematical Society*, 2005. To appear.

[173] M. Slater. Lagrange multipliers revisited: a contribution to non-linear programming. Cowles Commission Discussion Paper Math. 403, 1950.

[174] D.R. Smart. *Fixed Point Theorems.* Cambridge University Press, London, 1974.

[175] R.J. Stern and H. Wolkowicz. Indefinite trust region subproblems and nonsymmetric eigenvalue perturbations. *SIAM Journal on Optimization*, 5:286–313, 1995.

[176] R.E. Steuer. *Multiple Criteria Optimization: Theory, Computation and Application.* Wiley, New York, 1986.

[177] K.R. Stromberg. *An Introduction to Classical Real Analysis.* Wadsworth, Belmont, Cal., 1981.

[178] F.E. Su. Borsuk–Ulam implies Brouwer: a direct construction. *American Mathematical Monthly*, 109:855–859, 1997.

[179] C. Swartz. *An Introduction to Functional Analysis.* Marcel Dekker, New York, 1992.

[180] C.M. Theobald. An inequality for the trace of the product of two symmetric matrices. *Mathematical Proceedings of the Cambridge Philosophical Society*, 77:265–266, 1975.

[181] J. Van Tiel. *Convex Analysis: An Introductory Text.* Wiley, New York, 1984.

[182] H. Uzawa. The Kuhn–Tucker theorem in concave programming. In K.J. Arrow, L. Hurwicz, and H. Uzawa, editors, *Studies in Linear and Nonlinear Programming*, pages 32–37. Stanford University Press, Stanford, 1958.

[183] L. Vandenberghe, S. Boyd, and S.-P. Wu. Determinant maximization with linear matrix inequality constraints. *SIAM Journal on Matrix Analysis and Applications*, 19:499–533, 1998.

[184] J. von Neumann. Some matrix inequalities and metrization of matric-space. *Tomsk University Review*, 1:286–300, 1937. In: *Collected Works*, Pergamon, Oxford, 1962, Volume IV, 205-218.

[185] J. von Neumann and O. Morgenstern. *The Theory of Games and Economic Behaviour.* Princeton University Press, Princeton, N.J., 1948.

[186] H. Weyl. Elementare Theorie der konvexen Polyeder. *Commentarii Math. Helvetici*, 7:290–306, 1935.

[187] S.J. Wright. *Primal-Dual Interior-Point Methods.* SIAM, Philadelphia, 1997.

[188] K. Yosida. *Functional Analysis.* Springer-Verlag, Berlin, 1995.

[189] E. Zeidler. *Nonlinear Functional Analysis and its Applications I.* Springer-Verlag, New York, 1986.

Index

Page numbers in italics refer to principal references.